AF785921

# COMPUTER-AIDED MECHANICAL ASSEMBLY PLANNING

## THE KLUWER INTERNATIONAL SERIES IN ENGINEERING AND COMPUTER SCIENCE

ROBOTICS: VISION, MANIPULATION AND SENSORS

*Consulting Editor:* **Takeo Kanade**

***SHADOWS AND SILHOUETTES IN COMPUTER VISION,*** S. Shafer
ISBN: 0-89838-167-3
***PERCEPTUAL ORGANIZATION AND VISUAL RECOGNITION,*** D. Lowe
ISBN: 0-89838-172-X
***ROBOT DYNAMICS ALGORITHMS,*** F. Featherstone
ISBN: 0-89838-230-0
***THREE- DIMENSIONAL MACHINE VISION,*** T. Kanade (editor)
ISBN: 0-89838-188-6
***KINEMATIC MODELING, IDENTIFICATION AND CONTROL OF ROBOT MANIPULATORS,*** H.W. Stone
ISBN: 0-89838-237-8
***OBJECT RECOGNITION USING VISION AND TOUCH,*** P. Allen
ISBN: 0-89838-245-9
***INTEGRATION, COORDINATION AND CONTROL OF MULTI-SENSOR ROBOT SYSTEMS,*** H.F. Durrant-Whyte
ISBN: 0-89838-247-5
***MOTION UNDERSTANDING: Robot and Human Vision,*** W.N. Martin and J. K. Aggrawal (editors)
ISBN: 0-89838-258-0
***BAYESIAN MODELING OF UNCERTAINTY IN LOW-LEVEL VISION,*** R. Szeliski
ISBN: 0-7923-9039-3
***VISION AND NAVIGATION: THE CMU NAVLAB,*** C. Thorpe (editor)
ISBN: 0-7923-9068-7
***TASK-DIRECTED SENSOR FUSION AND PLANNING: A Computational Approach,*** G. D. Hager
ISBN: 0-7923-9108-X
***COMPUTER ANALYSIS OF VISUAL TEXTURES,*** F. Tomita and S. Tsuji
ISBN: 0-7923-9114-4
***DATA FUSION FOR SENSORY INFORMATION PROCESSING SYSTEMS,*** J. Clark and A. Yuille
ISBN: 0-7923-9120-9
***PARALLEL ARCHITECTURES AND PARALLEL ALGORITHMS FOR INTEGRATED VISION SYSTEMS,*** A.N. Choudhary, J. H. Patel
ISBN: 0-7923-9078-4
***ROBOT MOTION PLANNING,*** J. Latombe
ISBN: 0-7923-9129-2
***DYNAMIC ANALYSIS OF ROBOT MANIPULATORS: A Cartesian Tensor Approach,*** C.A Balafoutis, R.V. Patel
ISBN: 0-7923-9145-4
***PERTURBATION TECHNIQUES FOR FLEXIBLE MANIPULATORS:*** A. Fraser and R. W. Daniel
ISBN: 0-7923-9162-4

# COMPUTER-AIDED MECHANICAL ASSEMBLY PLANNING

*edited by*


**Luiz S. Homem de Mello**
*Jet Propulsion Laboratory*
*California Institute of Technology*

**Sukhan Lee**
*Department of Electrical Engineering Systems*
*University of Southern California*
*and*
*Jet Propulsion Laboratory*
*California Institute of Technology*



Springer Science+Business Media, LLC

**Library of Congress Cataloging-in-Publication Data**

Computer-aided mechanical assembly planning / edited by Luiz S. Homem de Mello, Sukhan Lee.
p. cm. -- (The Kluwer international series in engineering and computer science: SECS 148)
Includes bibliographical references and index.
ISBN 978-1-4613-6806-9 ISBN 978-1-4615-4038-0 (eBook)
DOI 10.1007/978-1-4615-4038-0
1. Computer-aided engineering. 1. Assembling machines. I. Homem de Mello, Luiz S. II. Lee, Sukhan. III. Series.
TA345.C6424 1991
670.42'7--dc20 91-21667
CIP

**Copyright** © 1991 Springer Science+Business Media New York
Originally published by Kluwer Academic Publishers,New York in 1991
Softcover reprint of the hardcover 1st edition 1991

All rights reserved. No part of this publication may be reproduced, stored in a retrieval system or transmitted in any form or by any means, mechanical, photo-copying, recording, or otherwise, without the prior written permission of the publisher, Springer Science+Business Media, LLC.

*Printed on acid-free paper.*

To my parents, Ligia and Fabio Homem de Mello.

To my parents, Anna and Hyung-Ki Lee.

# Contents

# Foreword

Some twenty years have elapsed since the first attempts at planning were made by researchers in artificial intelligence. These early programs concentrated on the development of plans for the solution of puzzles or toy problems, like the rearrangement of stacks of blocks. These early programs provided the foundation for the work described in this book, the automatic generation of plans for industrial assembly.

As one reads about the complex and sophisticated planners in the current generation, it is important to keep in mind that they are addressing real-world problems. Although these systems may become the "toy" systems of tomorrow, they are providing a solid foundation for future, more general and more advanced planning tools. As demonstrated by the papers in this book, the field of computer-aided mechanical assembly planning is maturing. It now may include:

- geometric descriptions of parts extracted from or compatible with CAD programs;
- constraints related to part interference and the use of tools;
- fixtures and jigs required for the assembly;
- the nature of connectors, matings and other relations between parts;
- number of turnovers required during the assembly;
- handling and gripping requirements for various parts;
- automatic identification of subassemblies.

This is not an exhaustive list, but it serves to illustrate the complexity of some of the issues which are discussed in this book. Such issues must be considered in the design of the modern planners, as they produce desirable assembly sequences and precedence relations for assembly.

As with other AI-based planning programs, the fundamental issues include knowledge representation and acquisition, search algorithms and inference techniques. Hence, several of the chapters of the book include discussions of modeling and representation of parts, liaisons between them and the processes which produce them. The basic approach to problem solving is systematic search in the problem space, which immediately raises the possibility of an explosion in the number of possible solutions. Hence, many current programs include editing provisions to eliminate undesirable branches of the search tree, or prune the tree in the process of developing the plans. Yet, there are situations in which the enumeration and examination of all feasible assembly orders has advantages, as demonstrated in some of the approaches discussed here. The amount of knowledge required for the intelligent determination of feasible assembly sequences and the subsequent editing and sorting is clearly enormous. Hence, some of the methods discussed in this book include interactive features, which enable users to assist in the process by means of question answering and other computer aids.

While it is evident that the automation of assembly planning is a maturing area of research, it is also evident that it is still in the research phase. Most of the authors of the papers in this book are associated with academia or with research centers. Computer-aided mechanical assembly planning is still largely an academic discipline. Its application in industry is still in the early stages. Among the reasons why its use in industry is not yet extensive are the following:

1. The computational efficiency of the today's planners limits them to assemblies with a relatively small number of parts. To handle products consisting of 60 or 80 parts, which occur in industry, current systems still need the help of a human operator.
2. The measures for the selection of desirable assembly sequences in existing planners are not yet wide enough. Cost, ease of assembly and robot gripping requirements are among the commonly used evaluation criteria. Fixturing and tooling requirements and the related issue of partial assembly turnovers are seldom discussed. Even less frequent are such practical production criteria as assembly line layout. Furthermore, many assembly processes incorporate various testing steps, which should be considered in the evaluation. Evaluation and rejection of candidate assembly sequences on the basis of such multiple criteria is clearly more difficult, but it would contribute to the perception of real world relevance of the assembly planners. Since the time associated with alternative plans and the related issues of tool change and fixture adjustment are of great industrial significance, it will also be important to integrate scheduling with planning in future systems.
3. There is a natural delay in the transfer of research knowledge from academia to industry. Industry is somewhat conservative in the introduction of new technology, which must be justified on the basis of both scientific and economic criteria.

The authors of the chapters in this book report on work that addresses these problems.

I indicated earlier that assembly planning is a maturing discipline. Perhaps it is more fair to say that it has reached adolescence. It has given up the toys of childhood, but it is not quite ready to assume all the responsibilities of the rough and tumble adult world of industry. This is perhaps the most exciting time to be associated with a new technology. Readers of this book will be ready for the developments to come.

**George A. Bekey**
Computer Science Department and
Center for Manufacturing and Automation Research
University of Southern California
Los Angeles, California

# COMPUTER-AIDED MECHANICAL ASSEMBLY PLANNING

# Chapter 1

# Introduction

**Luiz S. Homem de Mello and Sukhan Lee**

Intensifying competition in manufacturing has brought about redoubled pressure for cutting down costs, for improving product quality, and for minimizing the time from concept to production. These requirements, coupled with the ever growing complexity of products and of production systems, have contributed to a rising interest in *concurrent engineering*, or *simultaneous engineering*. These terms have been used to refer to the idea of integrating the design of a product and the design of its production system. Being able to take manufacturing considerations into account early in the development of a new artifact can greatly simplify its fabrication. Even a small change in the design of a product can have a large impact on the assembly alternatives. Where adopted, the concurrent engineering approach has led to more efficient production systems and therefore to lower costs compared to where design and production planning are separated. In addition, development times are shorter, and redesign due to manufacturing constraints is greatly reduced.

The introduction of the concurrent engineering approach has been facilitated by the recent progress in digital electronic technology. As computers become faster, more powerful, and less expensive, and as software engineering matures, industrial designers and engineers have had increased access to software tools that help them improve their productivity. Computer-aided design (CAD) programs, for example, are already well established and substantially improve the efficiency of the design process. Another field in which the use of computers is becoming widespread is process planning, that is, the generation of a sequence of machining cuts for the production of a part. Still another manufacturing domain in which computer aids are being introduced is the scheduling of factory resources.

This book focus on yet another area for software tools that has emerged more recently namely mechanical assembly planning. The use of computers for planning the assembly of mechanical products originated in the research on planning within artificial intelligence.

There are many reasons for the systematization and the computerization of assembly planning, some of which are listed below.

- Industrial designers will benefit from having a tool with which they can quickly assess their designs for ease of assembly.
- The planning and programming chores in manufacturing are time consuming and error prone. Moreover, the time spent in planning and programming may excessively delay the actual production. The automation of these chores expedites their execution, reduces their cost, and improves their quality.
- The tailoring of products for market niches is becoming more common. For small batches, the cost of manual planning and programming can weigh heavily in the total production cost.
- Although many experienced industrial engineers have a knack for devising efficient ways to assemble a given product, systematic procedures are necessary to guarantee that no good assembly plan has been overlooked. For complex products, the number of different assembly alternatives may be so large that even skillful engineers fail to notice many possibilities.
- In some cases, it is necessary to adapt the assembly process to different sets of machines. The need to produce different products in the same shop may lead to the choice of an assembly plan for a product that may not be the most efficient on ideal conditions, but uses the idle equipment. Likewise, when the same product is assembled in different sites, the plan that is more suitable to the available equipment may be different from one shop to the next. Automation allows the actual planning to be delayed until it is clear what machines will execute the assembly.
- In many applications of autonomous systems, it is impracticable to preprogram all tasks they might face. Such systems must have the ability to generate assembly or disassembly plans that fit the particular situation they encounter. Similarly, an opportunistic scheduler can be more effective if it is able to generate, in real time, the assembly plan that is more suitable for the order in which parts arrive or are picked from a bin.

A number of technical issues must be addressed for the automation of assembly planning. They include the following:

- **The representation of assemblies**

A computer representation of mechanical assemblies is necessary in order to automate the generation of assembly plans. The main issues here are deciding what information about assemblies is required, and how the information can be represented inside the computer.

Of course the shape of the parts and the geometric relationship between parts are crucial for assembly planning. Therefore, they must be represented, along with their tolerances, in any assembly model. But although the geometric aspects are very important, assembly models must also represent a number of other aspects such as the attachments that secure two parts together, or the chemical treatments (e.g., painting, lubrication) that must be applied after parts are joined.

A relational scheme seems ideal for representing assemblies since it can capture the geometric and mechanical relations between parts. However, for assemblies with large number of parts, a hierarchical scheme may be more efficient since many products are designed with natural hierarchies of subassemblies. In practice, a combination of the two schemes may attain the advantages of both.

- **The representation of assembly plans**

A computer system for assembly planning must have a way to represent the assembly plans it generates.

Several methodologies for representing assembly plans have been utilized. These include representations based on directed graphs, on AND/OR graphs, on establishment conditions, and on precedence relationships. A clear understanding of these alternative representations and of how one maps into the others is very important in developing an assembly planner. As later chapters will show, the ability to go back and forth from one representation scheme to another can lead to efficiency gains in the planning process.

In addition to representing the joining of parts or subassemblies, the representation schemes must also be able to represent the other operations such as the chemical treatments that must be applied after parts are put together.

- **The correctness and completeness of the planning process**

Clearly, to be an useful tool, an assembly planning system must only generate correct assembly plans. Furthermore, to solve problems that require optimization, such as the selection of the best assembly alternative, one must be able to traverse the space of all candidate solutions, regardless of the method used to solve the problem. It should be noted that the solution procedure does not need to go over all possibilities. What is important is that the method has the potential to generate all assembly plans.

- **The efficiency of the planning process**

Assembly planning is a computationally intensive task. Therefore, it is imperative that we seek new approaches that can reduce the computation required to generate assembly plans. Some approaches to improve the planning efficiency may sacrifice completeness. This happens, for example, when subassemblies are treated as units, in order to artificially reduce the number of parts. In these cases, it is important to ascertain that no good plan is being missed, that is, that the alternatives being pruned would not be among the preferred ones.

- **The selection of assembly plans**

The number of distinct feasible assembly plans can be very large even for assemblies made up of a small number of parts. Therefore, a complete enumeration of assembly plans is prohibitive in most real applications. Finding systematic ways to narrow down the alternatives is crucial for the automatic planning of assembly. Two kinds of approaches are currently being tried. One, more qualitative, is the development of rules that can be used to eliminate assembly plans that include difficult tasks or awkward intermediate subassemblies. Another approach, more quantitative, is to devise an evaluation function that computes the merit of assembly plans based, for example, on the cost of the resources needed to perform the assembly, on the total time required, and on the difficulty of execution. It seems likely that a combination of the qualitative and the quantitative approaches will attain the advantages of both.

- **The integration with CAD programs**

A mechanical assembly is a composition of interconnected parts. As mentioned above, more and more frequently the parts are being designed using CAD programs. Therefore, the shape of each part as well as other relevant information are already stored in computer databases. The assembly planning will be more efficient if those CAD databases can be directly input to the program that generates the assembly models.

- **The integration with task and motion planners**

Another result of the recent progress in digital electronics is the introduction of programmable robots in manufacturing. These machines can be adapted to execute different operations by changing their internal programs. Task and motion planners that will facilitate robot programming are currently being developed. With a view towards future integration, the output of assembly planners should be compatible with what is required by task and motion planners. Moreover, it is also desirable that assembly planners take into account the capabilities and limitations of task and motion planners.

All the above issues are active areas of research. The main goal of this book is to consolidate in one volume the main approaches to solving these problems. It has been divided into two parts: assembly modeling, and assembly planning. The next two sections present an overview of the book.

## 1.1 Assembly modeling

Part I contains four chapters which cover important issues in modeling assemblies.

Chapter 2 discusses the mathematical modeling of the geometric aspects of assemblies. Assembly models are defined in terms of configuration spaces whose elements correspond to collection of solid mechanical parts and their poses (i.e., positions and orientations.) A range of increasingly complex notions of assemblies is introduced. It includes the following:

1. rigid assemblies in which the relative positions of parts remains the same;
2. articulated mechanisms in which the relative positions of parts can change;
3. variational assemblies in which the shape of the parts can vary;
4. stochastic assemblies in which the relative positions of parts can vary.

Representation schemes previously developed for modeling individual parts can be readly extended for modeling the geometric aspects of these four notions of assemblies. Chapter 2 discusses these extensions. It also points to issues that are still open to research, such as the establishment of a sharp characterization for the subsets of configurations that correspond to physical assemblies.

Chapter 3 introduces a modular boundary class of models for solid objects. These models describe objects as the pairwise combination of face-adjacent parts. Compared with conventional boundary representation and constructive solid geometry, such models offer the putative advantages of locality of manipulation, capability of describing form features, and possibility of attaching tolerance information.

A specific model of that class, the Face-to-Face Composition (FFC) model, which was developed for an experimental geometric modeler, is presented. This model contains explicit information about interference among the components of an object. Juxtaposition and interference are represented in a hypergraph in which the nodes are component objects and the hyperarcs are connection and interference facets.

The data structure for the FFC mirrors a cellular decomposition of the modular object into non-overlapping, arbitrarily shaped cells. The FFC model can be constructed either by adding a single component at a time or by combining two composite objects represented by FFCs. An FFC can be readily evaluated to yield the boundary of the complete object.

An application of the FFC is the Production Graph, which represents alternative sequences of combining components. Valid sequences can be obtained using the validity checks developed for complete FFCs.

Chapter 4 presents a characterization of the spatial relationships between bodies in assemblies as trivial kinematic constraints. A kinematic constraint is defined as a set of displacements that can be expressed as a composition of cosets of Euclidean subgroups. A constraint is said to be trivial when it can be reduced to a single coset.

A graph of kinematic constraints is defined as a graph whose nodes correspond to workpieces and whose directed arcs are labeled with trivial kinematic constraints. The problem of how to find equivalent constraints between two bodies is addressed. By relying on the composition and intersection of subgroups, it is possible to carry out a topological analysis of the motion possibilities for a set of bodies linked by a set of trivial kinematic constraints. A basic algorithm for constraint propagation is presented. This algorithm computes the equivalent constraint between two bodies in a graph of kinematic constraints with arbitrary topology. An assembly example illustrates the algorithm's computation.

Chapter 5 formulates a mathematical programming approach to the solution of the problem of specifying the position of each part relative to the position of the other parts in an assembly.

The position of each part is specified based on geometric relationships between various features of the part and mating features of its neighboring parts. These feature relationships are treated as inequalities, and mathematical programming is used to find the optimal configuration of the parts.

The approach is amenable to both sequential and simultaneous strategies for computing the part positions. The computations are tractable and robust. Thus a variational assembly model can be constructed and evaluated at reasonable cost, and the assembly model will be compliant with part variations. This approach is particularly useful for solving problems in tolerancing.

## 1.2 Assembly planning

Part II contains ten chapters which address important issues in the systematization and computerization of mechanical assembly planning.

Chapter 6 discusses four of the most commonly used representations for assembly sequences. These are based on directed graphs, on AND/OR graphs, on establishment conditions and on precedence relationships. The correspondence between these representations as well as their correctness and completeness are established and are illustrated with two assembly examples.

Chapter 7 presents a basic algorithm for the generation of all mechanical assembly sequences for a given product. The algorithm employs a relational model of assemblies. In addition to the geometry of the assembly, this model includes a representation of the attachments that bind one part to another.

The problem of generating the assembly sequences is transformed into the problem of generating *disassembly* sequences in which the *disassembly* tasks are the inverse of feasible assembly tasks. This transformation leads to a decomposition approach in which the problem of disassembling one assembly is decomposed into distinct subproblems, each being to *disassemble* one subassembly. It is assumed that exactly two parts or subassemblies are joined at each time, and that whenever parts are joined forming a subassembly all contacts between the parts in that subassembly are established. The algorithm returns the AND/OR graph representation of assembly sequences. Bounds for the amount of computation involved are presented.

The correctness of the algorithm is based on the assumption that it is always possible to decide correctly whether or not two subassemblies can be joined, based on geometrical and physical criteria. An approach to compute this decision is presented. An experimental implementation for the class of products made up of polyhedral and cylindrical parts having planar or cylindrical contacts among themselves is described.

Chapter 8 presents a systematic method for the determination of assembly plans, described by assembly trees (or part trees). This method involves a recursive definition of the assembly process, a model of the end product defining all the actions which have to take place in the assembly process, and a formalization of assembly constraints. An assembly example illustrates the method.

The assembly process includes not only the mating and securing of parts but also all the other operations, referred to as complementary, such as inspection, test, cleaning and labeling. Accordingly, the model for the end product includes information about the required complementary operations.

The assembly constraints are divided into two classes: *operative constraints*, which define whether or not any candidate assembly operation is feasible; and *strategic constraints* which prune the awkward assembly plans. A resulting interactive software named LEGA has been implemented. LEGA is written in PROLOG and uses a database of constraints.

The strategic constraints are introduced in the product model. One kind of strategic constraint is to impose an intermediate subassembly. Another kind is to group parts that have to be assembled in sequence. These include stacks and ordered layers, two types of configurations that are common in practice.

Although there is some degree of subjectivity in the choice of the strategic constraints, their use have lead to large gains in planning efficiency. It is also possible to run LEGA with two or more distinct sets of strategic constraints and analyze the outcomes.

The operative constraints are of three types: geometric, stability and material. The latter corresponds to the availability of tools or other equipment and their capability to execute the assembly operations. The specification of stability and material constraints also involves some degree of subjectivity. Operative constraints are determined as the assembly plans are generated based on information supplied (interactively) by the user.

LEGA first tries to deduce whether or not a candidate operation is feasible from the constraints in its database. This kind of deductive inference is well suited to PROLOG. Optionally, a program connected to a CAD database is activated. If no deduction can be made automatically, then LEGA queries the user. When this occurs, new constraints are created and added to the database. Typically, the user is queried frequently at the beginning of the planning process and sporadically at the end.

LEGA has been effectively applied to products or subassemblies having up to 20 components.

Chapter 9 describes the GRASP assembly planner. The input to GRASP are three-dimensional models of parts and their locations. The output is an AND/OR graph representing the set of all geometrically feasible assembly sequences in which exactly one part is added at each assembly task, and it follows a straight line trajectory.

GRASP follows an approach similar to the algorithm described in chapter 7. It also transforms the problem of generating the assembly sequences into the problem of generating *disassembly* sequences in which the *disassembly* tasks are the inverse of feasible assembly tasks. GRASP, however, minimizes the geometric reasoning needed to test candidate assembly tasks. Whenever it does a geometric reasoning computation to find whether or not a part is movable, GRASP stores an expression encoding the conditions under which the given part would be movable. The subsequent analysis of candidate tasks will first try to deduce whether or not a part is movable from these expressions in order to avoid lengthy geometric reasoning computations.

Three types of conditions of increasing complexity are used in GRASP. The first, called *simple*, corresponds to the fact that if a part $p$ is movable in an assembly $A$, then it is also movable in any subassembly of $A$. The second, called *contact*, corresponds to the fact that if a part $p$ is not movable because it collides with one or more of the parts in $S$, then $p$ is not movable in any subassembly that includes all those obstructing parts. The third condition, called *local* is more elaborate. The parts of an assembly $A$ in contact with a given part $p$ are clustered in a way that all parts in a group constrain the freedom of $p$ in the same way. In a subassembly of $A$, $p$ will not be movable unless none of the parts in one such group are present.

Chapter 9 also discusses the computational complexity of GRASP and presents its performance for two assembly examples.

Chapter 10 describes an extension to GRASP aimed at eliminating the restriction that only one part be added in any assembly task. It presents a new algorithm for solving an important subproblem of the assembly planning problem, namely the generation of all the ways in which an assembly can be partitioned into two subassemblies.

Instead of generating all cut sets of the assembly's graph of connection, the algorithm presented in chapter 10 uses the geometry information to prune this search and to avoid the generation of many of the cut sets that do not correspond to feasible assembly tasks. An assembly example illustrates the approach. This new algorithm is shown to be sound and complete. Chapter 10 also presents an analysis of its complexity.

Chapter 11 describes the XAP/1 assembly planning system. It begins by comparing the version of the assembly planning problem addressed by the XAP/1 system to other systems along four major dimensions: range of operations allowed in the plans produced; degree of detail in which the plans are described; type of input data required; and degree of optimization done on plans.

The XAP/1 system is oriented toward plan optimization rather than toward generating all feasible plans. Due to this orientation, XAP/1 plans to somewhat greater detail than other planners. The plans it generates are such that only one part is moved at a time and no operation separates parts already joined. Those plans include not only the sequence in which parts are put together but also, for each part, the mating trajectory. In fact, plans produced by XAP/1 are a sequence of *insertion operations* each of which consists of inserting a part or subassembly into a fixture by following a specified trajectory.

XAP/1 generates plans by successively adding sequencing and trajectory assertions to a set until it describes only one plan. The geometric feasibility of the resulting plan is enforced by a single form of constraint. The search for an optimal plan is guided by advice from a set of plug-in criteria modules and an arbitration module. These criteria provide not only estimates of the quality of partially formed plans, they provide advice on which planning decisions should be considered next. Three criteria are discussed: *fixture complexity*, *directionality* and *manipulability*.

The chapter ends showing the performance of XAP/1 on some sample problems and discussing, briefly, extensions to the system's approach.

Chapter 12 presents BRAEN, a system that generates a disassembly sequence for a product from the boundary representations of its parts and other objects involved in the assembly process such as table-top and wall. The reverse of this sequence is the assembly plan. BRAEN assumes a single robot. The plans it produces are sequences of motions, each motion being a translation or a rotation of a component or multi-component subassembly.

In addition to planning at the level of motion specification, BRAEN is oriented towards generating a good sequence quickly rather than generating all feasible sequences. The use of a detailed geometric description for the objects enables the system to compute whether or not a motion is feasible.

The planning technique is centered on two modules. A freedom determination module uses an iterative surface subdivision technique to identify movable components and multi-component subassemblies. A disassembly module searches for a sequence of motions that will break one assembly into two subassemblies.

BRAEN uses three common sense techniques to enhance system performance. One technique deduces the feasibility or infeasibility of a motion in one configuration from the feasibility or infeasibility of the same motion in another configuration. Another technique involves trying first the motions that are more likely to be feasible. Yet another technique uses simple physics to model the effects of gravity and the stability of subassemblies.

Chapter 13 presents COPLANNER, an assembly planning system organized under the *Cooperative Problem Solver* paradigm. In this system, planning is carried out by the cooperation of several modules namely: the plan coordinator, the heuristic advisor, the geometric reasoner, the physical reasoner, the resource manager, and the blackboard.

COPLANNER operation is also based on a recursive decomposition of the assembly into subassemblies. In order to increase the planning efficiency, the system avoids the analysis of decompositions that do not correspond to feasible assembly tasks. This is achieved by clustering the parts that have to be assembled together. Chapter 13 introduces a systematic formulation to construct an *abstract liaison graph* representation of the assembly which merges sets of mutually inseparable parts, that is, those parts that must be assembled independently in any feasible assembly sequence.

The approach is then extended to cluster parts into *preferred* subassemblies based on a *weighed abstract liaison graph.* This graph is similar to the abstract liaison graph, but has weights assigned to its edges. These weights reflect the stability of part interconnection and the directional constraints of the motion that brings the two parts together. The degree of part aggregation can be adjusted by changing some heuristic coefficients. This kind of clustering sacrifices completeness since the sequences that interleave parts of different subassemblies cannot be generated. However, there are manufacturing gains in assembling these *preferred* subassemblies independently. The advantages include the better stability of the intermediate subassemblies, the less difficulty of the assembly tasks, and the greater cohesion of the parts in the *preferred* subassemblies.

COPLANNER has been implemented in Common Lisp and C on a Sun 260 workstation. The plans it generates include *assembly instructions* which schematically describe how to execute the joining of subassemblies.

Chapter 14 shows an assembly planning system that uses Design for Assembly (DFA) analysis to guide the generation of the preferred assembly plans.

Like the assembly planner described in chapter 13, the operation of this planning system is also based on a recursive decomposition of the assembly into subassemblies and on the use of an abstract liaison graph. In addition, the planning in this system incorporates the special processes, such as cleaning, testing or labeling, that must occur during the assembly. These special processes are taken into account in the assembly planning by the introduction of special precedence constraints. Furthermore, this planner can distinguish reversible from non reversible assembly tasks.

Chapter 14 also establishes methods of evaluating alternative assembly plans in terms of DFA criteria such as subassembly stability, directionality, subassembly poses, special process requirements, and parallelism in assembly. The number of fixtures, or holding devices, and the number of reorientations during assembly are identified through the analysis of stability and directionality. All these factors are used in defining cost and heuristic functions for an AO* search for an optimal plan.

Chapter 15 shows a system developed at the Charles Stark Draper Laboratory over the last five years. It is an integrated computer aid useful for assembly-line design and concurrent design of mechanical products.

First, a simple technique for generating assembly sequences on which the initial version of the system was based is presented. At the time of its development, the technique was a great help, and for many applications it was adequate and practical. The liaison diagram representation of assembly that was used as well as the technique itself still provide quick insight into product assembly. The simple technique can be invoked mentally and need not depend on computer aid. The potential role and form of computer aid, however, was immediately recognized and efforts toward devising computer aids were started at once.

The algorithms used in the current version of the system take advantage of the methodology presented in chapters 7 and 8, including: an efficient organization of interference questions based on cut-sets of all subassemblies; a disassembly (instead of assembly) paradigm to avoid the work of creating and discarding assembly dead ends; and the coding of part-interference data already at hand to screen subsequent part-interference questions and often infer answers, dramatically reducing question-count. The plan for this version of the system was to have the computer to work from a design solid-model data-base to answer the remaining necessary assembly interference questions. But consideration of the volume of computation needed combined with the great success in having an engineer familiar with the design answer the interference questions aided by simple screen characterizations of the subassemblies lead to the implementation of an interactive program for finding sequences.

The current system's algorithms and interactive programs for editing product assembly sequences are also described. Editing means and criteria are user-exercised and may be based on assembly-state and assembly-move issues; on assembly-line layout and topology issues; and on consideration of fixturing, orientation, and fixturing-change and reorientation counts. The on-line visual aids provided during generation and evaluation of sequences are illustrated with examples.

Various physical and economic criteria exist. These criteria evolved from and are related to the work on industrial assembly system design and product design. They include: pass-through of a particular assembly state; executing a particular partial sequence; avoiding a difficult assembly move; avoiding an awkward assembly state; choice of assembly line topology; minimizing nonproductive line tasks like refixturing and reorientation; minimizing various economic parameters. The criteria vary in editing power, need for an augmented information base, and ease of application. Knowledge of the power, information needs, ease of application, and the logical rules of sequence representation suggest the use of an application sequence strategy for the criteria.

# Part I

# Assembly Modeling

# Chapter 2

# Representations for assemblies

**Aristides A. G. Requicha and Timothy W. Whalen**

Mechanical, electrical and electronic products typically are *assemblies* of many component solid *parts*. The components of a product may be joined so as to form either (1) a *rigid* assembly, (2) an *articulated* collection of rigid bodies that may move relative to one another (often called a *mechanism*), or (3) a flexible, non-rigid assembly. No solid is perfectly rigid, and sometimes non-rigidity must be acknowledged explicitly, for example when two parts are press-fitted, or when one of the components is a spring.

The term *assembly* is commonly used in two senses, to denote either the *action* of joining several components, or the resulting *artifact*. This chapter is concerned primarily with assemblies as physical artifacts. We focus on two related issues: *what* information about assemblies must be captured, and *how* can such information be represented computationally in a form suitable for integrated, computer-aided systems that support the entire life-cycle of a product, from requirements analysis and design, through manufacturing and assembly, to field maintenance and disposition.

First we discuss mathematical models for assemblies, and introduce a range of increasingly complex notions of assembly. Next we consider schemes for representing assemblies computationally, and we show that one must be able to represent solid parts and their associated tolerances, plus mating relations between parts, and attributes of such relations. Part representations, mating relations, and tolerances are then discussed in more detail. The chapter concludes with a summary and discussion of open issues.

## 2.1 Mathematical models for assemblies

An assembly specification includes geometric as well as non-geometric information. Examples of the latter are the torques required to tighten the bolts in an engine block, and the characteristics of a welding join between two parts. In this section we ignore non-geometric aspects of assemblies, and we focus on geometry.

Individual components of an assembly can be modelled mathematically as $r$-sets, that is, as compact, regular, semi-algebraic subsets of $E^3$, the 3-D Euclidean space [31, 32]. But which mathematical objects correspond to assemblies? This is the major issue addressed in the remainder of this section.

A mathematical model for an *assembly instance* is a set of mathematical models for $n$ solids $S_i$, plus associated geometric transformations (i.e., mappings from $E^3$ to $E^3$) $T_i$ that define the solids' relative *poses* (i.e., locations and orientations). For non-rigid assemblies, for example those involving press-fit or flexible components, the $T_i$ correspond to suitable deformations concatenated with the rigid motions that establish the objects' poses. For simplicity, in the remainder of this section we consider only assemblies of rigid components, and we assume that there is an agreed coordinate frame in which all the poses are measured. Also, throughout the chapter we often refer to mathematical models for solids or assemblies simply as "solids" or "assemblies", when no confusion is likely to arise between abstractions (mathematical entities) and their physical counterparts.

Denote by $M_s$ the mathematical modelling space for solids, that is, the space of all $r$-sets in $E^3$. (Sometimes a smaller class of $r$-sets suffices as a modeling space, as we will see later.) A collection of $n$ solids $(S_1, S_2, \ldots, S_n)$ is an element $\sigma$ of the *solid configuration space* $C_s$, which is the direct product of $n$ copies of $M_s$. Similarly, a set of $n$ transformations is a point $\tau$ in the *transformation configuration space* $C_t$, which is the direct product of $n$ copies of $M_t$, the modeling space for transformations. For assemblies of rigid components the $T_i$ are rigid motions in $E^3$, $M_t$ is a 6-D space, and $C_t$ is the *pose configuration space*. An assembly instance $\alpha$ is a pair $(\sigma, \tau)$, that is, an element of the *assembly configuration space* $C_a = C_s \times C_t$. Equivalently, $\alpha$ is a $2n$-tuple $(S_1, S_2, \ldots, S_n, T_1, T_2, \ldots, T_n)$, and we often write $\alpha(\sigma, \tau)$ with $\sigma = (S_1, S_2, \ldots, S_n)$ and $\tau = (T_1, T_2, \ldots, T_n)$.

Intuitively, an assembly instance is simply a collection of parts in fixed poses. If there is no shape uncertainty associated with the parts, and these are attached rigidly to one another, then there is only one assembly instance. But in general there may be many instances associated with an assembly. For example, a shaft may rotate through an entire $(0, 2\pi)$ range with respect to a bushing. The poses of the parts in a mechanism vary continuously, and physical motions correspond to curves or higher-dimensional subsets of $C_t$. To cater to moving parts we define a mathematical model for a *nominal assembly* as a pair $(\sigma, \Theta)$, where $\sigma$ is a point of $C_s$, and $\Theta$ is a subset of $C_t$. Equivalently,

$$A(\sigma, \Theta) = \{\alpha(\sigma, \tau) | \tau \in \Theta\},$$

where $A$ denotes a nominal assembly and $\alpha$ one of its instances. We use the term "nominal" to emphasize that part variability is not taken into account – there is a given set of parts that may move in space in the manner prescribed by the poses in the set $\Theta$. (However, we do not imply that parts in the assembly must have the "nominal dimensions" defined in their tolerance specifications.) The pose of a part may depend in complicated ways on the poses of several other parts, and therefore it is not sufficient to define independent subsets of $M_t$ for each part. The set $\Theta$ captures the pose relationships for all the parts.

Much of the past research on assembly modeling has focused on nominal assemblies. But parts cannot be manufactured with perfect forms and dimensions, and the associated geometric uncertainties are important because they often determine whether an assembly is physically realizable, or whether an assembly plan will succeed. To take part variability into account we define a mathematical model for a *variational assembly* as a subset of $C_a$, or, equivalently, as the union of a set of nominal assemblies

$$A = \bigcup_{\sigma \in \Sigma} A(\sigma, \Theta(\sigma)),$$

where $\Sigma$ is a subset of $C_s$. In a general variational assembly many groups ($n$-tuples) $\sigma$ of parts are admissible, and for each group there is an associated set of poses corresponding to relative motions of the parts. The admissible motions may depend on the specific $\sigma$ being considered.

Typically, $\Sigma = V_1 \times V_2 \times \cdots \times V_n$, where each $V_i$ is a subset of $M_s$ called a *variational class* [33, 34] associated with a part. This means that each admissible part is a solid $S_i$ selected from its corresponding class $V_i$, independently of the selection of other parts in the assembly. Independent selection reflects the modern principle of part interchangeability. Note, however, that even in modern manufacturing practice sometimes parts must be "matched". This implies that interchangeability no longer applies at the part level, but rather at the level of certain subassemblies.

Not all $\alpha(\sigma, \tau)$ correspond to realizable assemblies. The following conditions must be satisfied.

- Non-interference — Two parts of the assembly in their specified poses must not occupy simultaneously a 3-dimensional region of space. This can be formulated mathematically by requiring that the regularized intersections [31, 32] (denoted by $\cap^*$) between all pairs of distinct parts be empty:

  $$\forall i, j (i \neq j) \Rightarrow S_i \cap^* S_j = \phi.$$

  (Regularized set operations are the topological closures of the interiors of their conventional counterparts [31, 32].) Note that standard, non-regularized intersection is not appropriate because parts in contact intersect over a region of their boundaries.

- Path-existence — It must be possible to move the parts continuously and without collision from a pose configuration in which they are sufficiently apart to the specified configuration. That is, there must exist a continuous trajectory $\tau(r), r \in [0, 1]$ in $C_t$ such that (1) $\alpha(\sigma, \tau(r))$ is a non-interfering assembly for all $r \in [0, 1]$, (2) $\alpha(\sigma, \tau(0))$ corresponds to a situation where the distances between the solids are large compared to the solids' dimensions, and (3) $\alpha(\sigma, \tau(1))$ is the specified assembly instance.

These conditions must be modified slightly when an assembly involves a press fit or another joining technique that causes a limited amount of interference. We will assume in the sequel that such modifications have been incorporated in the definitions of non-interference and path-existence when necessary.

Nominal and variational assemblies must not include instances that fail the non-interference or path-existence criteria. Path existence clearly implies non-interference, because the specified configuration $\alpha(\sigma, \tau(1))$ must be non-interfering. The converse is not true, as shown in Figure 2.1, which depicts a non-interfering assembly that does not satisfy the path-existence criterion. The shaft is a single part, and cannot be assembled to the bracket. Assembly instances that satisfy the path-existence condition (and hence both criteria) are called *geometrically feasible* or *geometrically realizable*. Nominal and variational assemblies also are called geometrically realizable if all of their associated instances are geometrically realizable. Later chapters of this book will show that an assembly instance may be geometrically feasible and yet fail other criteria. For example, it may be impossible to reach a component with the tool necessary to fasten it.

The mathematics of assembly modeling has not been fully worked out. We understand reasonably well the characteristics of $M_s$ and $M_t$, and hence of $C_a$. The pose configuration space $M_t$ typically is 6-D since general rigid bodies have 3 degrees of freedom of translational motion and 3 of rotational motion. Therefore $C_t$ is $6n$-dimensional. $M_s$ typically is not a finite-dimensional space. But if the solids in the represented domain can be described by a finite number

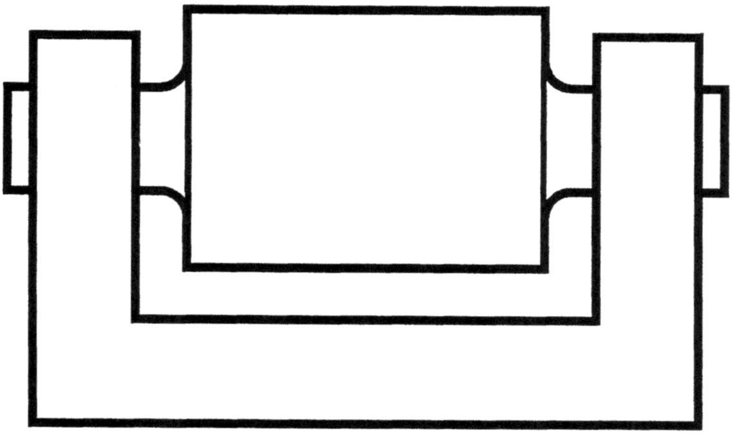

Figure 2.1: A non-interfering but geometrically infeasible assembly instance

of parameters, then $M_s$, and also $C_s$ and $C_a$, are finite-dimensional. For example, if all solids of interest are cuboids, the length, width and height completely define a solid, and $M_s$ is 3-D.

Some issues remain unresolved. For example, is any subset $\Theta$ of $C_t$ acceptable in an assembly model? And any $\Sigma$? We suspect that the answers may be "no". For example, not all subsets of $M_s$ are acceptable as variational classes. Some of them correspond to objects that are overconstrained, in the sense that portions of the objects' boundaries must be of perfect form or in a perfect relationship to others, and therefore are not manufacturable with physical processes, which have inherent uncertainties. A sharp characterization of variational classes is still evolving, but current thinking is that they must be regular sets in a topology related to that induced by the Hausdorff metric in $M_s$ [5]. Analogous results may apply to assembly modeling. We will ignore these issues in the sequel, and assume that feasibility is the only geometric condition that must be satisfied by mathematical models of assemblies.

The previous definitions are purely deterministic. But manufacturing and assembly processes have inherent uncertainties of a stochastic nature. Randomness can be introduced in our models by defining a *stochastic assembly configuration space* $\Omega_a$ that consists of $C_a$ with a probability density function $\Psi(\sigma, \tau)$ defined on it.

Random variations between parts sometimes compensate one another, leading to functionally acceptable assemblies of relatively imperfect components. It is often more economic to have loosely toleranced parts that sometimes (infrequently) cannot be assembled, than tightly toleranced parts guaranteed to always assemble. This implies that the geometric realizability requirements for stochastic assemblies should be relaxed. A stochastic assembly may contain infeasible instances, if the probability associated with such instances is sufficiently small.

The mathematics of stochastic assemblies is more complex and not as well developed as its deterministic counterpart.

## 2.2 Overview of assembly representation schemes

Let us focus initially on the geometric aspects of assembly representation. The notion of *representation scheme* and associated properties introduced in [32] for individual solids can be readily extended to assemblies by exploiting the mathematical models discussed in the previous section. A representation scheme for assemblies is simply a mathematical relation between the appropriate mathematical models and symbol structures called (computational) *representations.* (Here we are using the term "assembly" to encompass assembly instances, as well as variational and nominal assemblies.) The *domain* of a representation scheme is the set of mathematical models to which it applies. A representation is *valid* if it corresponds to at least one (geometrically feasible) assembly, and is *unambiguous* if it corresponds to only one assembly.

An assembly instance can be represented unambiguously by a collection of *solid models* (i.e., unambiguous representations for rigid solids) for its components, plus the associated geometric transformations that define the parts' poses. Solid modeling is a relatively mature technology (reviewed briefly in Section 2.3), and the representation of transformations by $4 \times 4$ matrices in homogeneous coordinates or by other means is well understood.

A representation for an assembly instance in terms of solid models and, say, $4 \times 4$ matrices of *numerical* elements is valid if the solid models and transformations are themselves valid, and if the corresponding configuration is geometrically realizable. Non-interference can be tested through pairwise regularized intersections between all of the components. Regularized intersections are provided in most of the modern solid modelers, although substantial amounts of computation are involved. Testing for path existence is much more complicated. We do not know of general algorithms capable of establishing that no path exists. In the current state of the art an assembly planner must be invoked. If it fails to find a path, one concludes that the assembly is likely to be unrealizable. Because extant assembly planners make a variety of restrictive assumptions (e.g., assembly paths must be straight lines, only one part is moved in each operation, and so on), planner failure does not guarantee geometric infeasibility. Success does ensure path existence.

Most of the commercially-available Computer Aided Design (CAD) systems provide facilities for defining assemblies through direct, explicit specification of poses. This approach has two major drawbacks:

- Assembly representations typically are constructed by human designers, and it is difficult to define explicitly the required transformations.
- A specific transformation defines a single point in pose configuration space. Therefore, the approach cannot describe articulated assemblies with moving parts.

A better approach is to define the poses indirectly, through *mating relations* between *surface features*, which are subsets of the parts boundaries. Mating relations establish geometric constraints between parts, and are closely related to the mechanical behavior of assemblies. Designers typically find mating relations a natural way of specifying assemblies. (Mating relations and similar concepts have been called in the literature "joints", "connections", "liaisons", "technologically and topologically related surfaces", and so forth.)

Indirect, constraint-based definition of assemblies raises delicate issues. For example: is a representation unambiguous? Whereas directly-specified poses obviously correspond to unambiguous representations, an indirect specification may correspond to a single pose configuration, to several, or to none at all. A *constraint satisfaction*, or *constraint evaluation* problem must be analyzed to determine if there are solutions, and if these are unique. Mating relations and constraint satisfaction are discussed in more detail in Section 2.4.

Variational, and even nominal, assemblies are complex mathematical objects that may involve complicated subsets of high-dimensional configuration spaces. How can such entities be represented computationally? Again, mating relations provide an answer. A mating relation may specify, for example, that two planar surfaces remain in contact. This constraint can be expressed in terms of a $4 \times 4$ matrix that defines the pose of one surface relative to the other. The matrix contains *symbolic* variables corresponding to the degrees of freedom not fixed by the constraint. In the example cited above three variables are needed, because there are two translational and one rotational degrees of freedom in a planar contact relation.

The validity of a nominal assembly representation is difficult to establish computationally, because it implies that all the corresponding assembly instances must be geometrically realizable. Even non-interference is difficult to test. If there is only one rigid body moving with respect to another, the motion is collision-free when the volume swept by the moving object does not intersect the other. (However, swept volumes are difficult to compute for complex, curved objects.) If several objects move simultaneously, ordinary swept-volume analysis is insufficient, and 4-D space-time sweeps or "extrusions" must be considered [7].

A variational assembly may be represented by a collection of *toleranced* parts connected through mating relations. Tolerancing (discussed in Section 2.5) is a method for representing variational classes through geometric constraints on part features. There are national (ANSI) and international (ISO) tolerancing standards. These are sometimes ambiguous, but roughly equivalent mathematical tolerancing theories have been proposed [33, 34], and ANSI has recently appointed committee Y14.5.1, charged with the task of defining mathematically the meaning of tolerance specifications.

When part variability is taken into account, it makes sense to consider mating relations that also involve geometric uncertainty. For example, specifying that a shaft and a bearing have two concentric cylindrical surfaces in contact is not an accurate description of the assembly for detailed analysis. Perfect contact would prevent the shaft from turning because of friction. Also, two imperfect cylinders in general can neither be in perfect contact nor be perfectly concentric. (What does concentricity mean for imperfect cylinders?) The functional requirements are for concentricity within some tolerance, and for a clearance (instead of contact) within some range. We do not know of standard means for representing geometric uncertainty between different parts in an assembly, but direct extensions of single-part tolerancing methods may be adequate.

Consider a variational assembly representation consisting of a set of variational classes, defined by toleranced solids, and of mating relations. Is the representation valid? For validity each possible combination of acceptable parts, that is, each $\sigma$ in $\Sigma = V_1 \times V_2 \times \cdots \times V_n$, must correspond to a geometrically realizable assembly for every pose configuration $\tau$ that satisfies the mating relations. Non-interference testing for a variational assembly is an exercise in *worst-case tolerance analysis*, discussed in Section 2.5. Path-existence testing involves assembly planning in the presence of geometric uncertainties.

Representations for stochastic and variational assemblies are similar. Stochastic assemblies require the additional specification of p.d.f.'s (probability density functions) to define $\Omega_a$. Typically, p.d.f.'s are associated to the characteristic parameters of each of the parts in the assembly. (This is not the most general approach possible, but it is the only one used in current practice, insofar as we know.) P.d.f. specification is done by selecting a specific statistical distribution (e.g., Gaussian) and assigning numeric values to its parameters (e.g., mean and standard deviation). Testing for (probable) geometric realizability involves *statistical tolerance analysis*, discussed in Section 2.5, and assembly planning under uncertainty.

In current industrial practice, a designer considers a critical requirement of an assembly (for example, a certain clearance between a pin and a hole) and converts it into tolerances associated with each of the mating parts through a process usually called *tolerance allocation* or *tolerance synthesis* [8]. The component tolerances are represented in engineering drawings or their electronic counterparts, but the critical assembly requirements usually are *not*. We believe that assembly representations should contain *both* assembly and component requirements, for the following reasons. The assembly requirements are insufficient. For example, a pin/hole clearance can be achieved by tolerancing the pin tightly and the hole loosely, or vice-versa, or by distributing the tolerances approximately equally between the two components. It is important to distinguish the three approaches because they have significant manufacturing and cost implications. The component tolerances also are insufficient. Without the explicit representation of the assembly-level requirements it is impossible to verify if they are indeed satisfied. Furthermore, should the component allocations need to be changed because of manufacturing or other life-cycle considerations, it is impossible to modify them automatically without knowledge of the assembly requirements.

If a representation contains both assembly-level tolerances (typically associated with mating relations), and component-level tolerances, it is important to keep the two sets logically separate. Together with untoleranced solid representations and mating relations, assembly-level tolerances define a variational assembly $\mathbf{A}_a$, whereas the component-level tolerances define another variational assembly $\mathbf{A}_c$. If $\mathbf{A}_a = \mathbf{A}_c$ the representation is redundant, and the assembly-level information is useful primarily when the design is modified, for example by changing the allocation of tolerances between individual components. But, in practice, the two sets of constraints often are *not* equivalent. A representation containing both assembly and component tolerances defines a variational assembly $\mathbf{A}_a \cap \mathbf{A}_c$. It is useful to introduce a notion related to validity, called *internal consistency*, to characterize assembly representations whose component-level constraints suffice to ensure that the assembly-level requirements are satisfied. A representation is *worst-case* internally consistent if $\mathbf{A}_a \supset \mathbf{A}_c$, and *statistically* internally consistent if the probability associated with $\mathbf{A}_c - \mathbf{A}_a$ is below a specified threshold. Internal consistency can be assessed by worst-case or statistical tolerance analysis, discussed in Section 2.5.

The concept of internal consistency may be enlarged so as to encompass other relationships between component-level and assembly-level data. For example, an assembly representation in which a square pin and a square hole mate through a kinematic revolute joint is internally inconsistent.

Not all assembly requirements are of a spatial nature, and even those which are geometric may not be expressible directly through mating relations. For example, one of the main requirements for a pick-and-place mechanism is that the end effector follow a specified trajectory, within a band of acceptable error. Another requirement is that the velocity have some specified range. This example shows that there is a fine line between assembly requirements and behavioral or functional characteristics of a product, and that it is not clear where the line should be drawn. We believe that all this information should be captured in the representation of a product, but not necessarily as part of what we are calling an assembly representation.

Let us turn now to non-geometric information that is directly relevant to assembling operations and cannot reasonably be inferred from other characteristics of the assembly. There are many examples: presence of adhesives or lubricants; welding data; fastener types; torques and forces required; special tools. We believe that all such information can be represented through *attributes* associated with mating relations.

We do not have mathematical models that encompass the non-geometric aspects of assemblies, and therefore formal definitions of unambiguity, validity, and so on, are not applicable. But an informal notion of validity is still useful. Establishing the validity of non-geometric data is a complex problem that may involve physical reasoning and an extensive base of experiential knowledge. For example, how are we to decide if the specification of a certain adhesive is valid?

## 2.3 Solid models and surface features

A solid model is an unambiguous computer representation for a physical solid object, modeled mathematically as an $r$-set [32]. Although many schemes exist for representing solids, the most useful are Constructive Solid Geometry (CSG) and Boundary Representation (BRep). Much has been written, rightly and wrongly, about the virtues of each of these schemes. We believe that both are important and have complementary characteristics. The modelers we build contain both.

Solids are represented in CSG by directed, rooted, acyclic graphs whose internal nodes correspond to regularized set operators or rigid motions, and whose terminal nodes correspond to primitive solids such as blocks, cylinders, or "sculptured", "free-form" solids. The primitives themselves typically are represented by a "type code" (for example, "block" or "cylinder") plus an $n$-tuple of parameters. For example, the parameters for a cylinder might be 7 real numbers, 2 defining the cylinder's size (i.e. height and radius) and the rest defining its position, with 3 corresponding to the coordinates of the center of a base, and the other 2 defining the direction of a vector aligned with the cylinder's axis.

The PADL-2 modeler [6] and some of the modern commercial systems can accommodate unevaluated, symbolic parameters for primitives and rigid motions. AM, an experimental assembly modeler under development at the University of Southern California's Programmable Automation Laboratory, admits as parameters arbitrary LISP expressions and functions. Symbolic parameters constitute a powerful representational capability. They can be used to establish constraints between objects' surfaces, to define object families, and to represent nominal assemblies through symbolic rigid-motion parameters. A specific instance of a solid in a parameterized family defined through CSG is constructed by binding numeric values to the symbolic variables, and evaluating the parameter expressions. Under very simple conditions (e.g., the size parameters for the primitives must be positive) the instantiated object is valid, that is, it has a corresponding $r$-set. This makes CSG-based parameterizations very attractive.

A BRep represents the topological boundary of an object through a graph whose nodes correspond to faces, edges and vertices, and whose arcs correspond to adjacency relations. BReps also can be parameterized, but this raises delicate problems. For example, certain combinations of parameter values may be inconsistent with the BRep structure. New faces, edges or vertices may have to be introduced or old ones deleted for the representations of certain object instances to be valid.

Mating relations and tolerance specifications are associated with constraints on subsets of the boundaries of solids. Most of these constraints apply to *surface features* of objects, although lower-dimensional entities are sometimes needed. A surface feature usually is a face, or a union of faces, and in rare cases it may be a subset or union of subsets of faces. For example, a flatness tolerance may apply only to a small region of a planar face because a mating part will contact the face only on that region. (Edge or vertex features can be defined in terms of surface features, and will be ignored in the sequel.)

BReps represent faces explicitly, and therefore can easily be extended to cater to surface features. But surface features also can be represented in CSG. The boundary of a CSG solid is a subset of the union of the boundaries of the primitives in the CSG representation. This implies that an object's (BRep) faces can always be associated with one or more *primitive faces.* Instead of representing a surface feature directly through a BRep node, we can represent it indirectly by the primitive face or faces that give rise to it. We need methods for representing faces of primitives, and for combining these, through a union operation, into larger features when necessary. A specific scheme, called VGraph, for representing surface features in terms of CSG and assigning them tolerances is discussed in [36], and has been implemented in an experimental version of PADL-2 and also in the AM system. Surface feature representations based on CSG are more complicated than their BRep counterparts, but offer advantages in dealing with *parameterized* families of objects.

## 2.4 Mating relations and constraint satisfaction

To support the computations needed to display assemblies, test them for interference, assess their stability, and for other applications, assembly instances must be represented by their component solids plus explicit, numerically-valued transformations with respect to a common or "lab" coordinate frame. A pose representation scheme based on $4 \times 4$ matrices, quaternions, or other methods, defines a set of *natural parameters* (e.g., Euler angles, rotation angles about the

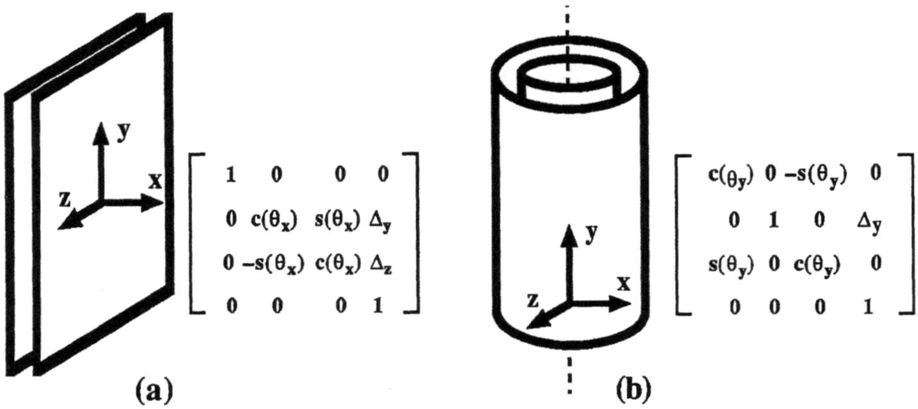


Figure 2.2: Coplanarity (a) and coaxiality (b) constraints and corresponding matrices with rotational (e.g. $\theta_x$) and translational (e.g. $\Delta_y$) free variables. ($c(\theta)$ and $s(\theta)$ denote $\cos\theta$ and $\sin\theta$.)

principal axes) that characterize unambiguously the poses of the components of an assembly in lab coordinates. But direct specification of natural pose parameters of parts or their surface features has drawbacks, as noted earlier. Indirect specification through relative distances, angles, and geometric constraints such as coplanarity, parallellism, and coaxiality is much more attractive. Some of these constraints (e.g., parallelism) are applied primarily to surface features of a single part, whereas others (e.g., coplanarity) correspond to mating relations between parts. All of these geometric constraints can be expressed as (often non-linear) *equations* on the natural pose parameters of surface features.

Mating relations between surface features can be described by *static* geometric constraints such as those just discussed. For example, "against" and "fits" conditions, which are equivalent to coplanarity and coaxiality for planar and cylindrical features, have been used in the RAPT system [1, 29] and in [21, 22]. Static constraints typically do not fix all of the degrees of freedom of a feature. For example, coaxiality between the cylindrical surfaces of a pin and a hole allows rotary motion about the axis, and translational motion along the axis direction. Static constraints between features of known geometry can be expressed as transformations with *symbolic* parameters that correspond to the degrees of freedom. Examples are provided in Figure 2.2. Each transformation maps a coordinate frame attached rigidly to one feature onto another frame attached to the other feature.

Mating relations also may be defined by *kinematic* constraints, which specify explicitly the desired relative motion between two features [19, 20, 27, 44]. For example, one might specify a translational or *prismatic* joint between a square bar and a square "hole". Kinematic constraints can be converted directly into symbolic-parameter relative transformations, and therefore are mathematically equivalent to their static counterparts. Nevertheless, we favor assembly representation schemes that support both static and kinematic constraints, because they help in capturing design intent, and therefore should make it easier to design an assembly or to reason about it. A kinematic constraint is closely related to the mechanical function and behavior a designer wants to achieve. In fact, a designer is likely to think first about the type of joint he or she wants to specify, and only later consider the detailed geometry of the surface features that "implement" the joint. Clearly, kinematic constraint information is available at the design stage, and can be easily captured if suitable user interfaces are provided. Note, however, that kinematic constraint specification raises the issue of consistency with features' geometry. For example, a rotary joint is incompatible with a square pin and hole geometry. The poses of components in a rigid assembly also can be defined through kinematic constraints, but a static-constraint specification is more natural, because no motion is intended.

We have been discussing *bi-directional* constraints. For example, coaxiality between a pin and a hole is a symmetric relation. If the pin's position were to change, the hole would have to move for the constraint to be maintained; similarly, a hole positional change would cause a corresponding change in the pin's pose. Alternatively, one can consider *uni-directional* constraints, which are akin to sequential operations, and sometimes are called *relative positioning* operations [13, 38, 45]. Uni-directional geometric constraints often can be captured by assigning to the pose parameters of a "target" feature the values of symbolic expressions involving the parameters of previously-defined "source" features. Figure 2.3 shows a very simple example. The left face of the small block $B$ can be constrained to be coplanar with the middle face of the L-shape $A$ if the position of $B$ is determined by evaluating the expression $c = a - b$. Observe that changes in $a$ or $b$ are correctly propagated to block $B$, and the constraint is enforced. However, a direct change in $c$ will not be reflected back to object $A$, and will produce a configuration that does not satisfy the coplanarity constraint. Expressions such as $c = a - b$ are not treated as equations, and it is not possible in this scheme to solve for $a$ given values for the other two parameters. Relative positioning via parameter expressions was implemented in the PADL-1 solid modeler through "distance chains" relating surfaces or half-spaces of objects defined by CSG [47], and is supported in PADL-2 [6] and some commercial modelers.

Unidirectional constraints are computationally convenient and surprisingly powerful, but have several drawbacks: (1) a sophisticated interface is needed to make the approach palatable to human users; (2) because constraints apply sequentially, previously-established relations may be broken unless special precautions are taken [19, 20, 38]; (3) complex constraints that correspond to systems of simultaneous equations are difficult, if not impossible to accommo-

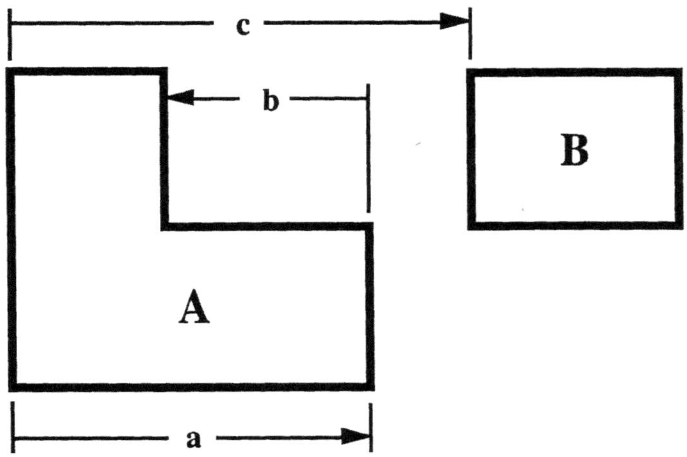


Figure 2.3: Uni-directional constraints and parameter expressions

date.

There is a substantial amount of additional literature on geometric constaints in the computer graphics area, from Sutherland's Sketchpad [41] to object-oriented approaches such as Borning's ThingLab [3]—see [38], which contains many references.

Consider now rigid assemblies defined through networks of mating relations. *Constraint satisfaction* methods analyze the networks to determine if they define a unique solution in pose configuration space, or if there are many solutions, or perhaps none, and to compute the solutions. The geometric constraints can be converted into a set of equations on the pose parameters of the surface features involved. An elegant technique for finding the relevant equations involves extracting cycles from the constraint network [1, 29]. In essence, constraint satisfaction amounts to studying the roots of a set of (nonlinear) equations.

Numerical solutions may be sought by using, for example, modified versions of Newton-Raphson iteration [21, 25, 37]. Modifications are needed because the number of equations often is larger than the number of unknowns, and redundant equations must be identified. Numerical solution of large systems of nonlinear equations raises several delicate issues, which include: (1) the process may fail to converge; (2) only one solution may be found when several exist; (3) the algorithm's behavior depends on the initial guess for the solution; and (4) the computation may be costly. Furthermore, if an assembly is articulated, the solutions contain higher-dimensional sets that correspond to the motions of the mechanism, and numerical equation solvers do not provide much useful information about the degrees of freedom of the assembly or about its motions.

An alternative approach consists of manipulating the equations symbolically [1, 29]. Casting the problem in algebraic terms enables powerful techniques to be deployed, for example Gröbner basis computation [16]. These techniques provide information about the entire set of solutions, its dimensionality, and so forth. Symbolic elimination and simplification methods produce results containing symbolic parameters that correspond to the degrees of freedom of an assembly. Unfortunately, Gröbner basis calculation and related symbolic algebra algorithms use rational arithmetic and are notoriously slow.

To mitigate some of the drawbacks of numerical and symbolic constraint satisfaction algorithms, a variety of heuristics and special-case short cuts have been proposed [1, 18, 29]. Sequential, uni-directional constraint satisfaction methods [38, 46] are computationally attractive, and the parameter expression approach is by far the fastest, since no equations are solved. But these methods have their own drawbacks, discussed earlier.

Recently, group theoretic methods have been applied to study the degrees of freedom of assemblies [26, 30, 43]. The key observation is that contact between two surfaces is maintained when the surfaces undergo rigid motions *only if the rigid motions leave the surfaces invariant.* For example, two cylinders in contact permit rotations around the axis and translations along the axis direction. These are precisely the rigid motions that map a cylinder onto itself, that is, the motions under which a cylinder is invariant—and therefore correspond to the *symmetries* of the cylinder. The symmetries of a feature are associated with a subgroup of the group of all rigid motions in Euclidean space. When parts are connected by several mating relations, their degrees of freedom may be computed by intersecting the corresponding symmetry groups. All the possible mating relations among a set of parts can be inferred by reasoning about symmetry groups. However, in our opinion, it is more reasonable to capture such relations at the design stage, since they are known to a designer even before the detailed geometry of the parts is specified.

In summary, assembly representations through mating relations give rise to networks of spatial constraints, and these are intimately associated with systems of non-linear equations, with all their inherent difficulties. Recent work by Kramer illustrates the current state of the art [19, 20]. Kramer combines symbolic and numerical methods, and converts bi-directional constraints into uni-directional relations for efficient solution. Finally, note that most of the research on geometric constraint satisfaction has been devoted to equality constraints (but see [28, 46]). Inequalities are important for dealing with geometric uncertainty.

## 2.5 Variational classes and tolerance analysis

Tolerances define permissible variations in the geometry of parts. A *variational class* is the set of solids that satisfies a given tolerance specification [33, 34]. (Note that other authors use these terms with a different meaning [45].) All the solids in a variational class should be "almost equal" in a suitable metric [5], functionally equivalent, and interchangeable in assembly operations.

Tolerance specifications amount to geometric constraints on the size, pose and form of subsets of a part's boundary. Typically they apply to surface features. The precise meaning of tolerance specifications is a topic of active research. There are two main approaches for defining tolerancing semantics:

- Shape and pose parameterization.
- Tolerance zone specification, which may be parametric or non-parametric.

We will explain these approaches with the help of a very simple 2-D example. Consider a planar quadrilateral polygon. If we assume that adjacent sides meet precisely at right angles, we have a perfect rectangle, which can be characterized completely by two parameters, its length $L$ and height $H$. We have thus defined a family of objects with an associated 2-D *parameter space*, which we can identify with $M_s$, the solid modeling space for this example. A pair $(L, H)$ is a point in this space and therefore it defines a specific rectangle instance. A tolerance specification corresponds to a subset of $M_s$. Typically, the subset is an interval $(L-\Delta L, L+\Delta L) \times (H-\Delta H, H+\Delta H)$. But other, more complex subsets may be defined indirectly, through constraints on entities that depend on the two parameters [45].

For a richer variational class we may relax the assumption of perfect orthogonality, and introduce four more parameters $\theta_1, \ldots, \theta_4$, which are the angles between adjacent sides. The "length" and "width" must be redefined as distances between specified vertices. A variational class now corresponds to a subset of a 6-D parameter space. This variational class includes quadrilaterals similar to that shown in Figure 2.4, which does not satisfy the earlier, perfect-orientation specification.

What we have done in both examples is to parameterize the poses of four straight lines, and constrain the poses through their associated parameters. The form of the lines is assumed perfectly straight. Perfect form is a reasonable first approximation, but a more refined tolerance specification must acknowledge that surfaces cannot be manufactured with perfect shapes. Imperfect form can be accommodated by using higher-order approximations. For example, we can replace the straight lines by second-degree curves (conics), and introduce additional parameters to define the conics. Alternatively a spline can be used. A tolerance specification still corresponds to a subset of parameter space, but the space's dimensionality has increased. Shape and pose parameterization is a

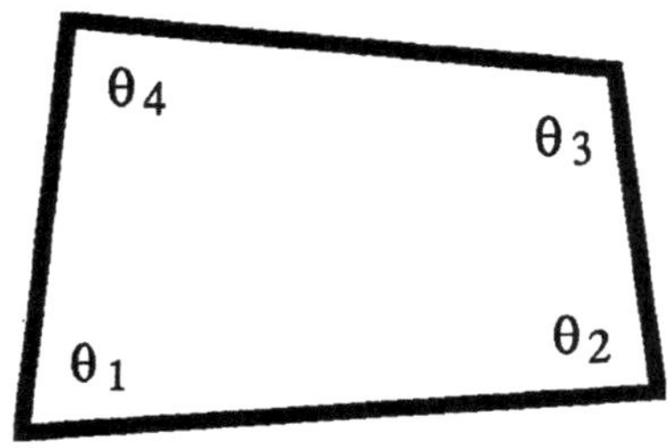


Figure 2.4: A quadrilateral with non-orthogonal sides

reasonable approach for perfect-form tolerancing, but its extension to imperfect form leads to a large number of parameters of dubious physical significance.

In contrast, the tolerance zone approach makes extremely mild assumptions about the nature of the surfaces involved. They are required only to "vary slowly" at the scale of the tolerance values specified [4, 33]. Tolerancing constraints in this approach are translated into set inclusion relations. Typically, a surface feature of an object is required to lie in a region of space called a tolerance zone. These zones may be constructed parametrically or non-parametrically [34]. We illustrate the two possibilities with an imperfect rectangle–see Figure 2.5. First we parameterize the rectangle by its length and height, as before, and specify an admissible range for each of the parameters. Next we construct the largest and smallest rectangles in the specified parameter range, and define a tolerance zone as their set difference, shown in Figure 2.5-a. Any object with a (slowly-varying) boundary in the tolerance zone is considered acceptable. In the non-parametric approach illustrated in Figure 2.5-b, we grow and shrink a perfect rectangle by specified amounts, and subtract the results to define the tolerance zones. The main distinction between parametric and non-parametric zones is the growing and shrinking method used. Instead of considering maximal and minimal values for parameters, expansion and contraction are achieved in the non-parametric approach through solid offsets [39], which are special cases of Minkowski operations or sweeps. The two approaches produce slightly different tolerance zones, as shown in the figure, but the differences do not seem to be practically important. Constructing tolerance zones for all the specifications used in practice is non-trivial, but an adequate theory is emerging [9, 17, 33, 40]. Note also that a tolerance zone specification may be converted into a set of constraints on parameters, *if* we assume that the

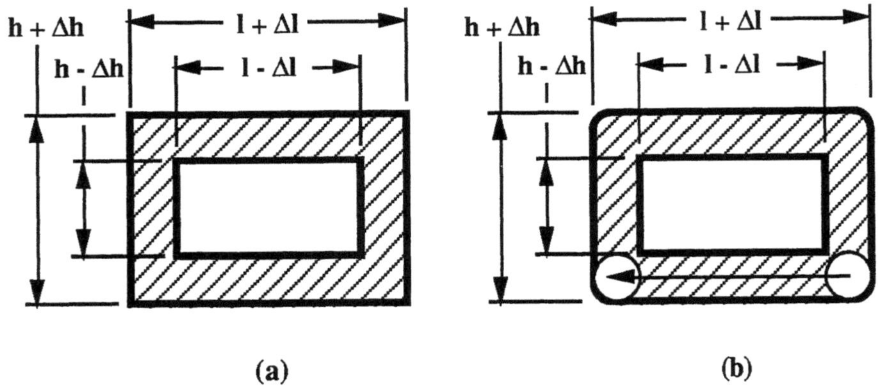


Figure 2.5: Parametric (a) and non-parametric (b) tolerance zones

surfaces of actual parts have specific, parameterized shapes.

Representations for variational classes depend on the specific approach adopted for tolerance semantics. The non-parametric theory requires representations for nominal solids, surface features, and attributes that constrain such features [36]. Shape parameterization, as well as parametric tolerance zones, have an additional requirement: feature and object representations must be parameterized.

Almost all of the extant tolerance analysis algorithms assume a semantics for tolerances in terms of shape and pose parameterizations. Typically, range or limit specifications are given for a few parameters or *dimensions* $d_i$, and the corresponding range is computed for a resulting dimension $d_r = f(d_i)$. Figure 2.6 shows a simple example. The resulting dimension is $d = a - (b + c)$. Given limits for $a, b$ and $c$, what are the corresponding limits for $d$? Observe that the relative location of the right face of the slot in the figure with respect to the left face of the slot can be defined either as the single distance $d$ or by the "chain" $-b + a - c$. (To construct this chain go left from the left face of the slot, then right to the rightmost face of the part, and then left to the right face of the slot.) In tolerancing jargon, two chains of dimensions associated with a feature constitute a *loop*. More generally, the given dimensions need not be aligned along a single direction, and a more complicated *vector loop* must be constructed to relate the relevant parameters. However, most systems deal only with linear loops of dimensions.

In most tolerance analysis programs the function $f$ that relates resulting and given dimensions must be specified by a user, either as a closed-form expression or by a procedure for computing $d_r$ from the other $d_i$. In some systems $f$ is

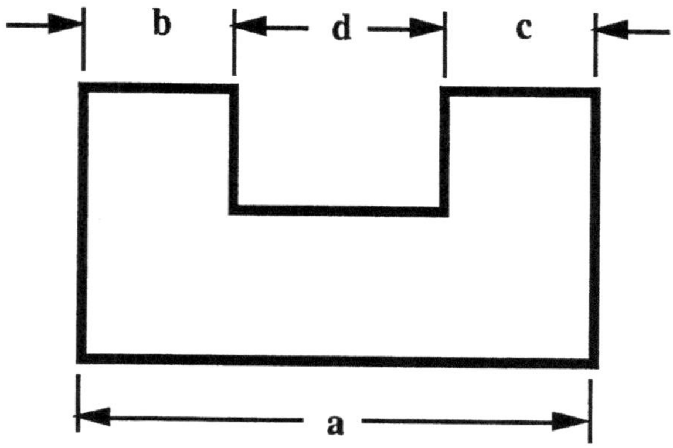


Figure 2.6: An indirectly-toleranced slot

defined implicitly by a simulation procedure that generates representations of parts in tolerance, "assembles" them, and "measures" the resulting dimensions [45].

Worst case analysis involves finding the minimal and maximal values for parameters, and therefore is an optimization problem. The equations that relate the resulting dimension to the given ones may be non-linear, when angular relationships are involved. Because the variations in the parameters are small, the equations often can be linearized, and the optimization carried out by linear programming [15, 42, 45].

Specific analysis problems must be formulated by human users so as to reflect critical assembly requirements, and the results of the analysis also must be interpreted by humans in most of the existing tolerance analysis programs. For example, the slot size in the example above may preclude assembly with a mating part, but such an inference is beyond the capabilities of typical industrial systems. Increasing automation is being demonstrated in research systems [45].

The same vector-loop considerations can be used for statistical tolerance analysis. Now the part dimensions are viewed as random variables. P.d.f.'s for the given dimensions are specified by a user, and the statistics of the resulting dimension are computed by using analytic statistical methods [2, 8, 23], or numerically, by Las Vegas techniques (which are the U.S. equivalent of old-world Monte Carlo analysis) [14, 45]. The traditional approach to tolerance analysis is well described in [2], which is a revised version of a report dating from the mid 1970's.

Judicious formulation of resulting dimensions coupled with worst-case or statistical tolerance analysis can go a long way towards establishing non-interference or internal consistency of assembly representations. In essence, traditional tolerance analysis seeks to show that constraints on components suffice to guarantee that assembly requirements are met, and therefore is closely related to internal consistency issues.

Tolerance analysis algorithms based on the tolerance zone approach are still at the research stage. The most interesting results thus far are reported in [10, 11, 12]. Work under way at the University of Southern California seeks to compute tolerance zones and verify clearance and fit conditions by using the ideas outlined in [35]. The tolerance zone approach deals naturally with imperfect form. Coupled with assembly representations that contain the critical assembly requirements, it is expected to lead to highly automated and powerful tolerance analysis systems.

## 2.6 Summary and open issues

This chapter introduced mathematical models for (the geometric aspects of) assemblies in terms of configuration spaces whose elements correspond to collections of solid mechanical parts and their poses (positions and orientations). Rigid assemblies were considered, as well as articulated mechanisms, with and without part variability, and with and without randomness. Sharp characterizations for the subsets of configuration space that correspond to physical assemblies are unknown.

Computational representations for assemblies proposed in the literature consist essentially of (toleranced) solid models for the component parts, mating relations between surface features, and attributes of such relations. Attributes establish geometric and non-geometric constraints on the assembly.

The validity and internal consistency of assembly representations raise a host of very complex problems, many of which require substantial mathematical and algorithmic development. These problems include interference calculations and path planning, to establish geometric realizability; constraint satisfaction, to find static poses and allowed motions of sets of parts connected by mating relations; and worst-case and statistical tolerance analysis, to check if part-level tolerance specifications ensure that assembly requirements are met.

Assembly representations should capture design requirements such as assembly clearances and desired motions. Some of this information can be inferred from the assembly geometry, but it seems more reasonable to capture requirements at the design stage, since they are known to the designer, than to re-create them later, in a process akin to reverse engineering. Computer aided *design* of assemblies was not considered in detail in this chapter but a good survey is available [24]. Two interesting issues raised by assembly design are the following.

1. Designers may proceed bottom-up, by constructing new part representations or using existing ones, and establishing mating relations and attributes. But they also may operate top-down. The first approach can be supported through relatively straightforward extensions of current solid modeling technology. The second, top-down approach is more challenging, because one must be able to ignore low-level details and establish relations between features of *incompletely specified* objects.

2. Assemblies are naturally decomposed by designers into subassemblies, typically through functional considerations. Subassemblies are not difficult to represent, and hierarchical structures can be combined with mating-relation graphs. However, the subassembly structure imposed by designers need not correspond to a desirable sequence of assembly tasks. In fact, some of the planners discussed in the following chapters assume a flat, non-hierarchical assembly structure and infer a suitable set of subassemblies associated with assembly operations.

The main conclusions of this chapter may be summarized as follows. The mathematical aspects of assembly modeling are not fully understood. Assembly representations through mating relations and attributes are reasonably well established, although most of the associated constraint satisfaction methods suffer from lack of generality, efficiency, or robustness (or all of the above). Properties of assembly representations such as validity and internal consistency involve a variety of complex, open issues. Algorithms for converting assembly representations into sequences of assembly operations are discussed in later chapters of this book.

# Acknowledgements

The authors were supported by the National Science Foundation under grants DMC-87-15404 and CDR-87-17322, by the industrial members of the Institute for Manufacturing and Automation Research (IMAR), and by the Industrial Associates of the Programmable Automation Laboratory, Institute for Robotics and Intelligent Systems (IRIS) of the University of Southern California.

# References

[1] A. P. Ambler and R. J. Popplestone, "Inferring the positions of bodies from specified spatial relationships", *Artificial Intelligence*, Vol. 6, No. 2, pp. 157-174, Summer 1975.

[2] Ø. Bjørke, *Computer-Aided Tolerancing*. New York: ASME Press, 2nd ed., 1989.

[3] A. Borning, "ThingLab – A constraint-oriented simulation laboratory", Ph.D. Dissertation, Dept. of Computer Science, Stanford University, July 1979.

[4] M. Boyer and N. F. Stewart, "Modelling spaces for toleranced objects", Département d'informatique et de recherche opérationnelle, Université de Montréal, July 1990.

[5] M. Boyer and N. F. Stewart, "Modelling spaces for toleranced objects: $\Re$-classes suitable for practical use", Département d'informatique et de recherche opérationnelle, Université de Montréal, November 1990.

[6] C. M. Brown, "PADL-2: A technical summary", *IEEE Computer Graphics and Applications*, Vol. 2, No. 2, pp. 69-84, March 1982.

[7] S. A. Cameron, "Modelling solids in motion", Ph.D. Dissertation, University of Edinburgh, 1984.

[8] K. W. Chase and W. H. Greenwood, "Design issues in mechanical tolerance analysis", *Manufacturing Review*, Vol. 1, No. 1, pp. 50-59, March 1988.

[9] F. Etesami, "Tolerance verification through manufactured part modeling", *Journal of Manufacturing Systems*, Vol. 7, No. 3, pp. 223-232, September 1988.

[10] A. Fleming, "Analysis of uncertainties in a structure of parts", *Proc. 9th Intl. Joint Conf. on Artificial Intelligence*, Los Angeles, CA, pp. 1113-1115, August 18-23, 1985.

[11] A. D. Fleming, "Analysis of uncertainties and geometric tolerances in assemblies of parts", Ph.D. Dissertation, Dept. of Artificial Intelligence, University of Edinburgh, 1987.

[12] A. Fleming, "Geometric relationships between toleranced features", *Artificial Intelligence*, Vol. 37, No. 1-3, pp. 403-412, December 1988.

[13] D. C. Gossard, R. P. Zuffante and H. Sakurai, "Representing dimensions, tolerances, and features in MCAE systems", *IEEE Computer Graphics & Applications*, Vol. 8, No. 2, pp. 51-59, March 1988.

[14] D. D. Grossman, "Monte Carlo simulation of tolerancing in discrete parts manufacturing and assembly", Computer Science Report Number STAN-CS-76-555, Stanford University, May 1976.

[15] P. Hoffman, "Analysis of tolerances and process inaccuracies in discrete part manufacturing", *Computer-Aided Design*, Vol. 14, No. 2, pp. 83-88, March 1982.

[16] C. M. Hoffmann, *Geometric and Solid Modeling.* San Mateo, CA: Morgan Kaufmann Publishers, 1989.

[17] R. Jayaraman and V. Srinivasan, " Geometric tolerancing: I. Virtual boundary requirements", *IBM Journal of Research and Development*, Vol. 33, No. 2, pp. 90-104, March 1989.

[18] H. Ko, "Empirical assembly planning: A learning approach", Ph.D. Dissertation, Dept. of Computer Science, University of Illinois at Urbana-Champaign, 1989.

[19] G. A. Kramer, "Solving geometric constraint systems", *Proc. 8th National Conf. on Artificial Intelligence*, Boston, MA, pp. 708-714, July 29-August 3, 1990.

[20] G. A. Kramer, "Geometric reasoning in the kinematic analysis of mechanisms", Ph.D. Dissertation (draft), University of Sussex, October 1990.

[21] K. Lee and G. Andrews, "Inference of the positions of components in an assembly: Part 2", *Computer Aided Design*, Vol. 17, No. 1, pp. 20-24, January/February 1985.

[22] K. Lee and D. C. Gossard, "A hierarchical data structure for representing assemblies: Part 1", *Computer Aided Design*, Vol. 17, No. 1, pp. 15-19, January/February 1985.

[23] W.-J. Lee and T. C. Woo, "Tolerances: Their analysis and synthesis", *Journal of Engineering for Industry*, Vol. 112, No. 2, pp. 113-121, May 1990.

[24] E. C. Libardi, J. R. Dixon and M. K. Simmons, "Computer environments for the design of mechanical assemblies: A research review", *Engineering with Computers*, Vol. 3, No. 3, pp. 121-136, Winter 1988.

[25] R. A. Light and D. C. Gossard, "Modification of geometric models through variational geometry", *Computer-Aided Design*, Vol. 14, No. 4, pp. 209-214, July 1982.

[26] Y. Liu and R. J. Popplestone, "Assembly feature-mating inference from solid models using symmetry groups", COINS Tech. Report 90-34, Computer and Information Science Dept., University of Massachusetts at Amherst, 1990.

[27] G. H. Morris and L. S. Haynes, "Robotic assembly by constraints", *Proc. 1987 IEEE Intl. Conf. on Robotics and Automation*, Raleigh, NC, pp. 1481-1486, March 31-April 3, 1987.

[28] G. Mullineaux, "Optimization scheme for assembling components", *Computer-Aided Design*, Vol. 19, No. 1, pp. 35-40, January/February 1987.

[29] R. J. Popplestone, A. P. Ambler and I. M. Bellos, "An interpreter for a language for describing assemblies", *Artificial Intelligence*, Vol. 14, No. 1, pp. 79-107, August 1980.

[30] R. J. Popplestone, Y. Liu and R. Weiss, "A group theoretic approach to assembly planning", *AI Magazine*, Vol. 11, No. 1, pp. 82-97, Spring 1990.

[31] A. A. G. Requicha, "Mathematical models of rigid solid objects", Tech. Memo. No. 28, Production Automation Project, Univ. of Rochester, November 1977.

[32] A. A. G. Requicha, "Representations for rigid solids: Theory, methods, and systems", *ACM Computing Surveys*, Vol. 12, No. 4, pp. 437-464, December 1980.

[33] A. A. G. Requicha, "Toward a theory of geometric tolerancing", *Intl. Journal of Robotics Research*, Vol. 2, No. 4, pp. 45-60, Winter 1983.

[34] A. A. G. Requicha, "Representation of tolerances in solid modeling: Issues and alternative approaches", in M. S. Pickett and J. W. Boyse, Eds., *Solid Modeling by Computers*. New York: Plenum Press, 1984, pp. 3-22.

[35] A. A. G. Requicha, "Solid modeling and its applications: Progress in tolerancing, inspection, and feature recognition", IRIS Tech. Report No. 259, Institute for Robotics and Intelligent Systems, University of Southern California, November 1989.

[36] A. A. G. Requicha and S. C. Chan, "Representation of geometric features, tolerances, and attributes in solid modelers based on constructive geometry", *IEEE Journal of Robotics and Automation*, Vol. RA-2, No. 3, pp. 156-186, September 1986.

[37] D. N. Rocheleau and K. Lee, "System for interactive assembly modelling", *Computer-Aided Design*, Vol. 19, No. 2, pp. 65-72, March 1987.

[38] J. R. Rossignac, "Constraints in constructive solid geometry", *Proc. 1986 Workshop on Interactive 3D Graphics*, Chapel Hill, NC, pp. 93-110, October 23-24, 1986.

[39] J. R. Rossignac and A. A. G. Requicha, "Offsetting operations in solid modelling", *Computer-Aided Geometric Design*, Vol. 3, No. 2, pp. 129-148, August 1986.

[40] V. Srinivasan and R. Jayaraman, "Geometric tolerancing: II. Conditional tolerances", *IBM Journal of Research and Development*, Vol. 33, No. 2, pp. 105-124, March 1989.

[41] I. E. Sutherland, "Sketchpad: A man-machine graphical communication system", Ph.D. Dissertation, Dept. of Electrical Engineering, Massachusetts Institute of Technology, January 1963.

[42] R. H. Taylor, "A synthesis of manipulation control programs from task level specifications", Ph. D. Dissertation, Dept. of Computer Science, Stanford University, July 1976.

[43] F. Thomas and C. Torras, "A group theoretic approach to the computation of symbolic part relations", *IEEE Journal of Robotics and Automation*, Vol. 4, No. 6, pp. 622-634, December 1988.

[44] R. B. Tilove, "Extending solid modeling systems for mechanism design and kinematic simulation", *IEEE Computer Graphics and Applications*, Vol. 3, No. 3., pp. 9-19, May/June 1983.

[45] J. U. Turner, "Tolerances in computer-aided geometric design", Ph.D. Dissertation, Dept. of Computer and System Engineering, Rensselaer Polytechnic Institute, May 1987.

[46] J. U. Turner, "Relative Positioning of parts in assemblies using mathematical programming", Tech. Report TR-89046, Rensselaer Polytechnic Institute, 1989.

[47] H. B. Voelcker, A. A. G. Requicha, E. E. Hartquist, W. B. Fisher, J. Metzger, R. B. Tilove, N. K. Birrell, W. A. Hunt, G. T. Armstrong, T. F. Check, R. Moote and J. McSweeney, "The PADL-1.0/2 system for defining and displaying solid objects", *Computer Graphics (Proc. Siggraph '87)*, Vol. 12, No. 3, pp. 257-263, August 1978.

# Chapter 3

# Representation of solid objects by a modular boundary model

**Leila De Floriani, Amitava Maulik, and George Nagy**

The geometric representation of man-made objects has always been considered essential for their design and construction. It is inconceivable that cathedrals, catapults, caravels and clockworks could have reached their level of perfection without the concurrent development of graphic tools as the *lingua franca* between designers, clients ("end-users"), and artisans. Drafting conventions were gradually refined and formalized according to the requirements of different disciplines (sheet metal, piping, trusses, part and assembly drawings, renderings). Till recently, "mechanical drawing" formed an important component of engineering and architectural education.

Early computer-aided drafting tools bore the same relation to engineering drawing as word processors to typing: they facilitated the preparation of neat, error-free prints, allowed storage and transmission in digital form, and speeded up immensely the updating of existing designs. Ancillary information, such as parts lists, machining directions, and surface finish, were kept in separate files and dimensions were treated as mere annotations. Electronic drawings contained the minimum amount of structural information: no other views could be displayed than those entered by the draftsman, and few additional properties (such as weight) could be calculated. This remained the case even after the introduction of wireframe models.

In the second phase of computerization, integral methods were developed to represent three-dimensional rigid objects themselves, rather than specific 2-D views of such objects. It was soon discovered that purely geometric coordinate information and surface equations left unresolved certain ambiguities: in particular, such location information is insufficient to differentiate the inside from the outside. This led to the introduction of data structures for representing simple topological concepts and relations. However, the integrity and consistency of 3-D representations could not be verified as easily as 2-D representations. It was therefore necessary to define transformations that, applied to a topologically valid object, guaranteed yielding another valid object. The computer models developed along these lines were sufficient to allow the manipulation and display of well-formed objects of arbitrary complexity, and form the basis of the current commercial systems.

The third phase of computerization, which includes the research reported here on modular boundary models, extended the above techniques to *families* of objects. The ability to combine objects has implications both in design, where the combination is conceptual, and in manufacturing, where the combination is physical. Useful combinations include not only juxtapositions, where objects do not spatially overlap, but combinations of interpenetrating components. In the latter case, voids (such as holes) are also considered objects, and multiple positive and negative objects may share the same space. One may equivalently consider a hole as the complement of the corresponding solid, or simply as a negative object. Naturally, the representation and verification of the validity of such modular objects becomes more complicated.

Another major limitation of classical CAD models has been their inability to describe form features and their relations. Pure solid modelers cannot be used for assembly and machining planning because they do not contain essential information that is most naturally associated with form features (i.e., tolerances and dimensions, materials, surface finish). *Modular boundary models* (MBMs) bridge the gap between the design and manufacturing phases through their ability to represent form features as model components [13,15]. Moreover, the MBM description of the object produced by the designer, in which components represent design features, can be locally modified (because of the modular nature of the model) to yield an MBM description in terms of manufacturing features.

Since modular boundary models are rooted in the boundary representation of monolithic objects, we first review alternative representation schemes for individual objects. Then, we introduce definitions for a class of modular boundary models for compound objects, and develop a specific model of that class, the

*Face-to-Face Composition* (FFC) model. The essential characteristics of this model are the explicit graph representation of the abutting and interpenetrating faces of the constituent object components, and the detailed representation of the resulting object in terms of a mutually exclusive, completely exhaustive, irregular cellular structure, called the *cellular model.*

Our high-level conceptual model (called the *FFC Graph*) allows the application of graph-theoretic tools to validity issues, including those that arise when the object is decomposed into its constituent components. In discussing validity issues, we find that the locality of information due to modularity assists greatly in establishing geometric as well as topological validity. A direct representation of the FFC graph in the form of adjacency relations is, however, too cumbersome for the complete low-level vertex, edge and facet information necessary for building and manipulating the model. This latter is thus relegated to the cellular representation.

Since boundary evaluation is a common requirement for any model, we first suggest an algorithm to evaluate the boundary of the final object. Then we describe a data structure appropriate for the cellular representation, which is an extension of Weiler's radial-edge structure [54], and show how the FFC model of an arbitrary compound object can be constructed. Finally, we introduce the *Production Graph* (based on the AND/OR assembly graph described in [14,25,48]), which shows the alternative sequences of material-removal and assembly operations for manufacturing the modeled object.

We conclude that modular boundary models fit in the long-established trend from human to programmatic computer utilization of models. First generation models could be used only for direct screen or plotter output. Wireframe models could be projected and, with some human intervention, generate numerical machine tool control code. Second-generation models were adequate for 3-D surface display under varying lighting and viewing conditions. The current generation of models is intended for the computer-integrated manufacturing paradigm, linking automated (feature-based) design, production engineering, and quality control.

However, before the ideas presented can be applied in an actual design and manufacturing environment, a number of additional problems must be solved. These include issues related to the numerical robustness of the various operations, the cost of the suggested data structure for objects typical of given applications, and the introduction of additional essential criteria for pruning the production graph to obtain feasible and economical manufacturing sequences. These issues are discussed briefly in the final section.

## 3.1 Object Representation Schemes

Several object representation schemes have been developed in the past for different applications. Interesting classifications and discussions of such schemes are given in [36,41,46]. Some object models are special-purpose, in the sense that they can effectively represent only special classes of objects. Examples of such models are sweep representations, generalized cylinders and cones, and blob models. General-purpose models are characterized by a larger descriptive power, and are those used in modern CAD/CAM systems.

The emphasis in object models for a CAD system used to be on the efficient performance of operations typical of a design environment. Examples of such operations were the creation of the object model, the visualization of the object on a graphic display, computation of integral properties, Boolean and interference computations. More recently, the emphasis has shifted to information for assembly and machining planning. In other words, the model should not only describe the geometry of a solid object, but be capable of representing information, like dimensions and tolerances, form features or surface finish, used in manufacturing. So, the new directions in research in solid modeling are towards more complete models which allow an explicit representation of tolerance and feature information.

Classical solid models for CAD/CAM applications can be broadly classified into *boundary*, *volumetric* and *hybrid* schemes. Boundary schemes describe an object in terms of the surfaces enclosing it. Volumetric schemes describe an object in terms of solid primitives covering its volume. Hybrid schemes combine the two approaches.

A *boundary representation* (B-Rep) of an object is a geometric and topological description of its boundary. The object boundary is segmented into a finite number of bounded subsets, called *faces*. Each face, in turn, is described by its bounding *edges* and *vertices*. Two other (non-primitive) elements are used to describe objects with multiply connected faces or internal cavities: the *loop* and the *shell*. A loop on a face $f$ is a closed chain of edges bounding $f$. A shell is defined as any maximal connected set of object faces. In a B-Rep a clear separation is made between geometry and topology. The geometric description consists of the shape and location in space of each of the primitive topological elements (vertices, edges and faces). Topological information is concerned with the adjacency relations between pairs of individual topological elements (25 relations, in total). Several data structures have been proposed to encode a B-Rep: the winged-edge structure [4], the symmetric structure [57], the face adjacency hypergraph [2]. They all store the five topological elements, but differ in the number and kind of relations they store. The radial edge structure, proposed and implemented by Weiler [54], is capable of describing also solids not bounded by two-manifold surfaces.

The descriptive power of a B-Rep depends on the surfaces used (planes, quadric or free-form surfaces). Boundary schemes can represent a wide variety of solid objects at arbitrary levels of detail. They are unambiguous, but generally not unique. Validity is quite difficult to establish. A topologically valid B-Rep can be constructed by the use of a limited set of primitive functions, called *Euler operators* [2,7,21,35,36]. Geometric validity must, however, be tested algorithmically.

Volumetric schemes can be classified into *decomposition* models, which describe an object as a collection of primitive objects combined with a single glueing operation, and *constructive* models, which describe an object as the Boolean combination of primitive point sets.

Decomposition schemes can be further subdivided into *object-based* and *space-based* schemes. The former describe an object as the combination of pairwise quasi-disjoint elementary cells whose union covers the object. Examples of such models are cell decompositions, like tetrahedralizations [5] mainly used for object reconstruction, and finite element meshes. Space-based schemes describe an object by subdividing the space into regular volume elements, called *voxels*. Examples are spatial enumerations and adaptive schemes, like the Octree or the Bintree [26,38,46,47,49,50]. Adaptive subdivision schemes achieve storage economy by combining neighboring voxels which are completely internal or completely external to the object. Space-based decomposition schemes provide only approximate object descriptions: the quality of the approximation is determined by a fixed resolution. Such representations are unambiguous and also unique, except for positional nonuniqueness: all space-based representations vary under rigid transformations. Octrees and Bintrees are especially interesting as auxiliary representations in a solid modeler, since Boolean operations and computation of integral properties can be done very efficiently on them. A general disadvantage of such schemes is the amount of storage required, even if pointerless linear representations have been developed [46].

Constructive schemes (*Constructive Solid Geometry*, CSG) [41,42,43] are a family of schemes for representing solids as Boolean combination of primitive components. A CSG model is described by a binary tree, called the *CSG tree*, in which internal nodes represent operators which can be either rigid motions or regularized set operators, while the terminal nodes are subsets of $E^3$. CSG schemes are unambiguous, but not unique. The validity of a CSG scheme can be checked at a purely syntactic level, provided that the primitives are bounded. The main disadvantages of a CSG are the difficulty in computing integral properties and extracting or describing information related to surfaces (surface finish, tolerances, etc.). Variants of the CSG tree have also been proposed. Wyvill and Kunii [58] devised the *CSG-DAG*, in which the Boolean operation allowed are set addition and set subtraction. This representation facilitates the construction of a spatial index, called a *PM-CSG tree*, on top of a CSG-DAG in order to speed up ray tracing on a constructive object description.

Hybrid models, as mentioned earlier, can be viewed as combinations of two different representation schemes. There are two major classes of hybrid schemes, *PM-Octrees* and *PM-CSG trees* (which combine an Octree with a boundary representation or a CSG model, respectively), and *Modular Boundary Models* (which combine a B-Rep with a constructive approach limited to a restricted set of Boolean operations). These latter will be discussed in the next section.

The major drawbacks of adaptive space-based decomposition models, like the Octree or the Bintree, are that they provide only approximate object descriptions, and contain a large number of nodes. Thus, several authors have proposed schemes that combine the Octree with a B-Rep by using the octree as a spatial index over the boundary description [3,8,9,12,20,23,39,46,47]. These schemes have different names, but their underlying principle is the same. Like an Octree, a *PM-Octree* is based on the recursive subdivision of a finite cubic universe containing the object into octants. Terminal nodes can be *full* or *void*, as in the octree, or they can be of type *face*, *edge* or *vertex*. Face nodes are crossed by a single object face, an edge node contains a portion of an edge together with the two faces incident on it, a vertex node contains exactly one vertex and portions of all the edges and faces incident on it. The main advantage of PM-Octrees is the simplicity of the algorithms for Boolean operations. Visualization and computation of integral properties can also be performed efficiently on such structures. PM-CSG trees are based on the same concept as the PM-Octree: the definition of a terminal node is changed to refer to a primitive object rather than to a boundary element [58].

Even by considering only the modeling operations performed in a CAD system, no representation scheme is uniformly best for all operations. So, many modeling systems maintain more than one object description. Octrees, and more recently, PM-Octrees, have been mainly used as a secondary representation to speed up Boolean operations and computation of integral properties. In many commercial systems, CSG has been used as an interface to the designer, while keeping a B-Rep as internal representation. Such modelers require conversion algorithms capable of translating data among the different schemes. A conversion algorithm must guarantee that the output is always correct and consistent with the input model. Also, ideally, the conversion should be completely invertible. This cannot happen when we convert from an exact model, like CSG or B-Rep, to any spatial enumeration model. A survey of conversion algorithms between representations can be found in [19].

Besides the "pure" solid models discussed above, which describe only the shape of the object, other object models containing also semantic information have been developed for machining and assembly. In this class, we include assembly models and feature-based models.

Generally speaking, assembly models describe an object as the composition of parts which must be combined to form the object, and the relative position and relations among the parts. They essentially differ in the kind of part-to-part relations they describe. Braid [6] defines a tree-like assembly structure, in which the terminal nodes are assembly components, described in a boundary form, and non-terminal nodes represent assembly operations. Assembly operations could be of type *collective* (i.e., placing components side by side), *conjunctive* or *disjunctive* (i.e., Boolean operations).

An early attempt at modeling the assembly of components was made in the AUTOPASS project [34]. The assembly model is a graph, in which the nodes are geometric objects (described by polyhedra) and the arcs represent four relations: *part-of*, *attachment*, *constraint* and *assembly-component*. More recently, an improved representation has been proposed by Lee and Gossard [32]. It is a hierarchical representation in which an assembly consists of subassemblies and components. Each component is described in a boundary form, and two mating relations are introduced: *against* (abutting planar faces) and *fits* (center lines of the two parts are collinear). This structure has been improved again by Ko and Lee [30] by adding additional mating conditions. Turner proposed an assembly model specifically developed for tolerance-based design [52].

Sanderson and Homem de Mello [48] discuss a set of algorithms and a relational scheme to generate a representation of all "feasible" assembly sequences. When given as input the pairwise relations (contacts, attachments, etc.) between the components, the scheme can handle assemblies which are constructed by combining two subassemblies at a time. The input relational model is converted to a graph. The cut-sets that correspond to feasible disassemblies are determined from the geometric feasibility, mechanical feasibility, and stability predicates defined by the authors. These predicates essentially test the possibility of local incremental translations, accessibility of the attachments, and gravitational stability. All the feasible disassemblies are then represented in the form of an AND/OR graph. The authors propose searching this graph for solution trees that represent either complete or partial assembly sequences. Partial sequences are used for the replacement of failed parts, whose identity cannot be predicted. We will make full use of these ideas in our model.

Lee and Shin [33] discuss a co-operative planning system with job-specific advisor routines to determine a partial-order graph for automatic generation of assembly sequences with a high degree of parallelism. The face-face contact between compatible surfaces of neighboring parts, and blocking relationships, where one part blocks the disassembly of another without actually touching it, are provided by the designer. The constraints imposed by these relationships are used by the *Geometric Reasoner* to determine the ranges of movements of each part. Then the *Heuristic Advisor* uses grouping heuristics and suggests tentative decompositions, which are checked for interference by the geometric reasoner. The *Plan Co-ordinator* takes the list of accepted decompositions and determines those tasks that can be accomplished most easily. Other advisors determine resources and their availability. This decomposition process is performed iteratively till the decompositions of all subassemblies are determined.

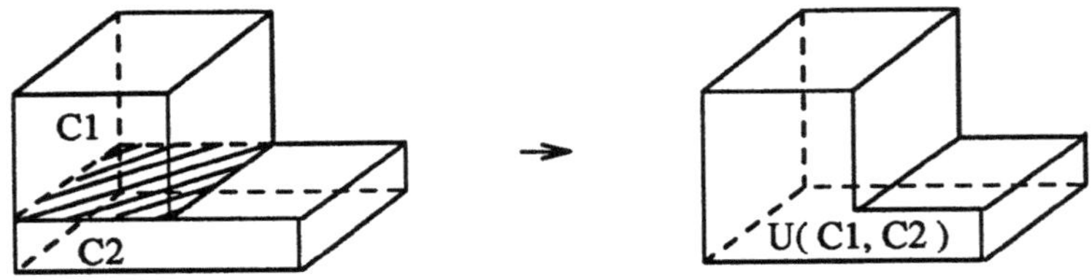


a) Disjoint Union of two cuboids $C_1$ and $C_2$.

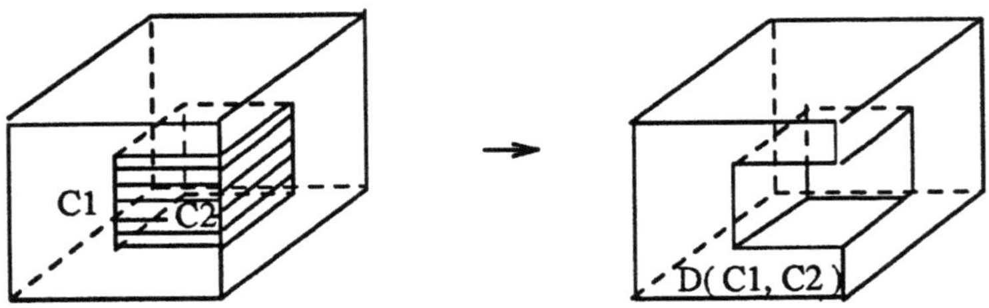


b) Restricted Difference of two cuboids $C_1$ and $C_2$.

Figure 3.1: Face-based operations on two cuboids $C_1$ and $C_2$.

U($C_i$,$C_j$), is the solid object defined as the set addition of $C_i$ and $C_j$. Given two components $C_i$ and $C_j$ such that $C_j$ is contained in $C_i$ and their boundaries intersect at some faces, then the Restricted Difference of $C_i$ and $C_j$, denoted D($C_i$, $C_j$), is the solid object defined as the set difference $C_i - C_j$. Thus, the two previous operators can be applied only to pairs of components such that their boundaries have a proper 2-D intersection. Such components are called *face-adjacent* components. This ensures that the results of these operations are objects in our domain. Figure 3.1 shows examples of the two operators using cuboidal components (the portions of the boundary that have 2-D intersection are shown hatched).

Note that the disjoint union operator is the *glue* operator defined in many boundary modelers [2,36], and also that the restricted difference operator can be interpreted as a glue operation applied to a positive and a negative object. A more convenient way of looking at an MBM, for the purpose of developing a boundary-based description, is to consider two kinds of components: *positive* and *negative* components corresponding to the actions of adding and removing material, respectively. Positive and negative components are then combined along faces by means of a single glue operator. The composition rules are:

(i) if $C_i$ and $C_j$ are both positive or both negative, then they must intersect only along their boundaries and have portions of faces in common.

(ii) if $C_i$ is positive and $C_j$ is negative, then $C_j$ must be contained in $C_i$, and they must intersect along their boundaries and share portions of faces.

A *Hierarchical Partial Order Graph* is formed whose terminal nodes represent simple parts and non-terminal nodes represent subassemblies along with the part relationships and parallelisms involved.

Most of the work on assembly reviewed above is based on *independent* representation of individual parts and *separate* data structures that store the relations between them. Repeated evaluations or cumbersome additions to the data structures may be required to efficiently perform incremental interference checks as new parts are added to the workplace or during tests for global detachability of parts from the assembly. But once two faces are specified and found to be in surface contact, the common portion of the two faces need not be treated during interference checks. This suggests deleting all the touching portions of the faces between different parts. Our modular description of the assembled object with parts as individual modules does exhibit this advantage.

Another important aspect in object models for manufacturing is the need for an explicit description of form features. In other words, as for the assembly, the object model should be expressed as the composition of parts which correspond to machining operations. Non-geometric information (which characterize the specific machining process) should be associated with form features. The problem of representing form features in an object model, or, equivalently, of developing a feature-based object representation scheme, has been addressed mainly in connection with the problem of extracting form features from the model produced by the designer (usually from a B-Rep) [15,16,22,24,27,28,56]. Henderson produces a feature-based representation, the feature graph, in which nodes are form features and arcs are relations among them. Woo uses a CSG model in which primitives are convex polyhedra extracted from a B-Rep by a recursive convex hull technique. A modular boundary model, which is a variation of the Hierarchical Face Adjacency Graph described in [13], is used in [15,16,22] to organize the extracted form features.

## 3.2 Modular Boundary Models

The class of solid objects we consider are those subsets of $E^3$ bounded by compact, orientable, two-manifold surfaces [1]. Modular Boundary Models are a family of object representation schemes which describe a solid object as the Boolean combination of parts defined by their boundary under a restricted set of Boolean operators.

Each part forming an MBM is called a *component*. A component is a solid object $C_i$ bounded by a compact orientable two-manifold surface (i.e., an object in our domain). A component is described through a boundary model. The Boolean operators in an MBM are a *disjoint union* and a *restricted set difference* operation. Given two components $C_i$ and $C_j$ such that their boundaries intersect only at some faces, the disjoint union of $C_i$ and $C_j$, denoted

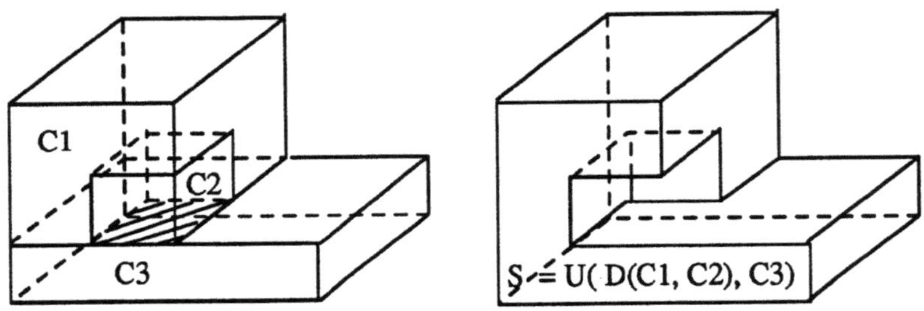


Figure 3.2: Cuboids $C_1$, $C_2$, and $C_3$ are all face-adjacent at the hatched connection face.

Positive and negative components are identified by the directions of their face normals: in a positive component they are directed outwards, in a negative component they are directed inwards.

Since an MBM is based on a Boolean combination of components represented in a boundary form, it could be considered as a hybrid boundary-CSG representation. The conceptual difference is that, due to the restricted operators allowed, an MBM describes the connection between face-adjacent components explicitly, while the two operators are implicit. Thus, the MBM is an unevaluated representation from which the reconstruction of the boundary of the object requires two-dimensional set operations only (to eliminate portions of faces).

Any face of a component which has a non-empty two-dimensional intersection with a face of at least one other component is called a *connection face.* The faces of the cuboids in Figure 3.1 to which the hatched portions belong are examples of connection faces. Note that a component may be face-adjacent at the same connection face to *several* components (see the example in Figure 3.2).

MBMs of the first generation, like the *Hierarchical Face Adjacency Hypergraph* (HFAH) [13] or the *Object Decomposition Graph* (ODG) [15], describe only face-adjacency relations between pairs of components in the form of a directed graph. In such a graph, the nodes correspond to the components defining the object decomposition. Each arc $(C_i, C_j)$ joining two face-adjacent components $C_i$ and $C_j$ represents the face-to-face relations between $C_i$ and $C_j$ and is labeled with the pairs of corresponding connection faces of $C_i$ and $C_j$. The orientation associated with the arc keeps track of the object construction sequence. Figure 3.3 shows an example of ODG. In the HFAH, the relations between object components is described as a tree, which restricts the class of modular decompositions which can be described. The HFAH has been developed and used [22] as a representation of object form features at different levels of resolution.

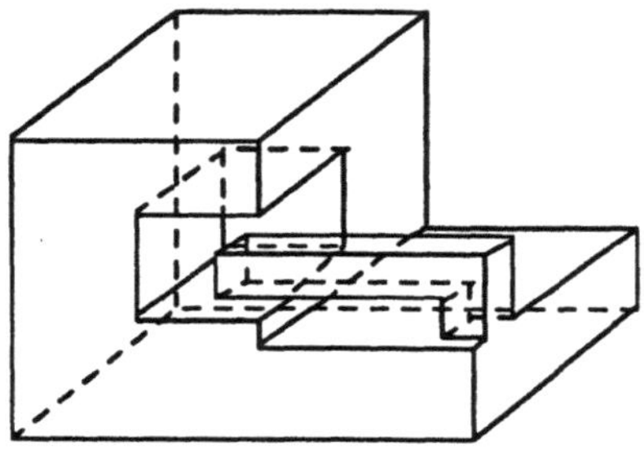

a) Object.

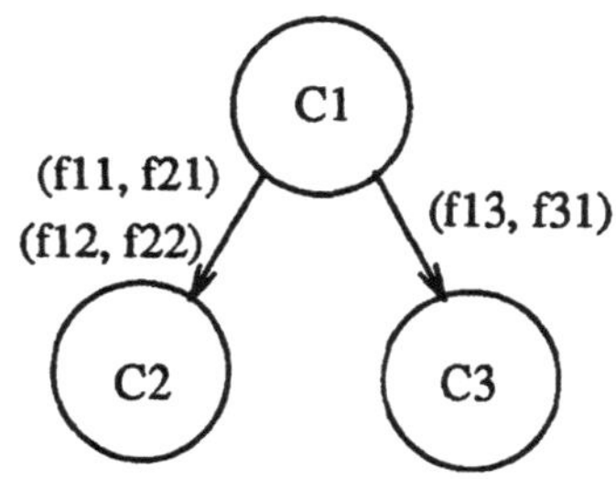


c) Object Decomposition Graph.

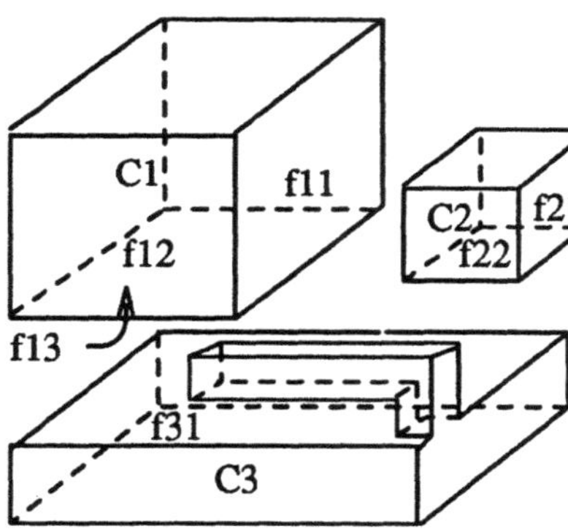


b) Exploded view of Modular Decomposition into components.

Figure 3.3: The Object Decomposition Graph and Modular Decomposition of an object.

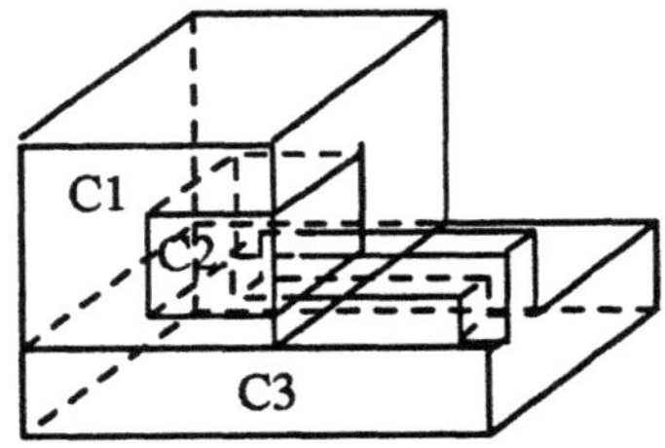


Figure 3.4: $C_1$ and $C_3$ have volumetric interference. $C_2$ and $C_3$ also have volumetric interference even though they do not share a connection.

## 3.3 The FFC Model

The components in an MBM can have spatial interference, and this does not only happen for a positive and negative component which are combined together (see the example of Figure 3.4). If we consider both the spatial interference among components and the partition of the component connection faces into subfaces shared with other face-adjacent components, we have a partition of the boundary of each component $C_i$ into portions of its original faces, that we call *facets*.

A facet of $C_i$ can be either:

(i) a maximal connected portion of the common intersection between one face of $C_i$ and one face of any face-adjacent component $C_j$ (and, thus, it is a subset of a connection face of $C_i$ and a connection face of $C_j$), or

(ii) a maximal connected portion of a face of $C_i$ defined by the intersection of such face with the boundary of any component $C_k$ having a volumetric interference with $C_i$.

A facet of a component $C_i$ is either a *connection facet*, when it is that subset that represents the common portion of a connection face, or it is a *boundary facet*. When two or more components are face-adjacent, only one connection facet is used to represent the common portion of their connection faces. Let $S$ be a solid object and $MD$ be a family of positive and negative components defining a modular decomposition of $S$ into face-adjacent components which, when combined through a glue operator, give $S$. The collection of the connection and boundary facets of all components of a modular decomposition $MD$ of an object defines a *fragmentation* $F$ of the union of the boundaries of its components. Figure 3.5 shows the fragmentation of the faces of the modular decomposition depicted in Figure 3.3.

Given the fragmentation $F$ defined by a modular decomposition $MD$ of $S$, each face $f_j$ in $F$ can be classified with respect to a component $C_i$ as follows:

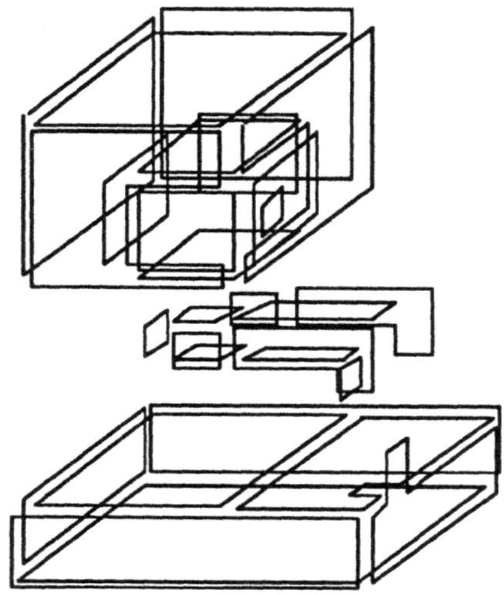

Figure 3.5: Exploded view showing facets in the fragmentation of the objects of Figure 3.3a.

(i) $f_j$ is a *connection* facet for $C_i$ (i.e., $f_j$ belongs to the boundary of $C_i$ and is shared by at least another component face-adjacent to $C_i$)

(ii) $f_j$ is a *boundary* facet for $C_i$ (i.e., $f_j$ belongs to the boundary of $C_i$ and is not shared by another component)

(iii) $f_j$ is *internal* to $C_i$ (i.e., $f_j$ belongs to the boundary of another component and is contained in $C_i$)

(iv) $f_j$ is *external* to $C_i$ (i.e., $f_j$ belongs to the boundary of another component and has no interference with $C_i$)

Figure 3.6 shows the classification of the facets of the modular decomposition of the object in Figure 3.2 according to the four categories listed above.

A connection facet is called *homogeneous* if it is a connection facet for an even number of components, otherwise it is called *non-homogeneous*. Figure 3.6 shows examples of homogeneous and non-homogeneous connection facets. Homogeneous connection facets do not belong to the boundary of the object. Thus, the evaluation of the boundary of $S$ from an MBM consists of eliminating the homogeneous connection facets.

The components in $MD$ and the fragmentation $F$ of the faces of the components in $MD$ with the above classification define a modular boundary model that we call the *Face-to-Face Composition* (FFC) model and denote $M = (MD,F)$. A high-level relational description of the FFC model is given by a hypergraph, called the *Face-to-Face Composition* (FFC) graph. In the FFC graph the nodes

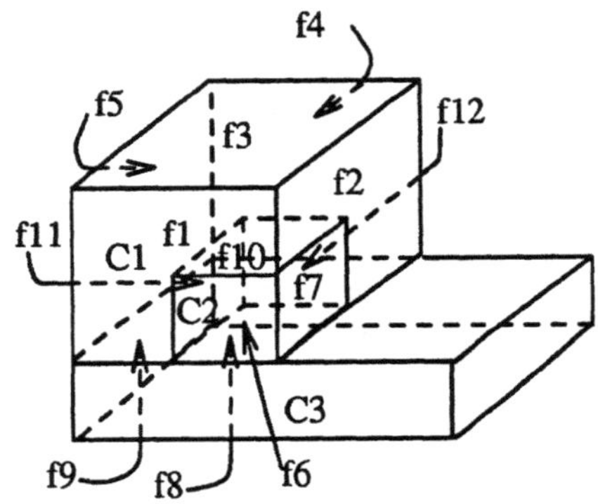


$f_1 - f_5$ are boundary facets of $C_1$.
$f_6 - f_9$ are connection facets of $C_1$.
$f_6$, $f_7$, and $f_9$ are homogeneous connection facets.
$f_8$ is a non-homogeneous connection facet.
$f_{10}$, $f_{11}$, and $f_{12}$ are internal facets to $C_1$.

Figure 3.6: Classifications of facets of components in Figure 3.2.

correspond to the components in *MD*, the hyperarcs to internal and connection facets. More formally, if $MD = \{C_1, C_2, \ldots, C_n\}$, and $F = \{f_1, f_2, \ldots, f_m\}$, then the FFC graph is a hypergraph $G = (N,A)$, where

(i) each node in $N$ corresponds to a component $C_i$ in *MD* (and thus can be identified with it);

(ii) each directed hyperarc $h$ in A is an ordered k-tuple $h = (C_{r1}, C_{r2}, \ldots, C_{rk})$, where $C_{rs}$, $s = 1, 2, \ldots, k$, are nodes of $G$. It corresponds to a facet $f_r$ in $F$ such that $f_r$ is a connection facet for at least two components in $h$ or an internal facet for at least one component in $h$, and $f_r$ is not external with respect to any component in $h$.

An attribute is associated with each hyperarc $h$ of $G$. It is an ordered k-tuple $h_a = (a_{r1}, a_{r2}, \ldots, a_{rk})$, where $a_{rs}$ is one of *internal, connection,* or *boundary.* Attribute $a_{rs}$ denotes the relation of facet $f_r$ with respect to component $C_{rs}$.

Figure 3.7 shows the FFC graph describing the modular decomposition depicted in Figure 3.3. Note that the external relation is not represented in the hyperarcs of the FFC graph. A facet which is only a boundary facet, and is not internal to any other component, is not described in the FFC graph. The FFC graph is a concise representation of the connection and interference relations among the components in the FFC model. Depending on its attribute, a hyperarc $h$ in the FFC graph can be classified as a *connection arc,* if all the attributes in $h_a$ are of type connection, as an *interference arc,* if at least one attribute

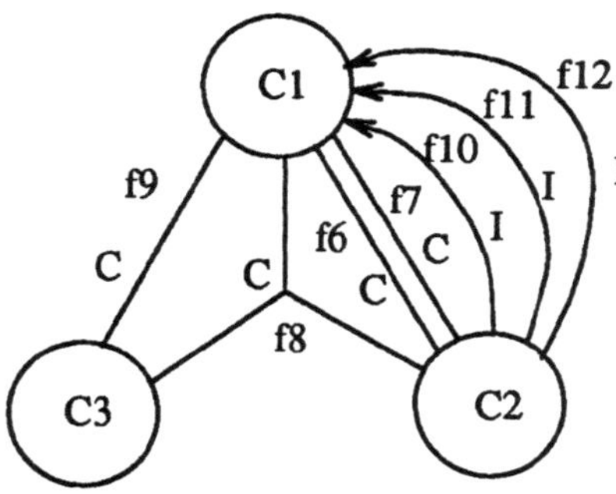


Figure 3.7: FFC graph for the modular decomposition shown in Figure 3.2 and Figure 3.6. ($C$ and $I$ denote connection and interference arcs respectively.)

in $h_a$ is of type internal, and there is no connection attribute in $h_a$, or as a *mixed arc*, when $h_a$ contains attributes of both type connection and internal. The spanning subgraph of the FFC graph G containing all the connection and mixed arcs of $G$ is called the *connection subgraph* of G).

We can obtain global interference information from the FFC graph. Two components $C_i$ and $C_j$ have a proper volumetric interference (i.e., their intersection is a 3-D subset of their volumes), if there exists a hyperarc $h$ incident in both of them such that the attribute associated with only one of them is of type internal. A component $C_i$ is contained in another component $C_j$ if

(i) there exists one hyperarc incident on $C_j$ for each facet of $C_i$,

(ii) a connection or an internal facet in $C_j$ corresponds to each connection facet in $C_i$,

(iii) and an internal facet in $C_j$ corresponds to each boundary facet in $C_i$ (see Figure 3.7).

## 3.4 Validity issues in the FFC model

The classification of the facets in a fragmentation induced by a modular decomposition of a solid object can be used to verify validity issues related to an FFC model. We assume that the single FFC components are single-shell solid objects bounded by two-manifold surfaces, and they are described by a valid boundary model. Topological validity is guaranteed by Euler operators, while geometric validity is checked algorithmically. We want to ensure that the boundary model of an object obtained by evaluating its FFC model is both topologically and geometrically valid.

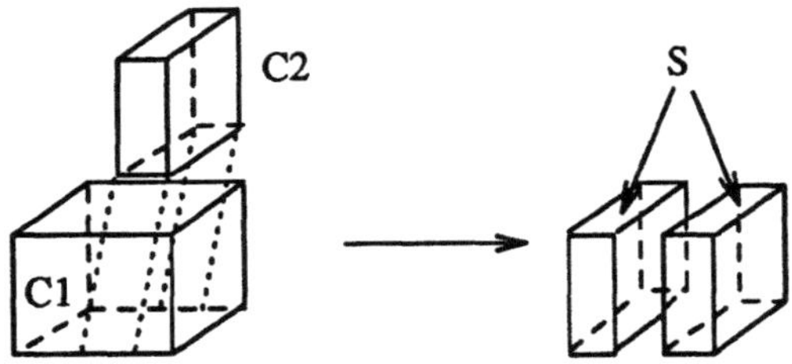


a) Resulting object $S$ has multiple shells.

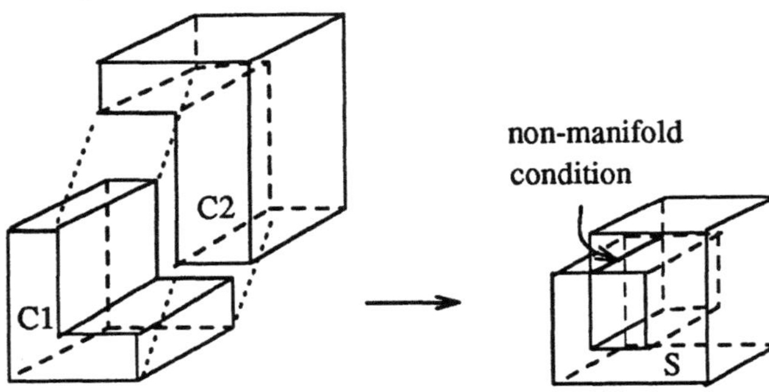


b) $S$ has a non-manifold condition at an edge.

Figure 3.8: Objects outside the domain of the FFC model as currently defined.

One problem is that we may obtain a representation of an object outside our domain by combining components which are in the domain: for instance, we can have an object composed of more than one shell (see Figure 3.8a), or an object bounded by a non-manifold surface (see Figure 3.8b). For now, we check this algorithmically, but it would be possible to extend our domain. Moreover, the evaluated representation of $S$ must describe an object without self-intersecting parts, and thus all the volumetric interferences must be eliminated by a suitable sequence of glue operations. A simple case illustrating this point occurs in Figure 3.3 if only $C_1$ and $C_3$ are combined to form $S$. This can clearly be checked by building the evaluated representation of $S$ and then looking for self-intersecting portions. This information, however, is embedded in the relations between the components in the modular decomposition $MD$ and the facets of fragmentation $F$. Intuitively, the evaluated representation can describe a solid object only if the facets lying on the boundary have one extra positive, or one extra negative, component (depending on whether the $MD$ describes a positive or a negative object) on one side and no material on the other side.

The definition introduced is a static definition of validity which does not take into account whether $S$ can be constructed through a sequence of glue operations by starting from the given modular decomposition. In other words, an

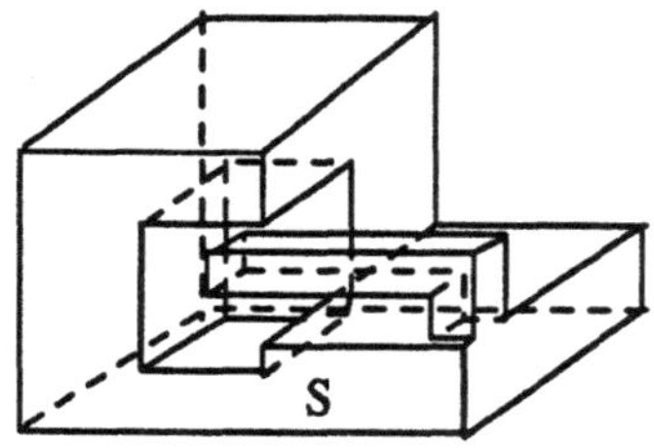


Figure 3.9: If $C_2$ is lowered into $C_3$, then a valid object that cannot be built from the modular decomposition in Figure 3.3b.

FFC model of an object can be valid, but not constructible (see the example of Figure 3.9). This is not a real problem since such a model could not be actually constructed with validity checks being performed at each update.

Necessary and sufficient conditions for validity have been proven in [10]. The main result for validity of the FFC representation of positive objects is the following (an analogous result holds for negative objects). An FFC model is a *valid representation* of an object if and only if each facet $f_j$ in $F$ satisfies one of two conditions on $n_j$, the difference between the number of positive and negative components in $MD$ for which $f_j$ is an internal facet:

(i) if $f_j$ is a homogeneous connection facet, then $n_j$ is equal to $-n$ or $1-n$, where $n$ is the difference between the number of positive and negative components on either side of $f_j$ for which it is a connection facet;

(ii) if $f_j$ is either a boundary or a non-homogeneous connection facet, then $n_j$ is equal to $-r$, where $r$ is the minimum between the differences of the number of positive and negative components for which $f_j$ is a connection facet, computed on each side of $f_j$.

The previous result allows a validity check which involves only boundary and connection facets. From the classification of the facets in a fragmentation $F$ introduced in the previous section, it follows that the boundary of $S$ is formed by the non-homogeneous connection facets and by the boundary facets of the fragmentation. The homogeneous connection facets correspond to the sets of facets of $F$ inside and outside $S$.

## 3.5 Evaluating an FFC Model

The problem of converting an FFC model into a boundary representation is termed *boundary evaluation* (by analogy with the well-known problem of converting from CSG to B-Rep). The conversion problem is important for compatibility with other modeling systems based on B-Rep, and for being able to use visualization techniques developed for boundary descriptions. Further reserch will go in the direction of designing specific algorithms for displaying objects described by an FFC model without the need of evaluating the model.

While the evaluation of a CSG representation is a difficult task because a large amount of information about the boundary of the object is only implicitly encoded in a CSG (see [44,45,51]), the evaluation of an FFC model is quite easy since the information about the object boundary is clearly represented in the FFC. The evaluation task is even easier if we do not require that the descriptions produced at intermediate steps be valid.

The evaluation of an FFC model $M$ of an object S, described by a modular decomposition $MD$ and by a fragmentation $F$, consists of iteratively applying a union operation to pairs of face-adjacent components. At each step, two face-adjacent components $C_i$ and $C_j$ are joined together along their common connection facets. If a connection facet $f_j$ is homogeneous, it is eliminated as soon as any two components sharing it are merged. If a connection facet is non-homogeneous, it remains as part of the boundary of $S$. The evaluation process can thus be regarded as the elimination of the homogeneous connection facets from $F$. Note that the resulting boundary description of $S$ might contain facets belonging to the same surface. In a planar-faced object environment, a segmentation of the boundary of $S$ into maximal connected faces could be produced by applying a face-growing algorithm to the segmentation produced by the evaluation process. Examples of these two situations are shown in Figure 3.10. For clarity the face growing algorithm has been used on all figures illustrating the object.

The boundary evaluation algorithm outlined above can produce invalid representations at intermediate steps. To be sure that we produce valid intermediate results, the algorithm must follow the *design sequence*, i.e., the sequence used by the designer to create the object.

The boundary evaluation algorithm of an FFC model can be expressed as a merging algorithm applied to the connection subgraph of its FFC graph. Performing the union of two components $C_i$ and $C_j$ is equivalent to merging the corresponding two nodes in the connection graph, and also in the set of extreme nodes of hyperarcs incident on both nodes. A hyperarc $h$ is eliminated when all its extreme nodes have been merged together. Figure 3.11 illustrates this process on the connection subgraph of the FFC-graph shown in Figure 3.7.

The boundary evaluation algorithm can also be applied in local modifications of the FFC model, for instance, for forming composite components from elementary ones. When the boundary evaluation is applied to a single pair of face-adjacent components, we want to be sure that the boundary description of the resulting component is valid. Validity checks on such descriptions can be performed through the local checks discussed in the previous section.

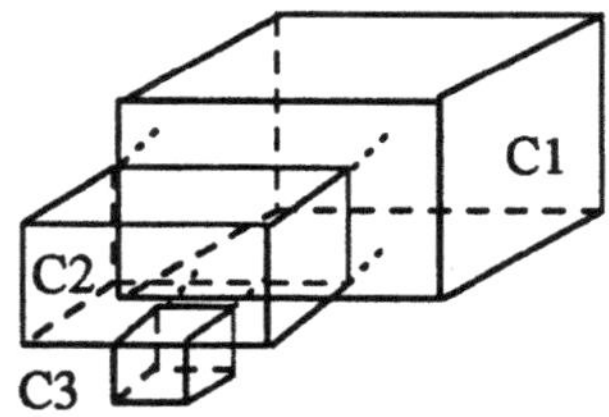


a) Exploded view of the modular decomposition of $S$.

b) Intermediate object $S'$. c) Final object $S$.

Proceeding from $S'$ to $S$ requires:
i) Deletion of the homogeneous connection facet (shown hatched).
ii)Merging at edge $e$ the two facets $f_1$ and $f_2$, that belong to the same plane surface.

Figure 3.10: Face-growing on an object with two positive ($C_1$ and $C_3$) and one negative ($C_2$) components combined at their front faces.

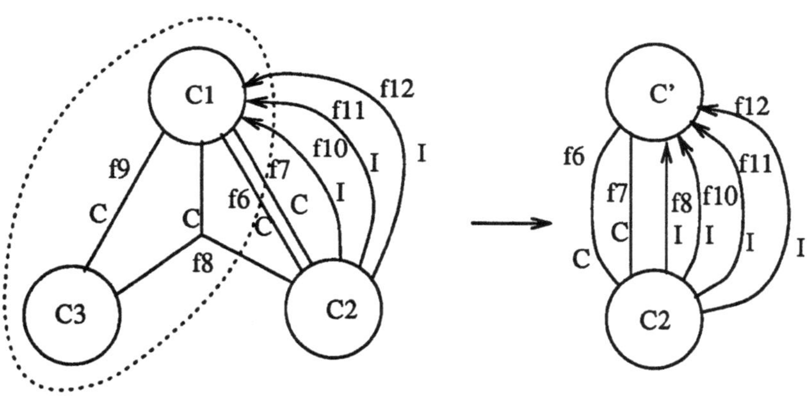


Figure 3.11: The hyperarc corresponding to facet $f_9$ is eliminated when components $C_1$ and $C_3$ (Figures 3.2, 3.6, 3.7) are merged, and $f_8$ becomes an internal facet.

Boundary evaluation is an irreversible process since once two components have been joined the FFC model does no longer contains information representing the two original components. Provisions for *undo* operators must be made.

## 3.6 Cellular representation of the FFC Model

An FFC model of an object described by a modular decomposition $MD$ can be represented as a partitioning of the portion of 3-D space defined by the union of the components in $MD$ into pairwise quasi-disjoint 3-D cells. This is a special type of cellular decomposition [41], since the cells can be empty or full and their union covers the union of the components in $MD$. Here, $S$ is covered by the union of only the full cells.

Given the modular decomposition $MD = \{C_1, C_2, \ldots, C_n\}$, the connection and interference information among the components of $MD$, represented by a fragmentation $F$, defines a *cellular decomposition* $CD = \{c_1, c_2, \ldots, c_p\}$ of the $\mathrm{U}C_i$ into cells which satisfy the following properties:

(i) Each cell $c_i$ is a subset of at least one component in $MD$.

(ii) The interiors of the cells are pairwise disjoint.

(iii) The union of the cells in $CD$ is the same as the union of the components in $MD$ (each considered as a positive component); this union is called *volume occupied* by $MD$ and denoted $V$.

(iv) Each facet $f_j$ in $F$ is either a common facet to two cells or a facet bounding $V$.

(v) Each vertex and each edge of the cellular representation must be common to all the cells adjacent to it.

(vi) A cell is either *empty* if it describes empty space, or *full*, if it is a part of the object volume.

Thus, a cellular representation of the FFC model is partially an object-based (like tetrahedralizations, finite element meshes, etc.) and partially a space-based (like octrees or bintrees) decomposition. If we consider only the full cells, $CD$ reduces to a cellular decomposition. Unlike space-based and object-based decompositions, the cells in the cellular FFC representation have a more complex shape. A cell can be any simply-connected solid object (cells cannot have internal cavities). Figure 3.12 shows an exploded view of the cells in the cellular representation of $S$ shown earlier in Figure 3.3.

We encode the FFC model by using a data structure for its cellular representation plus a binary matrix describing the component-facet and facet-component relations. The component-cell and cell-component relations can be obtained from the previous ones provided that we store the cell-facet and facet-cell relations in the data structure for the cellular representation.

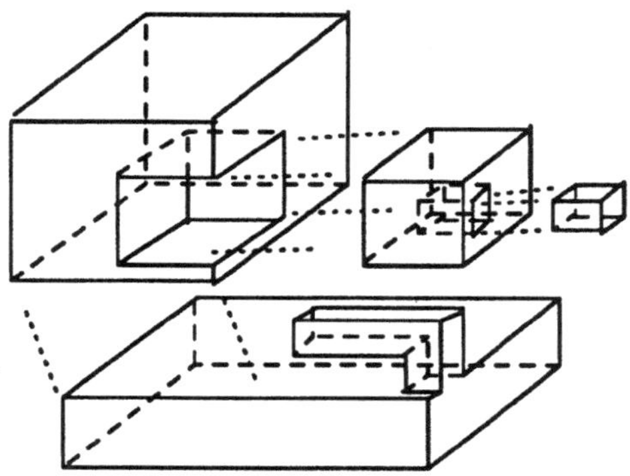

Figure 3.12: Exploded view of cells covering the union of the components in Figure 3.3.

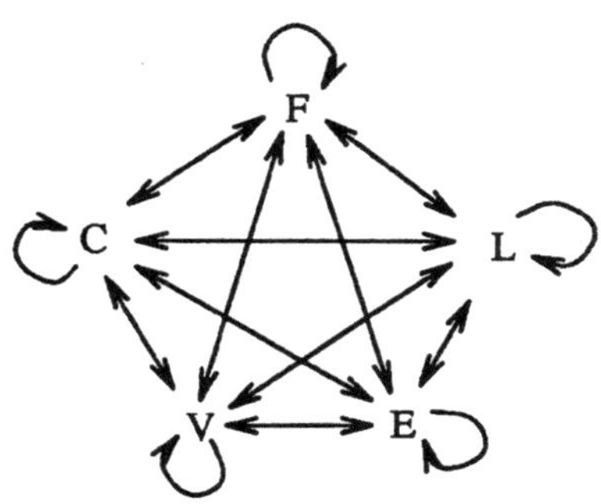


Figure 3.13: There are 25 possible relations between elements in a cellular decomposition.

A cellular decomposition *CD* is defined by five topological elements, namely, *cells*, *facets*, *loops*, *edges* and *vertices*. We can define 25 pairwise adjacency relations between each ordered pair of elements (arrow diagram in Figure 3.13), and 16 adjacency relations among the elements in a single cell.

The topological elements required to represent the two manifold boundary of a single cell and their inter-relationships are shown in Figure 3.14a. The relations satisfy the definition of the symmetric data structure proposed by Woo and is sufficient to describe any arbitrarily complex two-manifold surface [57].

The six stored adjacency relations (arrow diagram in Figure 3.14b) are defined as follows:

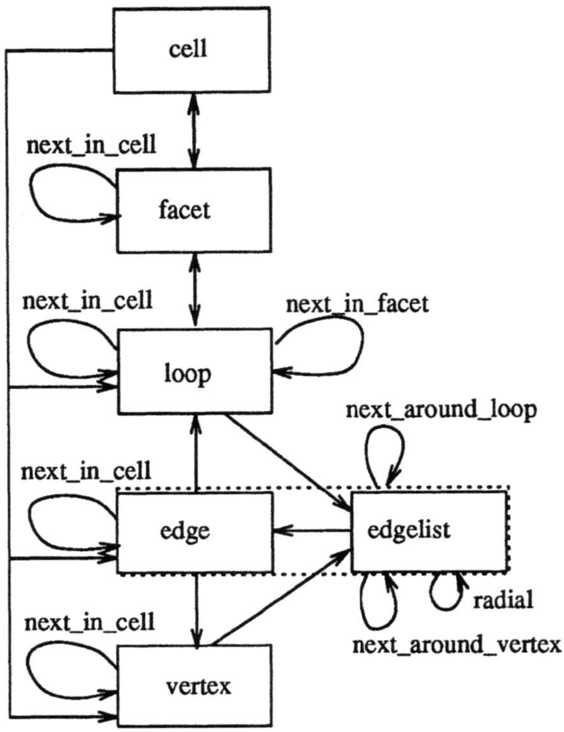


a) Relationships between elements in the Symmetric data structure.

$$F \longleftrightarrow L \longleftrightarrow E \longleftrightarrow V$$

b) There are six relations stored in a cell.

Figure 3.14: The Symmetric data structure.

(i) Facet-Loop (FL): loops belonging to a given facet $f$.

(ii) Loop-Facet (LF): facets (at most two) containing a given loop $l$.

(iii) Loop-Edge (LE): ordered list of edges forming a given loop $l$.

(iv) Edge-Loop (EL): loops to which a given edge $e$ belongs.

(v) Edge-Vertex (EV): extreme vertices of a given edge $e$.

(vi) Vertex-Edge (VE): ordered list of the edges incident in a given vertex $v$.

Note that the loop-facet, edge-loop and edge-vertex relations are *constant*, in the sense that they involve a constant number of elements, while the remaining three are *variable* relations. It can be proven that such relations are sufficient to characterize the boundary of a single-shell solid object without errors or ambiguities [37]. Moreover, the remaining ten relations can be retrieved from the six stored ones in a number of operations proportional to the number of elements involved in each relation.

The proposed data structure represents our cellular decomposition using the symmetric data structure for each individual cell. The topological elements and their inter-relationships are shown in Figure 3.15. It may be easily observed that the cell-elements maintain all their relationships in the symmetric data structure. A separate set of *object elements* have been defined as a framework for the identical cell elements and to store the geometry of facets and vertices. Adjacency relations between object elements of different types are not stored explicitly. They are derived from the corresponding cell elements.

Figure 3.16 shows the edgelists around a cell-loop and a cell-vertex in the symmetric data structure. It may be noted that the edgelists around a cell-vertex and cell-loop are traversed in a counter-clockwise direction when viewed from outside the solid. Since the same edgelists are used around cell-loops and cell-vertices a convention is used for the orientations of the edgelists with respect to the cell-loops and the cell-vertices. If an edgelist belongs to the lists of cell-vertex $cv_1$ and cell-loop $cl_1$ then the cell-edge starts from $cv_1$ when traversing $cl_1$. Figures 3.17, 3.18, and 3.19 show the extensions to the cellular data structure. The relations between the cell-elements remain the same. A cell has to be hypothetically isolated from the cellular object and then viewed from outside for the orientation conventions to be verified. Note that each facet, edge, vertex and loop is represented in each cell to which it belongs and also as an element of the decomposition.

## 3.7 Building an FFC Model

An FFC model of a solid object can either be constructed by combining a single component to an existing model (which is initially null) or by combining two separate FFC models. The first method can be considered as a specialization

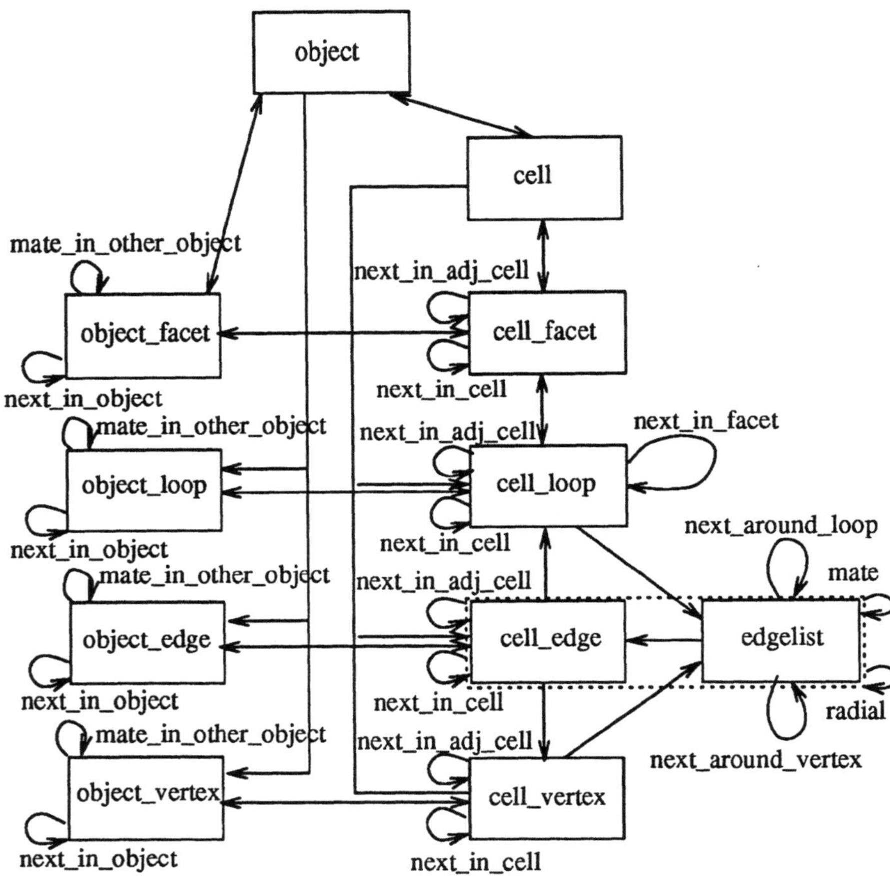


Figure 3.15: Relationships between elements in the cellular data structure.

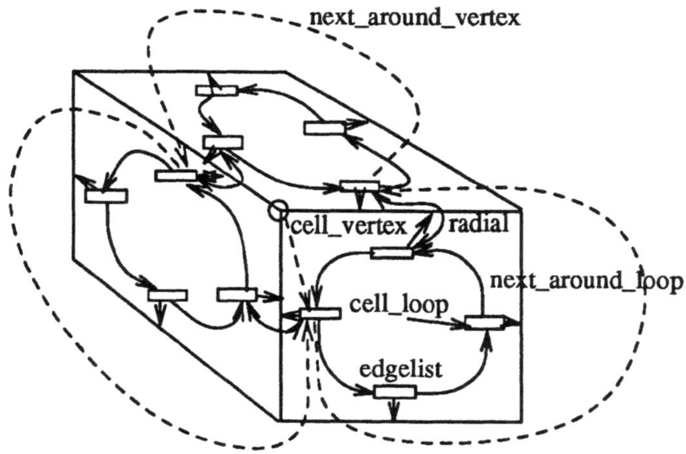


Figure 3.16: Edgelists around a cell loop and around a cell vertex in the Symmetric data structure.

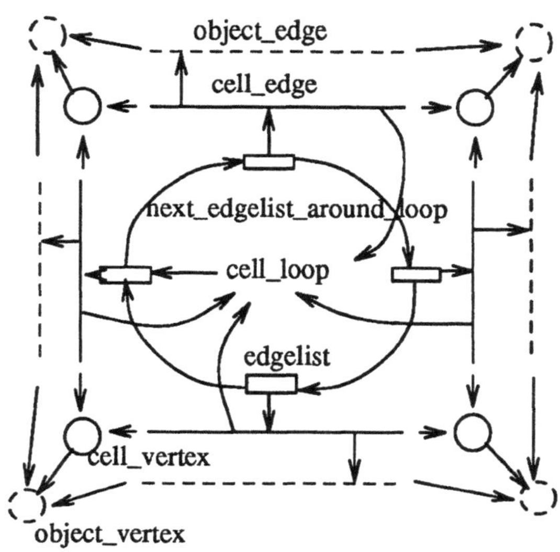


Figure 3.17: Plan view of a cell-loop as a list of cell edges.

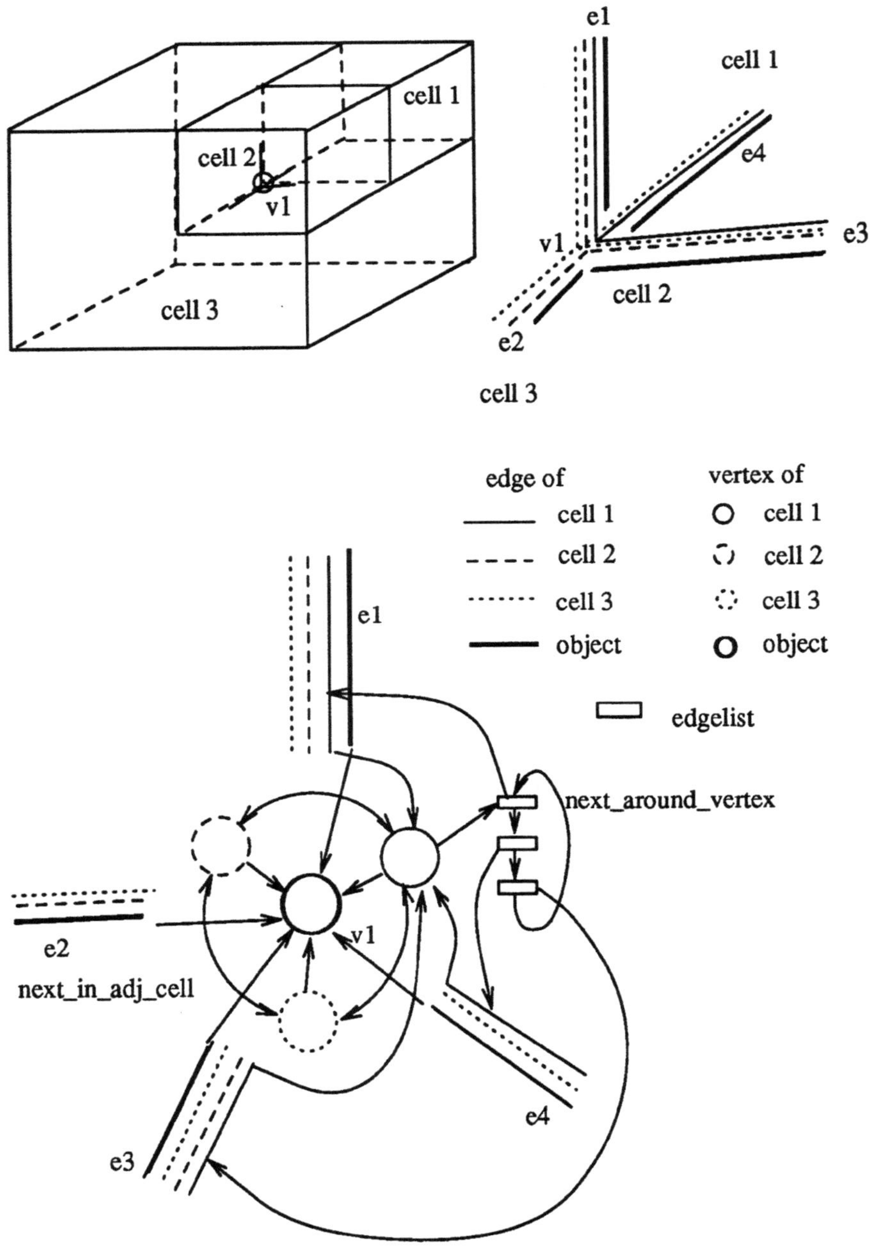


Figure 3.18: View of a cell-vertex as a list of cell-edges in the cellular data structure. (Only the adjacencies of a cell-vertex of cell 1 are shown in full.)

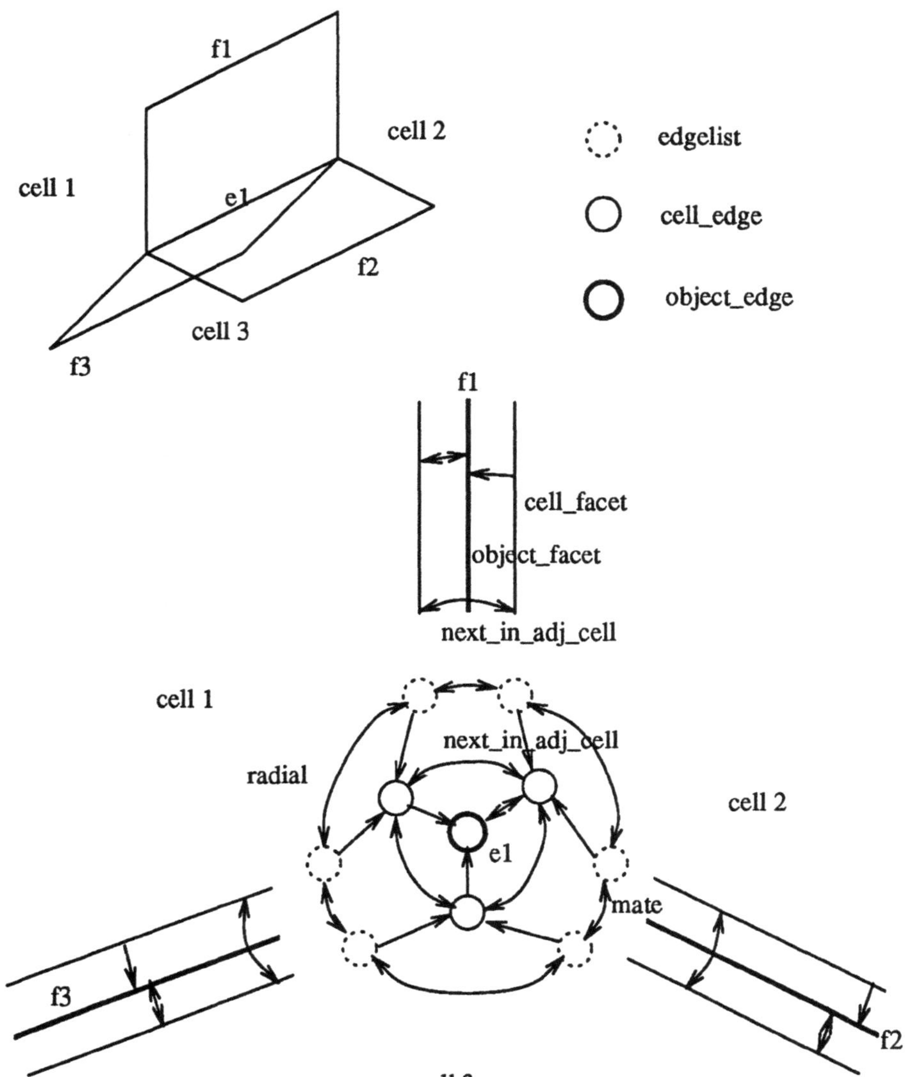


Figure 3.19: Cross-section of three object-facets sharing a common object-edge and the associated cell elements. (Only the relationships between similar elements have been labeled.)

of the second, since a single component is a special case of an FFC model. For now, we have restricted our attention to FFC model construction by the first method, and this operation, that we call COMPOSE, is currently being implemented in our experimental geometric modeler based on the FFC model.

The steps required to perform COMPOSE are first described followed by specialized Euler operators defined to create and manipulate the cellular data structure during COMPOSE.

### 3.7.1 Adding a component to an FFC model: COMPOSE

An FFC model can be constructed by a sequence of COMPOSE operations applied to an existing model (which is initially empty). A COMPOSE operation consists of adding a new component $C_i$ to an existing FFC model $M$. The requirements are that either the new component share portions of faces with the boundary of the object $S$ represented by $M$, and have no volumetric interference with $S$ (if $S$ and $C_i$ are both positive or negative), or $C_i$ be contained in $S$ (if $S$ and $C_i$ have opposite sign).

In terms of the FFC graph, a COMPOSE is equivalent to adding a new node $C_i$ and a set of connection and interference hyperarcs describing the connection and interference relations of the new components with the existing ones. The addition of a component $C_i$ can modify the connection facets of other components by splitting existing connection facets. Figure 3.20 shows the modification of existing facets and the FFC graph when adding a component.

The cellular representation of the FFC model is especially useful to enhance the efficiency of the construction of the FFC model. Each time a new component is added, it is intersected only with a restricted number of cells. Also, once interferences and connections are computed, we have to check the validity of the resulting object by checking the parity of the modified and new facets (see section 3.4). This also ensures that, at the completion of a sequence of COMPOSE operations, we have a valid object and a valid sequence of pairwise face-to-face compositions of parts which produce valid components at each intermediate step.

The inputs to the COMPOSE algorithm are as follows:

(i) The component-facet relations.

(ii) The cellular data structure of the FFC model.

(iii) The parity counters associated with each facet.

(iv) A description of the component being added.

(v) At least one pair of faces, with a face from the component and a face of a component in the FFC model, should be specified to abut after the specified transformations.

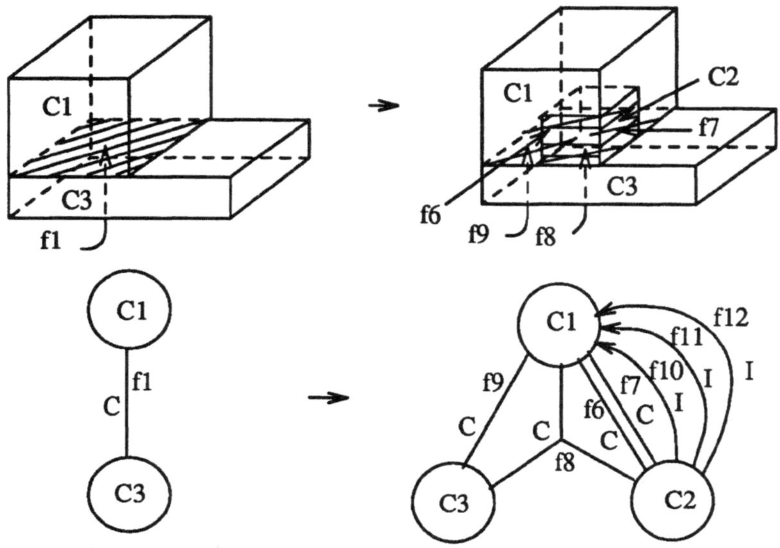


Figure 3.20: Some facets and the FFC-graph are modified when $C_2$ is added. See Figures 3.6 and 3.7. (Arc labels $C$ indicate connection facets.)

The outputs of the algorithm are as follows:

(i) The new component-facet relations.

(ii) The new cellular data structure.

(iii) The result of the validity check on the model.

The COMPOSE algorithm consists of the following steps:

(i) Convert component description to the symmetric data structure of a single cell using Euler operators.

(ii) Starting from the connection facets specified detect the cells from the current FFC model that need to be intersected with the new cell.

(iii) Intersect cells using Euler operators.

(iv) Update parity counters and perform validity checks.

(v) Update the component-facet relations.

### 3.7.2 Euler operators to manipulate the cellular representation.

The COMPOSE operation needs to create a new cell from the component and to manipulate the cellular data structure by creating and destroying topological elements in it. The operators to accomplish this should have correctness properties similar to that of the widely used Euler operators [2,21,35,53].

In a cellular decomposition the boundary of each cell is represented by a boundary model for a single shell. Utilizing this property the operators can maintain the correct topology of individual cells. Another property of the cellular decomposition is that any two cell-elements sharing an object-element must have the same geometry. Therefore, adding a cell-edge to a cell-loop requires adding a cell-edge to the other cell-loop sharing the same object-loop. Similarly, splitting a single cell-edge requires splitting all the cell-edges sharing the corresponding object-edge. Thus, the same operations are performed on several cells. Since their effect on the individual cells is the same as that of the traditional Euler operators, we have retained the nomenclature and added a prefix (CD) to the name of each operator.

The set of Euler operators defined by Mantyla [35] were implemented at first to manipulate the symmetric data structure representing the boundary of a single cell. These operators were extended to manipulate a cellular decomposition. The names and actions of the operators and a few simple examples are shown in Figures 3.21 and 3.22.

## 3.8 The Production Graph

The FFC model of an object $S$ can be built by pairwise combination, through the glue operation defined in section 3.2, of FFC models of simpler parts. This also means that the evaluation of an FFC model can be performed by merging pairs of object parts. We do not allow the construction of an FFC model by combination of more than two parts at a time. If an FFC model $M$ of an object $S$ can be built by adding a single component to an existing FFC model (initially a single component) at each step, then $S$ is called ***linearly constructible.*** The object depicted in Figure 3.3 is an example of a linearly constructible object.

Given a modular decomposition MD associated with an FFC model $M$ of an object $S$, each sequence of pairwise combinations of face-adjacent parts, which produces an evaluated boundary description of $S$, is called a ***composition sequence.*** A composition sequence can be described as a binary tree, called an ***evaluation tree***, in which the root represents the boundary model of $S$, the leaves represent the components in *MD*, and the intermediate nodes contain valid descriptions of parts of $S$. Figure 3.23 shows the evaluation tree for the object in Figure 3.3. A composition sequence for the object is obtained by a

| OPERATOR | ACTION |
|---|---|
| CD_M_CFLV | Make cell, face, loop, vertex |
| CD_K_CFLV | Kill cell, face, loop, vertex |
| CD_M_EV | Make edge, vertex |
| CD_K_EV | Kill edge, vertex |
| CD_M_EF | Make edge, face |
| CD_K_EF | Kill edge, face |
| CD_M_EKL | Make edge, Kill loop |
| CD_K_EML | Kill edge, Make loop |
| CD_M_F | Make face, Kill loop hole, or Make face, cell |
| CD_K_F | Kill face, Make loop hole, or Kill face, cell |
| CD_S_E | Split edge, Make vertex |
| CD_J_E | Join edge, Kill vertex |

Figure 3.21: Euler operators for the cellular decomposition.

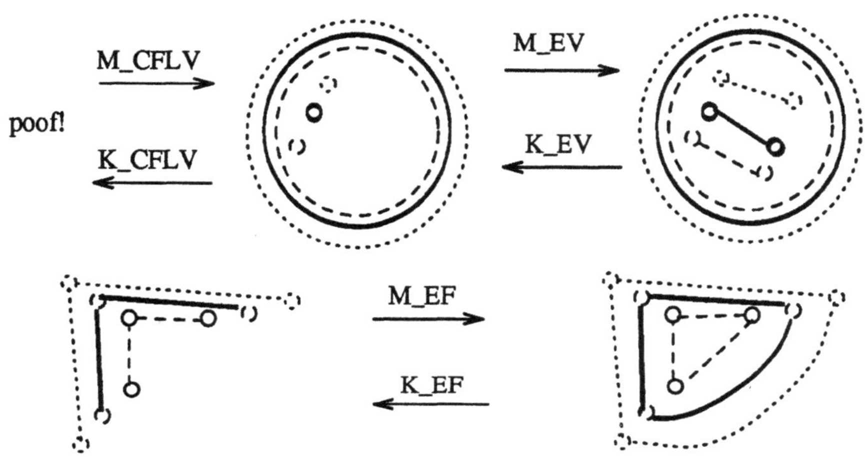


Figure 3.22: Examples of the actions of the first six Euler operators on a cellular decomposition.

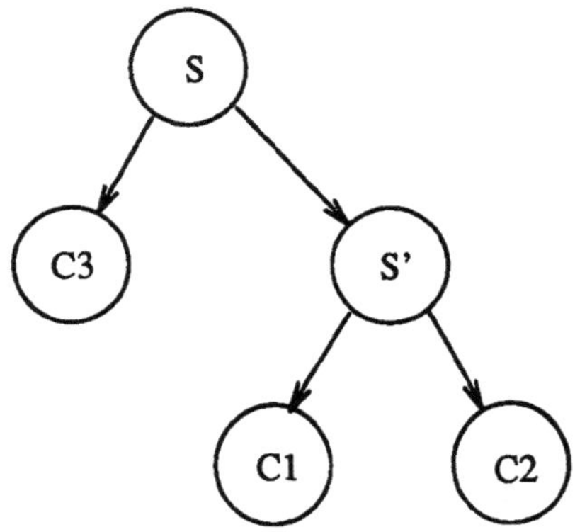


Figure 3.23: Evaluation tree for the object ($S$) depicted in Figure 3.3.

*postorder* traversal of the evaluation tree. The *preorder* traversal gives a *decomposition sequence*, i.e., a sequence of valid decompositions of S which produce the components in $MD$.

An evaluation tree is generated during the construction process. Such a tree reduces to a list when we use the operator COMPOSE described in section 3.7.1. The evaluation tree can be used to evaluate the boundary of an object from its FFC model, and this ensures that we obtain a valid representation at each intermediate step. There are many ways of constructing an object from its FFC model that are different from the one chosen by the designer. All feasible evaluation trees are described by the AND/OR graph that we term the *production graph*. OR nodes represent alternative composition/decomposition sequences, AND nodes the combination of two face-adjacent parts. Figure 3.24b shows the production graph of the object depicted in Figure 3.24a assuming that the object of Figure 3.3a is placed on a large slab $C_4$.

The production graph is a further development of the assembly AND/OR graph proposed by Sanderson and Homem de Mello [25,48]. It does not include all possible component combinations, but represents only those that produce a valid representation of a part at each intermediate step. Given a production graph, a particular evaluation tree can be extracted by traversing the graph from the OR node describing S and selecting one arc incident from each OR node and both arcs incident from each AND node traversed.

The production graph must be computed by selecting all decomposition sequences that produce feasible intermediate results. A decomposition sequence can be derived by recursively splitting $S$ into two valid parts. To produce a decomposition into valid representations of the parts, we must check the validity of such representations by applying the results in section 3.4.

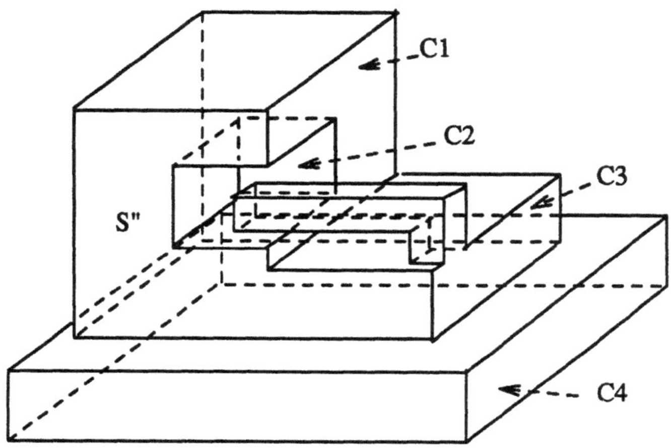


a) Object $S''$ obtained by placing $S$ of Figure 3 on a slab $C_4$.
(The dashed arrows identify the contribution of each component.)

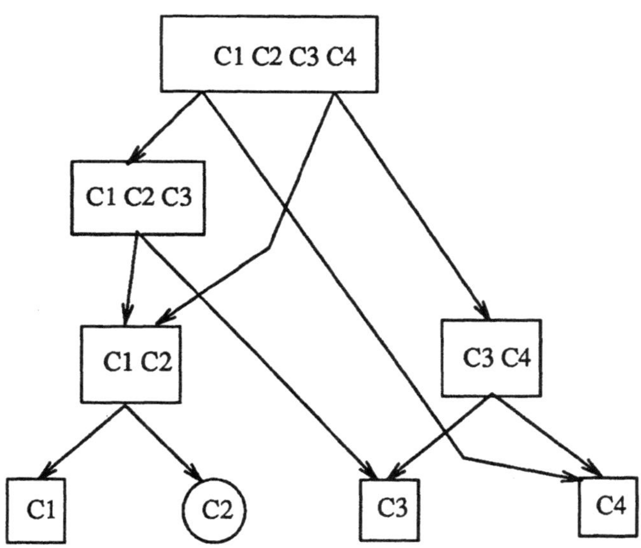


b) Production graph of $S''$.

Figure 3.24: An object and its production graph.

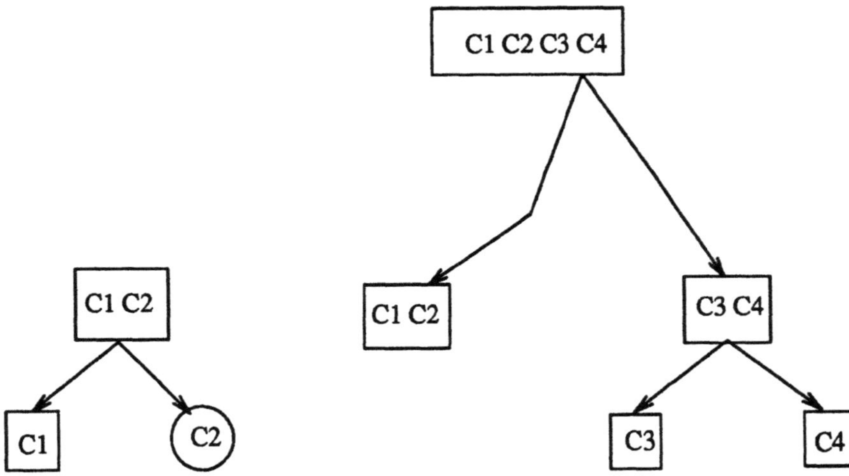


a) Cleavage graph of $S''$. b)Assembly graph of $S''$.

Figure 3.25: The cleavage and assembly graphs of the object $S''$.

Feasible assembly and machining sequences are represented in the production graph as feasible composition sequences. In the production graph, an AND node can describe addition of two parts with the same sign, and is called an *assembly node*, or the combination of a positive and a negative part, and is called a *cleavage node*. Assembly nodes describe an assembly operation, cleavage nodes a machining (material removal) operation. Any subgraph of the production graph entirely composed of assembly nodes is called an *assembly graph*. Similarly, any subgraph made only of cleavage nodes is termed a *cleavage graph*. Figures 3.25a and 3.25b show assembly and cleavage subgraphs of the production graph of Figure 3.24b.

In those situations when it is possible to combine all negative components with the positive ones that contain them, the production graph reduces to an assembly graph, and thus it describes all feasible assembly sequences that can be obtained by starting from such parts. In this case, we will not have any internal facet in the fragmentation defining the FFC model, and the FFC graph will reduce to a connection graph in which the hyperarcs describe mating relations among components to be assembled together. In general, however, the assembly and cleavage graphs cannot be separated in the production graph, since material removal operations are usually interspersed with assembly operations in a production sequence.

## 3.9 Concluding Remarks

We have attempted to lay the groundwork for the development of third-generation solid modeling systems. The FFC model differs from previous models in the high-level, graph-theoretic representation of juxtaposition and interference relations among the faces of overlapping, arbitrary components. These components themselves constitute individual objects that are represented by their boundaries and are guaranteed to be topologically valid through the use of Euler operators. The proposed data structure for the resulting object contains sufficient information to establish the adjacency relationship among the cells, and also their relationship to the original object components. An algorithmic method is available to construct compound objects from constituent objects that are themselves composed of individual components and are represented by FFCs. The recovery of the boundary of the compound object from the cellular model is quite straightforward. A possible application of the model is exhibited by the production graph, which shows alternative sequences of machining and assembling the modeled object.

A number of issues remain to be solved. Foremost among them is the development of robust geometric algorithms capable of dealing with almost coincident vertices, almost collinear edges, and almost coplanar faces resulting from numerical approximation.

The storage efficiency of the proposed data structure could undoubtedly be improved, since edges and facets are represented more than once. A direction to investigate is extension of the data structures developed for tetrahedral decompositions [18].

In the production graph, feasible decomposition sequences could be generated more efficiently by exploiting graph-theoretic properties of the conditions developed in [10]. Further, more realistic constraints on machining and assembly need to be developed in terms of specific manufacturing environments, such as tool-path and robot-arm geometries.

So far, only the algorithm for composing a single component with an FFC model has been developed in detail. Although feasibility is obvious, considerable work remains to be done to fill in the steps for combining two arbitrary FFC models.

It is worth investigating combining modular boundary representation with a PM-Octree in a solid modeler. Boolean operations and the computation of mass properties can be accomplished efficiently on PM-Octree representations. Furthermore, space-decomposition representations are easily generated from multiple camera views of actual objects [40].

We are actively looking at each of these problems and continuing implementation of the FFC model for polyhedral geometries.

## Acknowledgments

We gratefully acknowledge support under National Science Foundation grant IRI-8704718 and INT-8714578, NATO Collaborative Research Grant 0498/87, and the New York State Science and Technology Council through the RPI Center for Advanced Technology. We are pleased to acknowledge Elisabetta Bruzzone's essential contribution to the analysis of the validity of the FFC model, under the sponsorship of the CNR (Italy).

## References

[1] M. Agoston. *Algebraic Topology*, Marcel Dekker, New York, 1976.

[2] S. Ansaldi, L. De Floriani, B. Falcidieno, Geometric modeling of solid objects by using a face adjacency graph representation, *Computer Graphics*, 19(3):131-139, July 1985.

[3] D. Ayala, P. Brunet, R. Juan, I. Navazo, Object representation by means of non minimal division quadtrees and octrees, *ACM Transaction on Graphics*, 4(1):41-59, January 1985.

[4] B. G. Baumgart, Winged-edge polyhedron representation, *Technical Report* STAN-CS-320, Computer Science Department, Stanford University, Stanford, CA, 1974.

[5] J. D. Boissonnat, Geometric structures for three-dimensional shape representation, *ACM Transactions on Graphics*, 3(4):266-286, April 1984.

[6] I. C. Braid, On storing and changing shape information, *CAD Group Document #97*, University of Cambridge, Computer Laboratory, Cambridge, December 1977.

[7] I. C. Braid, R. C. Hillyard, I. A. Stroud, Stepwise construction of polyhedra in geometric modeling, in *Mathematical Models in Computer Graphics and Design*, K. W. Brodlie, ed., Academic Press, New York, pages 123-141, 1980.

[8] P. Brunet, I. Navazo, Geometric modeling using exact octree representation of polyhedral objects, *Proceedings EUROGRAPHICS'85*, pages 159-169, 1985.

[9] P. Brunet, I. Navazo, Solid representation and operation using extended octrees, *ACM Transaction on Graphics*, 8, 1989.

[10] E. Bruzzone, Validity issues in the Face-to-Face Composition model, *Technical Report*, 89-025, Rensselaer Polytechnic Institute, Troy, NY, October 1989.

[11] E. Bruzzone, A. Maulik, A cellular data structure for solid objects in the Face-to-Face Composition model, *Technical Report*, 89-026, Rensselaer Polytechnic Institute, Troy, NY, October 1989.

[12] I. Carlbom, I. Chakravarty, D. Vanderschel, A hierarchical data structure for representing spatial decomposition of 3-D objects, *IEEE Computer Graphics and Applications*, 5(4):24-31, April 1985.

[13] L. De Floriani, B. Falcidieno, A hierarchical boundary model for solid object representation, *ACM Transactions on Graphics*, 7(1):42-60, 1988.

[14] L. De Floriani, A. Maulik, G. Nagy, Manipulating a modular boundary model with a face-based graph structure, *Geometric Modeling for Product Engineering. IFIP WG 5.2*, M. J. Wozny, J. U. Turner, K. Preiss, Eds., North-Holland, pages 131-143, 1989.

[15] L. De Floriani, Feature extraction from boundary models of three-dimensional objects, *IEEE Transactions on Pattern Analysis and Machine Intelligence*, 11(8):798-798, August 1989.

[16] L. De Floriani, E. Bruzzone, Building a feature-based object description from a boundary model, *Computer Aided Design*, December 1989.

[17] L. De Floriani, G. Nagy, A graph-based model for face-to-face assembly, *Proceedings IEEE Int. Conference on Robotics and Automation*, Scottdale, pages 75-78, 1989.

[18] L. De Floriani, A pyramidal data structure for triangle-based surface description, *IEEE Computer Graphics and Applications*, 9(2):67-78, March 1989.

[19] L. De Floriani, E. Puppo, Representation and conversion issues in solid modeling, *Progress in Computer Graphics* (to appear).

[20] M. J. Durst, T. L. Kunii, Integrated polytrees: a generalized model for the integration of spatial decomposition and boundary representation, *Theory and Practice of Geometric Modeling*, W. Straßer, H. P. Seidel, Eds., Springer-Verlag, pages 329-348, 1989.

[21] C. M. Eastman, K. Weiler, Geometric modeling using Euler operators, *Proceedings of the First Conference on Computer Graphics and CAD/-CAM Systems*, Cambridge, Ma, pages 248-259, May 1979.

[22] B. Falcidieno, F. Giannini, Automatic recognition and representation of shape-based features in a geometric modeling system, *Computer Vision, Graphics and Image Processing*, 48(1):93-123, October 1989.

[23] K. Fujimura, T. L. Kunii, A hierarchical space indexing method, *Proceedings of Computer Graphics'85*, Tokyo, TI-4:1-14, 1985.

[24] M. R. Henderson, Extraction of feature information from three-dimensional CAD data, *PhD Thesis*, Purdue University, 1984.

[25] L. S. Homem de Mello, A. Sanderson, AND/OR graph representation of assembly plans, *AAAI-86 Proceedings of the Fifth National Conference on Artificial Intelligence*, American Association for Artificial Intelligence, Morgan Kaufmann Publishers, 1986.

[26] C. L. Jackins, S. L. Tanimoto, Octrees and their use in representing three-dimensional objects, *Computer Graphics and Image Processing*, 14(3):249-270, November 1980.

[27] G. E. Jared, Shape features in geometric modeling, *Solid Modeling by Computers: from Theory to Applications*, Plenum, New York, 1984.

[28] S. Joshi, T. Chary, Graph-based heuristics for recognition of mechanical features from a 3-D solid model, *Computer Aided Design*, 20(2), 1988.

[29] M. Karasick, On the representation and manipulation of rigid solids, *PhD thesis*, School of Computer Science, McGill University, Montreal, 1988.

[30] H. Ko, K. Lee, Automatic assembling procedure generation from mating conditions, *Computer Aided Design*, 19(1):3-10, 1987.

[31] Y. T. Lee, A. A. G. Requicha, Algorithms for computing the volume and other integral properties of solids. II. A family of algorithms based on representation conversion and cellular approximation, *Communications of the ACM*, 25(9):642-650, September 1982.

[32] K. Lee, D. Gossard, A hierarchical data structure for representing assemblies: Part 1., *Computer Aided Design*, 17(1):20-24, 1985.

[33] S. Lee, Y. G. Shin, Automatic construction of assembly partial-order graphs, *Proceedings RPI Conference on Computer Integrated Manufacturing*, pages 383-392, 1988.

[34] L. I. Liberman, M. A. Wesley, AUTOPASS: an automatic programming system for computer, *IBM Journal of Research and Development*, 21, pages 321-333, July 1977.

[35] M. Mäntylä, R. Sulonen, GWB: a solid modeler with Euler operators, *IEEE Computer Graphics and Applications*, 2(7):17-31, September 1982.

[36] M. Mäntylä, *An Introduction to Solid Modeling*, Computer Science Press, Rockville, MD, 1987.

[37] A. Maulik, A graph-based approach to solid modeling, *PhD Thesis Proposal*, Department of Electrical, Computer, and Systems Engineering, Rensselaer Polytechnic Institute, Troy, NY, October 1988.

[38] D. Meagher, Octree encoding: a new technique for the representation, the manipulation, and display of arbitrary 3-d objects by computer, *Technical Report*, Department of Electrical, Computer, and Systems Engineering, IPL-TR-80-111, Rensselaer Polytechnic Institute, Troy, NY, October 1980.

[39] I. Navazo, Extended octree representation of general solids with plane faces: model structure and algorithms, *Computers & Graphics*, 13(1):5-16, 1989.

[40] H. Noborio, S. Fukada, S. Arimoto, Construction of the octree approximating three-dimensional objects by using multiple views, *IEEE Transactions on Pattern Analysis and Machine Intelligence*, 10(6):769-781, November 1988.

[41] A. A. G. Requicha, Representations of rigid solids: theory, methods, and systems, *ACM Computing Surveys*, 12(4):437-464, December 1980.

[42] A. A. G. Requicha, H. B. Voelcker, Solid modeling: a historical summary and contemporary assessment, *IEEE Computer Graphics and Applications*, 2(2):9-24, March 1982.

[43] A. A. G. Requicha, H. B. Voelcker, Solid modeling: current status and research directions, *IEEE Computer Graphics and Applications*, 3(7):25-37, October 1983.

[44] A. A. G. Requicha, H. B. Voelcker, Boolean operation in solid modeling: boundary evaluation and merging algorithms, *Proceedings IEEE*, 73(1):30-44, January 1985.

[45] J. R. Rossignac, H. B. Voelcker, Active zones in CSG for accelerating boundary evaluation, redundancy elimination, interference detection, and shading algorithms, *ACM Transactions on Graphics*, 8(1):51-87, January 1988.

[46] H. Samet, *The Design and Analysis of Spatial Data Structures*, Addison-Wesley, Reading, MA, 1990.

[47] H. Samet, *Applications of Spatial Data Structures*, Addison-Wesley, Reading, MA, 1990.

[48] A. Sanderson, L. S. Homem de Mello, Automatic generation of mechanical assembly sequences, *Geometric Modeling for Product Engineering. IFIP WG 5.2*, M. J. Wozny, J. U. Turner, K. Preiss, Eds., North-Holland, pages 461-482, 1989.

[49] M. Tamminen, H. Samet, Efficient octree conversion by connectivity labeling, *Computer Graphics*, 18(3):43-51, July 1984.

[50] M. Tamminen, O. Karonen, M. Mäntylä, Ray-casting and block model conversion using a spatial index, *CAD Journal*, 16(4), 1984.

[51] R. B. Tilove, Set membership classification: a unified approach to geometric intersection problems, *IEEE Transactions on Computers*, C-29(10):874-883, October 1980.

[52] J. U. Turner, Tolerances in Computer-Aided Geometric Design, *Ph.D. dissertation*, Department of Electrical, Computer, and Systems Engineering, Rensselaer Polytechnic Institute, Troy, NY, May 1987.

[53] K. Weiler, Edge-based data structures for solid modeling in a curved-surface environment, *IEEE Computer Graphics and Applications*, 5(1):21-40, January 1985.

[54] K. Weiler, Topological structures for geometric modeling, *Ph.D. dissertation*, Department of Electrical, Computer, and Systems Engineering, Rensselaer Polytechnic Institute, Troy, NY, August 1986.

[55] P. W. Wilson, M. Pratt, Requirements for support of form features in a solid modeling system, *Technical Report*, Geometric Modeling Project, CAM-I, 1985.

[56] T. C. Woo, Feature extraction by volume decomposition, *Proceedings Conference on CAD/CAM in Mechanical Engineering*, MIT, Cambridge, MA, March 1982.

[57] T. C. Woo, A combinatorial analysis of boundary data structure schemata, *IEEE Computer Graphics and Applications*, 5(3):19-27, March 1985.

[58] G. Wyvill, T. L. Kunii A functional model for constructive solid geometry *The Visual Computer*, 1(1):3-14, July1985.

# Chapter 4

# Graphs of kinematic constraints

**Federico Thomas**

When a set of kinematic constraints are imposed between several rigid bodies, finding out the set of configurations that satisfy all these constraints is a matter of special interest. The problem is not new and has been discussed, not only in Kinematics, but also in the design of object level robot programming languages for assembly tasks.

This chapter deals with the problem of finding out how constrained movements, or kinematic constraints, are propagated and how some workpieces in an assembly reduce their degrees of freedom after this propagation, and how inconsistencies between constraint movements can be found. Special attention is paid to those problems which can be solved using a simple topological analysis derived from the Theory of Continuous Groups of Transformations.

Part of the material presented herein has already appeared in [16]. Here important points have been clarified and some modifications have been introduced. Also, an important part of this chapter is devoted to the propagation of kinematic constraints using part of the material appeared in [17].

This chapter is structured as follows. Section 4.1 shows the important role of kinematic constraints in the assembly domain. Section 4.2 provides all basic theory about kinematic constraints needed in this chapter. Section 4.3 essentially deals with the basic operations to be carried out on a graph of kinematic constraints, namely composition, intersection and star-polygon transform. Section 4.4 introduces a basic algorithm for constraint propagation, which avoids the application of the star-polygon transform when obtaining the equivalent constraint between any two bodies in a graph of kinematic constraints with arbitrary topology. Section 4.5 presents an example and, finally, Section 4.6 gives a brief summary of the main points in this chapter.

## 4.1 The role of kinematic constraints in the assembly domain

In the assembly domain, it does not suffice to make the workpiece models produced by a CAD system available in the programming environment, but a description of the way the different pieces should be fitted together is also required. This description can be provided in full detail by either the designer or the programmer, or rather be automatically inferred, at least in part, from constraints derived from both the shapes of the workpieces involved in the assembly, after trying to find matings of complementary subparts between them, and the mechanics of the assembly operations themselves.

Matings of complementary subparts of different workpieces have a direct translation into constrained movements, or kinematic constraints. In general, this translation assumes that the legal motion for compatible pairs of predefined subparts, or features, is provided by the user and thus already known. Alternatively, it would be possible to infer legal motions directly from geometric models of predefined features. This problem has been reduced to find local symmetries [9] or, when working with polyhedral workpieces, to find cycles of edges [18]. The recognition and extraction of expected patterns of geometry and topology, corresponding to particular engineering functionality, as described in [20], will play an important role in this area in a near future.

In the assembly domain, kinematic constraints are not only relevant when mating complementary subparts, but also when specifying relative locations between workpieces, specially when using an interactive graphics system. Let us look at a simple example. In order to specify the location of the block with reference to the box in fig. 4.1, we impose that faces $P_1$ and $P_2$ of the block be against $P_4$ and $P_3$ of the box, respectively. Then, we might ask: Is there any configuration satisfying both constraints? In other words, are they consistent? If the answer is yes, how many degrees of freedom remain between the block and the box? Which are the values of the constrained degrees of freedom? It will be shown that a directed graph of kinematic constraints, that is, a graph whose nodes correspond to workpieces and whose arcs are labeled with a set of

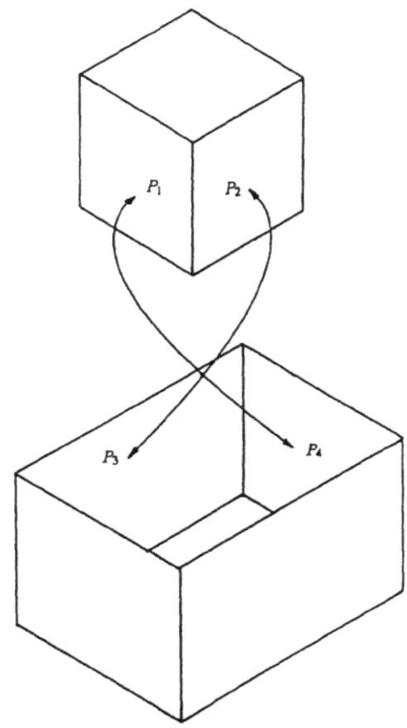


Figure 4.1: Specifying the location of a block with reference to a box using a set of kinematic constraints.

legal transformations linking the coordinate reference frame of the corresponding workpieces, is a proper data structure to represent these problems.

If we want to deal with graphs of kinematic constraints with arbitrary topology and constraints, then we must be able to find a solution to any inverse kinematic problem. Nevertheless, no general satisfactory solution, convenient for practical use, has been found for the general inverse kinematic problem. This problem is highly complicated because of its non-linearity, non uniqueness of the solution and existence of singularities. Fortunately, most kinematic graphs arising in the assembly domain are *quite* simple, since most planes and axis of symmetry of the involved geometric features are parallel and orthogonal in the final assembly.

The automatic manipulation of kinematic constraints has attracted a lot of attention not only in Kinematics, but also in the design of object level robot programming languages, such as RAPT [14] or LM-Geo [12]. Several algebraic symbolic approaches have emerged, among which we will mention a system of rewriting rules [14] and a table look-up procedure [8].

Algebraic symbolic (as opposed to a numerical) methods for dealing with kinematic constraints can shed light on basic aspects of the problem. For example, as it is shown in [17], the way they propagate provide useful information on the sequence of assembly.

The algebraic symbolic method used by the RAPT interpreter can be factored into a solution for the rotation which will determine angles, followed by the formation of real equations involving variables representing linear displacements and sines and cosines of the angle variable, which, in general, are difficult to deal with.

The approach presented herein distinguishes between *topological* and *geometrical* analysis of a set of kinematic constraints. The described topological analysis, well suited for the assembly domain, is derived from the Theory of Continuous Groups of Transformations, and it was essentially devised by Hervé in [7] for obtaining the number of degrees of freedom in mechanisms (see [1] for a revision). This analysis takes advantage of the fact that the legal relative motions resulting from mating two complementary subparts, such as pegs and holes or grooves and tongues, constitute cosets of subgroups of the Euclidean group, leading to a procedure based on a set of look-up tables.

## 4.2 The Euclidean group and kinematic constraints

It is well known that a rigid body in 3-dimensional space has 6 degrees of freedom, and, given a reference frame, any displacement can be obtained by a pure rotation about the origin followed by a pure translation.

The set of all displacements of a rigid body, with the composition operation, is isomorphic to the Special Euclidean group $SE(3)$. The decomposition $SE(3) = \Re^3 \times SO(3)$ shows the aforementioned fact that for any $\mathbf{D} \in SE(3)$,

$$\mathbf{D} = \mathbf{Trans}(\mathbf{v})\mathbf{Rot}(\mathbf{u}, \theta),$$

where $\mathbf{Trans}(\mathbf{v})$ is a translation along the vector $\mathbf{v} \in \Re^3$ and $\mathbf{Rot}(\mathbf{u}, \theta) \in SO(3)$ is a rotation of angle $\theta$ about the axis $\mathbf{u}$. Rotations about the axes $\mathbf{x}$, $\mathbf{y}$ and $\mathbf{z}$ are denoted by $\mathbf{Twix}$, $\mathbf{Twiy}$ and $\mathbf{Twiz}$, respectively. An arbitrary rotation can be written, using Euler's decomposition, as:

$$\mathbf{Rot}(\mathbf{u}, \theta) = \mathbf{Twix}(\varphi)\mathbf{Twiz}(\phi)\mathbf{Twix}(\psi).$$

A rotation can also be expressed using only the $\mathbf{Twix}$ operator and constant rotations as follows:

$$\mathbf{Rot}(\mathbf{u}, \theta) = \mathbf{Twix}(\alpha)\ \mathbf{XTOY}\ \mathbf{Twix}(\beta)\ \mathbf{XTOY}\ \mathbf{Twix}(\gamma)$$

where the constant rotation $\mathbf{XTOY}$ is defined as $\mathbf{Twiz}(\pi/2)$.

While the results presented below do not depend on a particular representation of $SE(3)$, we will use the well known 4×4-matrix representation of homogeneous transforms [13], which has become fashionable because its simplicity. Let us see a brief overview to this representation (see [2] for alternative representations such as screw coordinates, quaternions, dual numbers, etc.)

### 4.2.1 Homogeneous transformations. An overview

The representation of objects in an $n$-dimensional space using homogeneous coordinates needs a space of dimension $n+1$ from which the original space is recovered by projection. For example, the vector $\mathbf{v} = x_1\mathbf{i} + y_1\mathbf{j} + z_1\mathbf{k}$, where $\mathbf{i}, \mathbf{j}, \mathbf{k}$ are unit vectors along the Cartesian coordinate axes, is represented using homogeneous coordinates as a column vector:

$$\mathbf{v} = \begin{pmatrix} x \\ y \\ z \\ t \end{pmatrix}$$

so that

$$\begin{aligned} x_1 &= x/t \\ y_1 &= y/t \\ z_1 &= z/t \end{aligned}$$

Henceforth we will normalize $t = 1$.

A transformation $\mathbf{H}$ is a $4 \times 4$-matrix so that, the image of a given point $\mathbf{v}$ under this transformation is represented by the matrix product $\mathbf{u} = \mathbf{H}\mathbf{v}$.

#### Translations

A transformation $\mathbf{H}$ representing a translation by a vector $\mathbf{d} = a\mathbf{i} + b\mathbf{j} + c\mathbf{k}$ will be:

$$\mathbf{H} = \mathbf{Trans}\ (\mathbf{d}) = \mathbf{Trans}\ (a, b, c) = \begin{pmatrix} 1 & 0 & 0 & a \\ 0 & 1 & 0 & b \\ 0 & 0 & 1 & c \\ 0 & 0 & 0 & 1 \end{pmatrix}$$

Thus, given a vector $\mathbf{v} = (x, y, z, 1)^t$, its image $\mathbf{u}$ under $\mathbf{H}$ will be

$$\mathbf{u} = \mathbf{H}\ \mathbf{v} = \begin{pmatrix} x+a \\ y+b \\ z+c \\ 1 \end{pmatrix}$$

It is easy to prove that the set of all translations constitutes a group under the matrix product operation, which will be denoted by $T$.

### Rotations

The transformations representing rotations about the $x$, $y$, and $z$ axes by angles $\psi$, $\theta$ or $\phi$, respectively, are:

$$\mathbf{Rotx}(\psi) = \begin{pmatrix} 1 & 0 & 0 & 0 \\ 0 & \cos\psi & -\sin\psi & 0 \\ 0 & \sin\psi & \cos\psi & 0 \\ 0 & 0 & 0 & 1 \end{pmatrix}$$

$$\mathbf{Roty}(\theta) = \begin{pmatrix} \cos\theta & 0 & -\sin\theta & 0 \\ 0 & 1 & 0 & 0 \\ \sin\theta & 0 & \cos\theta & 0 \\ 0 & 0 & 0 & 1 \end{pmatrix}$$

$$\mathbf{Rotz}(\phi) = \begin{pmatrix} \cos\phi & -\sin\phi & 0 & 0 \\ \sin\phi & \cos\phi & 0 & 0 \\ 0 & 0 & 1 & 0 \\ 0 & 0 & 0 & 1 \end{pmatrix}$$

Each element $ij$ of the $3 \times 3$ upper left submatrix is equal to the cosine of the angle between the $i$-axis of the original coordinate frame and the $j$-axis of the rotated one.

These matrices, as well as their products, are orthogonal matrices with determinant equal to $+1$. They also constitute a group under matrix multiplication which will be denoted by $S_o$.

### Displacements

The transformations representing rotations and translations can be multiplied, and the resulting matrices are said to describe *displacements*.

The following properties must be emphasized:

- *Decomposition of a displacement.* Every displacement $\mathbf{H}$ can be decomposed into the product of a translation and a rotation, so that

$$\mathbf{H} = \mathbf{Trans}(\mathbf{d})\,\hat{\mathbf{H}} = \mathbf{Trans}(a,b,c)\,\hat{\mathbf{H}} \quad , \quad \forall \mathbf{H} \in SE(3)$$

  where $\hat{\mathbf{H}}$ is the rotation component of the displacement $\mathbf{H}$ or, in other words, is the matrix resulting from setting the first three elements – $a$, $b$ and $c$ – of the last column of $\mathbf{H}$ to zero.

- *Composition of n displacements.*

$$\mathbf{H}_1 \cdots \mathbf{H}_i \cdots \mathbf{H}_n = \textbf{Trans}\ (\mathbf{d}_1)\ \hat{\mathbf{H}}_1 \cdots\ \textbf{Trans}\ (\mathbf{d}_n)\ \hat{\mathbf{H}}_n =$$

$$\textbf{Trans}\ (\mathbf{d}_1 + \hat{\mathbf{H}}_1\mathbf{d}_2 + ... + \hat{\mathbf{H}}_1\hat{\mathbf{H}}_2 \cdots \hat{\mathbf{H}}_{n-1}\mathbf{d}_n)\ \hat{\mathbf{H}}_1\ \hat{\mathbf{H}}_2 \cdots\ \hat{\mathbf{H}}_n\ ,$$

$$\forall \mathbf{H}_1 \cdots \mathbf{H}_n \in SE(3)$$

If a transformation is postmultiplied by another transformation, the latter is applied with respect to the transformed frame described by the former. Conversely, if a transformation is premultiplied by another one, the latter is applied with respect to the reference frame [13]. Other authors [14], in using the transposes of the above defined transformations, adhere to the inverse rule.

- *Inverse displacement.* Because of the properties of orthogonal matrices, the inverse displacement of $\mathbf{H}$ is:

$$\mathbf{H}^{-1} = \hat{\mathbf{H}}^t\ \textbf{Trans}\ (-a, -b, -c)\ ,\quad \forall \mathbf{H} \in SE(3)$$

where $\hat{\mathbf{H}}^t$ denotes the transpose matrix of $\hat{\mathbf{H}}$.

A given displacement has been denoted using a upper case bold letter. Hereafter, sets of displacements, possibly subgroups, will be denote using just an upper case letter.

### 4.2.2 Subgroups of the group of displacements

It is well known that a *group* is a set of elements closed under an associative operation with an identity and inverse elements, as is the group $SE(3)$ of displacements. A *subgroup* $S \subset SE(3)$ is a subset of $SE(3)$ which is itself a group under the same operation. The composition of elements of $SE(3)$ can be extended to the composition of elements and subgroups. If $S \subseteq SE(3)$ and $\mathbf{D} \in SE(3)$, then the *right coset* $S \cdot \mathbf{D}$ is the set $\{\mathbf{H} \cdot \mathbf{D} \mid \mathbf{H} \in S\}$. The *left coset* $\mathbf{D} \cdot S$ and the *two-sided coset* $\mathbf{D}_1 \cdot S \cdot \mathbf{D}_2$ can be similarly defined. More generally, the composition of two subgroups $S_1 \cdot S_2$ is defined as $\{\mathbf{D}_1 \cdot \mathbf{D}_2 \mid \mathbf{D}_1 \in S_1, \mathbf{D}_2 \in S_2\}$.

**Definition 1 (Conjugation classes of subgroups of SE(3))** *Every such class is an equivalence class with respect to the relation:*

$$S_1 \sim S_2 \Leftrightarrow \exists \mathbf{D} \in SE(3)\ \mid\ S_2 = \mathbf{D}S_1\mathbf{D}^{-1}$$

*$S_1$ and $S_2$ being subgroups of $SE(3)$.*

Table 4.1: Classification of the subgroups of *SE(3)* into conjugation classes

| Dimension (d.o.f.) | Notation | Conjugation class and associated lower pair | Geometric elements of definition | Canonical subgroup |
|---|---|---|---|---|
| 0 | I | Identity displacement | | I |
| 1 | $T_{\mathbf{v}}$ | Rectilinear translation (P) Prismatic | A direction of translation given by a vector **v** | {**Trans** (x,0,0) \| x $\in \Re$} |
| | $R_{\mathbf{u}}$ | Rotation around an axis (R) Revolution | An axis of revolution **u** | {**Twix** ($\psi$) \| $\psi \in (-\pi, +\pi]$} |
| | $H_{\mathbf{u},p}$ | Helicoidal movement (H) Screw | An axis of revolution **u** and a thread pitch $p$ | {**Trans** (x,0,0)**Twix**($p$x) \| x $\in \Re$ , $p$ = constant } |
| 2 | $T_P$ | Planar translation | A plane $P$ | {**Trans** (0,y,z) \| x, y $\in \Re$} |
| | $C_{\mathbf{u}}$ | Lock movement (C) Cylindrical | An axis **u** | {**Trans** (x,0,0) **Twix** ($\psi$) \| x $\in \Re, \psi \in (-\pi, +\pi]$} |
| 3 | $T$ | Spatial translation | | {**Trans** (x,y,z) \| x, y, z $\in \Re$} |
| | $G_P$ | Planar sliding (E) Plane | A plane $P$ | {**Trans** (0,y,z) **Twix** ($\psi$) \| y, z $\in \Re, \psi \in (-\pi, +\pi]$} |
| | $S_o$ | Spheric rotation (S) Spherical | A point $o$ in the space | {**Twix** ($\psi$) **XTOY Twix**($\xi$) **XTOY Twix** ($\eta$) \| $\psi, \xi, \eta \in (-\pi, +\pi]$} |
| | $Y_{\mathbf{v},p}$ | Translating screw | A direction of revolution **v** and a thread pitch $p$ | {**Trans** (x,y,z) **Twix**($p$x) \| x, y, z $\in \Re$ , $p$ = constant } |
| 4 | $X_{\mathbf{v}}$ | Translating gimbal | A direction of revolution **v** | {**Trans** (x,y,z) **Twix**($\psi$) x, y, z $\in \Re, \psi \in (-\pi, +\pi]$} |

There exists infinite subgroups of $SE(3)$, but they can be classified into a finite number of conjugation classes. This suggest that we can represent each conjugation class by a *canonical subgroup*, so that all subgroups of the same class can be expressed as a conjugate of it.

An exhaustive classification of the continuous subgroups of $SE(3)$ into conjugation classes can be carried out using classic methods of analysis of finite dimension continuous groups [3]. A list of the classes thus obtained and a canonical subgroup for each of them is shown in table 4.1 (adapted from [7]). Note that all lower pairs are included in this classification. Let us recall that a lower pair exists when one element is coupled to the other via a wrapping action and contact takes place along a surface.

The notation used for these conjugation classes appears in the second column of table 4.1. Each class can be characterized by a set of geometric elements of definition which appear as subindices in the notation of the class. A geometric element of definition of a given subgroup is an affine space of $\Re^3$ of dimension 0, 1 or 2 (a point, line or a plane) which characterize the subgroup. A scalar is also required to characterize the $H_{\mathbf{u},p}$ and $Y_{\mathbf{v},p}$ subgroups. An instance of this elements leads to a subgroup belonging to the class. Instances will be denoted using numerical subindices. For example, $T_P$ denotes the conjugation class of planar translations and $T_{P_1}$ denotes a given subgroup belonging to this class.

The canonical subgroups are chosen in such a way that their geometric elements of definition satisfy the following conditions:

- if it is a point, it coincides with the origin of the reference frame;
- if it a line, it passes through the the origin of the reference frame and the $x$ axis is aligned with it; and
- if it is plane, it passes through the origin of the reference frame and the $x$ axis is orthogonal to it.

The election of canonical subgroups is thus arbitrary. If $S_i$ is a canonical subgroup, it will be denoted $\hat{S}_i$. Given a subgroup $S_1$, $(S_1)^G$ denotes the canonical subgroup in the same class.

The *degree of freedom* of a kinematic chain is defined as the necessary and sufficient number of variables that define uniquely the position and orientation of all the workpieces involved. The *dimension* of one of the foregoing subgroups is defined as the degree of freedom of the constrained motion it allows. A set of variables is thus associated with every subgroup. The dimension of a subgroup is indicated as $dim(\cdot)$, where $(\cdot)$ denotes one of those subgroups. Obviously, $dim(SE(3)) = 6$.

When the geometric elements of definition of two different subgroups satisfy some kind of spatial relationship – such as parallelism, collinearity or perpen-

Table 4.2: Conditions of inclusion of one subgroup of $SE(3)$ into another

$\mathbf{u}_0 \bot P_0 \Rightarrow \mathbf{u}_0$ perpendicular to $P_0$
$\mathbf{u}_1 \bowtie \mathbf{u}_0 \Rightarrow \mathbf{u}_1$ and $\mathbf{u}_0$ collinear
$\mathbf{u}_0 \parallel \mathbf{v}_0 \Rightarrow \mathbf{u}_0$ and $\mathbf{v}_0$ parallel

| | $T_{P_0}$ | $C_{\mathbf{u}_1}$ | $T$ | $G_{P_1}$ | $S_{o_0}$ | $Y_{\mathbf{v}_0,p_1}$ | $X_{\mathbf{v}_1}$ |
|---|---|---|---|---|---|---|---|
| $T_{\mathbf{u}_0}$ | $\mathbf{u}_0 \parallel P_0$ | $\mathbf{u}_0 \parallel \mathbf{u}_1$ | $\forall \mathbf{u}_0$ | $\mathbf{u}_0 \parallel P_1$ | | $\mathbf{u}_0 \bot \mathbf{v}_0$ | $\forall \mathbf{u}_0$ |
| $R_{\mathbf{u}_0}$ | | $\mathbf{u}_0 \bowtie \mathbf{u}_1$ | | $\mathbf{u}_0 \bot P_1$ | $o_0 \in \mathbf{u}_0$ | | $\mathbf{u}_0 \parallel \mathbf{v}_1$ |
| $H_{\mathbf{u}_0,p_0}$ | | $\mathbf{u}_0 \bowtie \mathbf{u}_1$ | | | | $\mathbf{u}_0 = \mathbf{v}_0, p_0 = p_1$ | $\mathbf{u}_0 \parallel \mathbf{v}_1$ |
| $T_{P_0}$ | | | $\forall P_0$ | $P_0 \parallel P_1$ | | $P_0 \bot \mathbf{v}_0$ | $\forall P_0, \forall \mathbf{v}_1$ |
| $C_{\mathbf{u}_0}$ | | | | | | | $\mathbf{u}_0 \parallel \mathbf{v}_1$ |
| $T$ | | | | | | | $\forall \mathbf{v}_1$ |
| $G_{P_0}$ | | | | | | | $P_0 \bot \mathbf{v}_1$ |
| $Y_{\mathbf{v}_0,p_0}$ | | | | | | | $\mathbf{u}_0 \parallel \mathbf{v}_1$ |
| $X_{\mathbf{v}_0}$ | | | | | | | |

dicularity –, one may become subgroup of the other. The conditions of inclusion of one subgroup into another appear in table 4.2 (adapted from [7]).

Now, we can introduce a formal definition of kinematic constraint.

**Definition 2 (Constraints and linking displacements)** *A constraint R is a set of displacements which can be expressed as a composition of cosets of canonical subgroups. That is,*

$$R = \mathbf{L}_0 \hat{S}_1 \mathbf{L}_1 \hat{S}_2 \mathbf{L}_2 \cdots \mathbf{L}_{n-1} \hat{S}_n \mathbf{L}_n \tag{4.1}$$

*where* $\mathbf{L}_1, \ldots, \mathbf{L}_{n-1}$ *are defined as linking displacements. A constraint is said to be* trivial *when it can be reduced to a single coset.*

The interest of most mechanisms is to provide a constrained motion which cannot be expressed as a constraint in the way it has been defined here. Nevertheless, we are not interested in analyzing mechanisms, but reasoning about constrained motions in the assembly domain.

Hereafter, we will assume that our constraints are trivial. In this particular case, if $R_i = \mathbf{L}_0\hat{S}_i\mathbf{L}_1$, then $R_i^G$ will denote the canonical subgroup $\hat{S}_i$, thus extending the notation introduced for subgroups.

Constraints will be denoted by $R_i$, where $i$ is a subindex that identifies it. If $R_i$ is the set of legal transformations from the reference frame of $\mathcal{B}_1$ to the reference frame of $\mathcal{B}_2$, $R_i^{-1}$ denotes the set of legal transformations in the way around, i.e. from $\mathcal{B}_2$ to $\mathcal{B}_1$. Note that $R_i^G = (R_i^{-1})^G$ for all $R_i$.

A constraint $R_i$ has the variables and geometric elements of definition inherited from $\hat{S}_i$. Given a reference frame, the subgroup with the same geometric elements of definition as a given constraint $R_i$ will be called its associated subgroup, which will be denoted by $R_i^A$. Obviously, $R_i^A \sim R_i^G$.

# 4.3 Operations on a graph of kinematic constraints

A directed graph of kinematic constraints – or GR graph, for short – is defined as a graph whose nodes correspond to workpieces and whose directed arcs are labeled with constraints. The two basic operations on a graph of kinematic constraints are composition and intersection of constraints. The former (fig. 4.2a) involves finding the constraint between bodies $\mathcal{B}_1$ and $\mathcal{B}_3$ that results from composing the constraint between $\mathcal{B}_1$ and $\mathcal{B}_2$ – say $R_i$ – with that between $\mathcal{B}_2$ and $\mathcal{B}_3$ – say $R_j$ –, which will be denoted by $R_i \cdot R_j$. The latter operation (fig. 4.2b) permits combining two given constraints, $R_i$ and $R_j$, between the same two workpieces into a single resulting constraint, which will be denoted by $R_i \cap R_j$. Let us analyze both operations in terms of composition and intersection of subgroups.

## 4.3.1 Composition

Let us assume a universe of three bodies – $\mathcal{B}_1$, $\mathcal{B}_2$ and $\mathcal{B}_3$ – linked by two trivial constraints

$$R_{12} = \mathbf{A}_1\hat{S}_1\mathbf{A}_2$$

$$R_{23} = \mathbf{B}_1\hat{S}_2\mathbf{B}_2.$$

Then, the equivalent constraint between bodies $\mathcal{B}_1$ and $\mathcal{B}_3$, that results from composing $R_{12}$ and $R_{23}$, is:

$$R_{13} = R_{12}R_{23} = \mathbf{A}_1\hat{S}_1\mathbf{A}_2\mathbf{B}_1\hat{S}_2\mathbf{B}_2 = \mathbf{A}_1\mathbf{A}_2S_1S_2\mathbf{B}_1\mathbf{B}_2 \tag{4.2}$$

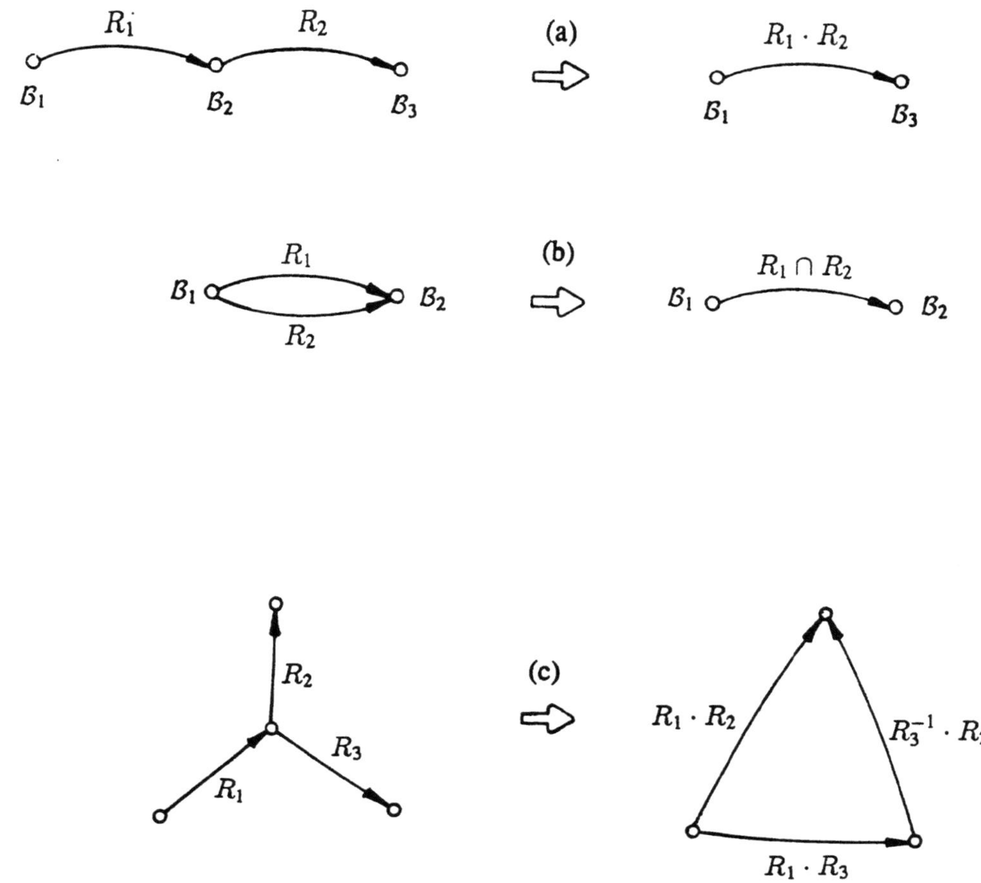


Figure 4.2: Operations on a graph of kinematic constraints: (a) composition; (b) intersection; and (c) star-polygon transform.

where

$$S_1 \sim \hat{S}_1 \text{ and } S_2 \sim \hat{S}_2$$

We will denote $(R_{12}R_{23})^C = S_1 S_2$ according to 4.2.

Thus, the problem of composing two trivial constraints can be reduced to the problem of composing two subgroups, and the outcome of the composition of two continuous subgroups of $SE(3)$ can be tabulated as shown in table 4.3 (adapted from [7]).

Clearly, the composition of two trivial constraints needs not be a trivial constraint itself, and the only information we need to find it out is the spatial relationships between their geometric elements of definition of both constraints.

When we compose two constraints expressed in terms of canonical subgroups, the linking displacement ($\mathbf{A}_2\mathbf{B}_1$ in (4.2)) captures the information about the spatial relationship between their geometric elements of definition. Taking advantage of this fact, we can check the linking displacement to find whether the composition of two trivial constraints can be reduced to a trivial constraint.

### 4.3.2 Intersection

If body $\mathcal{B}_3$, still in the same example above, is rigidly linked to $\mathcal{B}_1$ forming a closed kinematic chain, the intermediate body $\mathcal{B}_2$ will only have the possibilities of motion given by $R_{12} \cap R_{23}^{-1}$.

We can write,

$$R_{12} \cap R_{23}^{-1} = \mathbf{A}_1\hat{S}_1\mathbf{A}_2 \cap \mathbf{B}_2^{-1}\hat{S}_2\mathbf{B}_1^{-1} = (S_1' \cap S_2'\mathbf{B}_2^{-1}\mathbf{B}_1^{-1}\mathbf{A}_2^{-1}\mathbf{A}_1^{-1})\mathbf{A}_1\mathbf{A}_2 =$$

$$= (S_1' \cap S_2'\mathbf{C})\mathbf{A}_1\mathbf{A}_2$$

If $S_1'$ and $S_2'$ are subgroups of $SE(3)$, then $(S_1' \cap S_2'\mathbf{C})$ is either null or is a coset of $S_1' \cap S_2'$ (proposition 2 of [15]). Then, we have

$$R_{12} \cap R_{23}^{-1} = (S_1' \cap S_2')\mathbf{D}\mathbf{A}_1\mathbf{A}_2 \quad \text{iff} \quad R_{12} \cap R_{23}^{-1} \neq \emptyset \tag{4.3}$$

where

$$\mathbf{D} = \mathbf{E}\mathbf{C}, \ \mathbf{E} \in S_2', \ \mathbf{D} \in S_1', \tag{4.4}$$

$S_1'$ being a conjugate subgroup of $\hat{S}_1$, and $S_2'$ of $\hat{S}_2$.

We will denote $(R_{12} \cap R_{23}^{-1})^I = S_1' \cap S_2'$ according to 4.3 and 4.4.

Table 4.3: Intersection and regular representation for the composition of all pairs of subgroups of $SE(3)$ whose intersection is different from the identity displacement or one is not subgroup of the other

| Groups to be composed | Conditions on the geometric elements | Conditions on the linking displacement | Intersection | Regular representation |
|---|---|---|---|---|
| $T_{P_0}\cdot T_{P_1}$ | | | $T_{\mathbf{v}_0}$<br>$\mathbf{v}_0 = P_0 \cap P_1$ | $T$ |
| $T_{P_0}\cdot G_{P_1}$ | | | $T_{\mathbf{v}_0}$<br>$\mathbf{v}_0 = P_0 \cap P_1$ | $X_{\mathbf{v}_0}$<br>$\mathbf{v}_0 \perp P_1$ |
| $G_{P_0}\cdot G_{P_1}$ | | | $T_{\mathbf{v}_0}$<br>$\mathbf{v}_0 = P_0 \cap P_1$ | $R_{\mathbf{u}_0}\cdot T\cdot R_{\mathbf{u}_1}$<br>$\mathbf{u}_0 \perp P_0, \mathbf{u}_1 \perp P_1$ |
| $Y_{\mathbf{v}_0,p_0}\cdot T_{P_0}$ | $\mathbf{v}_0 \not\perp P_0$ | $l_{11} \neq \pm 1$ | $T_{\mathbf{v}_1}$<br>$\mathbf{v}_1 \parallel P_0,\ \mathbf{v}_1 \perp \mathbf{v}_0$ | $X_{\mathbf{v}_0}$ |
| $Y_{\mathbf{v}_0,p_0}\cdot G_{P_0}$ | $\mathbf{v}_0 \not\perp P_0$ | $l_{11} \neq \pm 1$ | $T_{\mathbf{v}_1}$<br>$\mathbf{v}_1 \parallel P_0,\ \mathbf{v}_1 \perp \mathbf{v}_0$ | $X_{\mathbf{v}_0}\cdot R_{\mathbf{u}_0}$<br>$\mathbf{u}_0 \perp P_0$ |
| $Y_{\mathbf{v}_0,p_0}\cdot Y_{\mathbf{v}_1,p_1}$ | $\mathbf{v}_0 \not\parallel \mathbf{v}_1$ | $l_{11} \neq \pm 1$ | $T_{\mathbf{v}_2}$<br>$\mathbf{v}_2 \perp \mathbf{v}_0,\ \mathbf{v}_2 \perp \mathbf{v}_1$ | $R_{\mathbf{u}_0}\cdot T\cdot R_{\mathbf{u}_1}$<br>$\mathbf{u}_0 \parallel \mathbf{v}_0,\ \mathbf{u}_1 \parallel \mathbf{v}_1$ |
| $Y_{\mathbf{v}_0,p_0}\cdot C_{\mathbf{u}_0}$ | $\mathbf{u}_0 \perp \mathbf{v}_0$ | $l_{11} = 0$ | $T_{\mathbf{v}_0}$ | $Y_{\mathbf{v}_0,p_0}\cdot R_{\mathbf{u}_0}$ |
| $C_{\mathbf{u}_0}\cdot C_{\mathbf{u}_1}$ | $\mathbf{u}_0 \parallel \mathbf{u}_1$ | $l_{11} = \pm 1$<br>$l_{24} \neq 0$ or $l_{34} \neq 0$ | $T_{\mathbf{u}_0}$ | $C_{\mathbf{u}_0}\cdot R_{\mathbf{u}_1}$ |
| $T_{P_0}\cdot C_{\mathbf{u}_0}$ | $\mathbf{u}_0 \parallel P_0$ | $l_{11} = 0$ | $T_{\mathbf{u}_0}$ | $T_{P_0}\cdot R_{\mathbf{u}_0}$ |
| $T\cdot C_{\mathbf{u}_0}$ | | | $T_{\mathbf{u}_0}$ | $X_{\mathbf{v}_0}$<br>$\mathbf{v}_0 \parallel \mathbf{u}_1$ |
| $G_{P_0}\cdot C_{\mathbf{u}_0}$ | $\mathbf{u}_0 \parallel P$ | $l_{11} = 0$ | $T_{\mathbf{u}_0}$ | $G_{P_0}\cdot R_{\mathbf{u}_0}$ |
| $X_{\mathbf{v}_0}\cdot C_{\mathbf{u}_0}$ | $\mathbf{u}_0 \not\parallel \mathbf{v}_0$ | $l_{11} \neq \pm 1$ | $T_{\mathbf{u}_0}$ | $X_{\mathbf{v}_0}\cdot R_{\mathbf{u}_0}$ |
| $Y_{\mathbf{v}_0,p_0}\cdot C_{\mathbf{u}_0}$ | $\mathbf{u}_0 \parallel \mathbf{v}_0$ | $l_{11} = \pm 1$ | $H_{\mathbf{v}_0,p_0}$ | $X_{\mathbf{v}_0}$ |
| $G_{P_0}\cdot C_{\mathbf{u}_0}$ | $\mathbf{u}_0 \perp P_0$ | $l_{11} = \pm 1$ | $R_{\mathbf{u}_0}$ | $X_{\mathbf{v}_0}$<br>$\mathbf{v}_0 \perp P_0$ |
| $S_{o_0}\cdot C_{\mathbf{u}_0}$ | $o_0 \in$ axis $\mathbf{u}_0$ | $l_{11} = \pm 1$<br>$l_{24} = 0$<br>$l_{34} = 0$ | $R_{\mathbf{u}_0}$ | $S_{o_0}\cdot T_{\mathbf{u}_0}$ |
| $S_{o_0}\cdot G_{P_0}$ | | | $R_{\mathbf{u}_0}$<br>$o_0 \in$ axis $\mathbf{u}_0$<br>$\mathbf{u}_0 \perp P_0$ | $S_{o_0}\cdot T_{P_0}$ |
| $S_{o_0}\cdot X_{\mathbf{v}_0}$ | | | $R_{\mathbf{u}_0}$<br>$o_0 \in$ axis $\mathbf{u}_0$<br>$\mathbf{u}_0 \parallel \mathbf{v}_0$ | $SE(3)$ |
| $S_{o_0}\cdot S_{o_1}$ | | | $R_{\mathbf{u}_0}$<br>$\mathbf{u}_0 = \frac{\overline{o_0 o_1}}{\lvert o_0 o_1\rvert}$ | $S_{o_0}\cdot R_{\mathbf{u}_0}\cdot R_{\mathbf{u}_1}$<br>$o_1 \in$ axis $\mathbf{u}_0$<br>$o_1 \in$ axis $\mathbf{u}_1$ |
| $Y_{\mathbf{v}_0,p_0}\cdot Y_{\mathbf{v}_1,p_1}$<br>$(p_0 \neq p_1)$ | $\mathbf{v}_0 \parallel \mathbf{v}_1$ | $l_{11} = \pm 1$ | $T_{P_0}$ $P_0 \perp \mathbf{v}_0$ | $X_{\mathbf{v}_0}$ |
| $Y_{\mathbf{v}_0,p_0}\cdot X_{\mathbf{v}_1}$ | $\mathbf{v}_0 \not\parallel \mathbf{v}_1$ | $l_{11} \neq \pm 1$ | $T_{P_0}$ $P_0 \perp \mathbf{v}_0$ | $X_{\mathbf{v}_0}\cdot R_{\mathbf{u}_1}$<br>$\mathbf{u}_1 \parallel \mathbf{v}_1$ |
| $G_{P_0}\cdot Y_{\mathbf{u}_1,p_0}$ | $\mathbf{v}_0 \perp P_0$ | $l_{11} = \pm 1$ | $T_{P_0}$ | $X_{\mathbf{v}_0}$ |
| $G_{P_0}\cdot X_{\mathbf{v}_0}$ | $\mathbf{v}_0 \not\perp P_0$ | $l_{11} \neq \pm 1$ | $T_{P_0}$ | $R_{\mathbf{u}_0}\cdot T\cdot R_{\mathbf{u}_1}$<br>$\mathbf{u}_0 \perp P_0,\ \mathbf{u}_1 \parallel \mathbf{v}_0$ |
| $G_{P_0}\cdot T$ | | | $T_{P_0}$ | $X_{\mathbf{v}_0}$ $\mathbf{v}_0 \perp P_0$ |
| $Y_{\mathbf{v}_0,p_0}\cdot T$ | | | $T_{P_0}$ | $X_{\mathbf{v}_0}$ |
| $X_{\mathbf{v}_0}\cdot X_{\mathbf{v}_1}$ | $\mathbf{v}_0 \not\parallel \mathbf{v}_1$ | $l_{11} \neq \pm 1$ | $T$ | $R_{\mathbf{u}_0}\cdot T\cdot R_{\mathbf{u}_1}$<br>$\mathbf{u}_0 \parallel \mathbf{v}_0,\ \mathbf{u}_1 \parallel \mathbf{v}_1$ |

Note that, although the intersection of two subgroups is at least the identity displacement, the intersection of two constraints may be the empty set.

When the intersection of two constraints is null, i.e. it is not possible to find a set of displacements satisfying (4.4), it implies that both kinematic constraints can not be simultaneously satisfied. This situation can not be detected through the intersection of subgroups. Roughly speaking, if we state our problems of kinematic constraints purely in terms of compositions and intersections of subgroups, we will be unable to detect inconsistencies. As it has been pointed out in [1], Group Theory provides the means for a *topological analysis* of the behavior of a set of bodies linked by a set of kinematic constraints, but a *geometric analysis* is required if we care about dimensions.

Thus, the problem of intersecting two constraints, say $\mathbf{A}_1\hat{S}_1\mathbf{A}_2$ and $\mathbf{B}_2^{-1}\hat{S}_2\mathbf{B}_1^{-1}$, can be expressed, if their intersection is different from null, in terms of the intersection of two subgroups, and the information about the spatial relationships between their geometric elements of definition can be obtained either from $\mathbf{A}_2 \cdot \mathbf{B}_1$ or $\mathbf{B}_2 \cdot \mathbf{A}_1$. Obviously, both information must be consistent.

The outcomes of the composition and intersection of two continuous subgroups of $SE(3)$, for all those cases in which the intersection is different from the identity displacement or one subgroup is a subgroup of the other, have also been tabulated in table 4.3. $l_{ij}$ denotes the element $(i, j)$ of the $4 \times 4$ homogeneous transformation representation for the linking displacement.

See [9] for deeper prospects on the intersection of, possibly not continuous, subgroups of $SE(3)$.

As a summary, we can say that: (a) the composition of two trivial constraints is sometimes a trivial constraint; (b) the intersection of two trivial constraints is a trivial constraint or null; and (c) the intersection of two non-trivial constraints is not necessarily a constraint, as defined here.

**Definition 3 (Independence and inconsistency)** *Two trivial constraints, $R_1$ and $R_2$, are said to be independent iff $(R_1 \cap R_2)^I$ is the identity displacement, and they are said to be inconsistent iff $R_1 \cap R_2$ is the empty set.*

Let us suppose that we want to find out the dimension of $R_{13} = R_{12} \cdot R_{23}$ or, in other words, the number of d.o.f. of the body $\mathcal{B}_3$ with respect to $\mathcal{B}_1$. It can be stated that:

$$dim(R_{13}) = dim(R_{12}) + dim(R_{23}) - dim(R_{12} \cap R_{23})$$

This formula can be extended to the composition of $n$ constraints, leading to a variation of the *Chebyshev-Grübler-Kutzbach* formula

$$dim(R_{1,n+1}) = \sum_{1}^{n} dim(R_{i,i+1}) - \sum_{l=2}^{n} dim(R_{1,l} \cap R_{1,l+1}) \qquad (4.5)$$

where

$$R_{ij} = \prod_{l=i}^{j-1} R_{l,l+1}$$

There are many examples of kinematic chains whose degree of freedom cannot be determined from its sole topology, i.e., they are elusive to the application of 4.5 [1, page 86].

**Definition 4 (Regular representation)** *The composition of two trivial constraints, $R_3 = R_1 R_2$ provide a regular representation for $R_3$ iff $(R_1 \cap R_2)^I = \mathbf{I}$. Then, $dim(R_3) = dim(R_1) + dim(R_2)$*

Notice that regular representations are not unique.

### 4.3.3 Examples

Firstly, let us analyze the composition of two constraints whose associated subgroups are $G_{P_0}$ and $X_{\mathbf{u}_0}$. This composition can be expressed as:

$$\begin{aligned} R_c &= \mathbf{A}\ \hat{S}_1\ \mathbf{L}\ \hat{S}_2\ \mathbf{B} \\ &= \mathbf{A}\ \mathbf{Trans}(0, y, z)\ \mathbf{Twix}(\theta)\ \mathbf{L}\ \mathbf{Trans}(x', y', z')\ \mathbf{Twix}(\psi)\ \mathbf{B} \end{aligned}$$

where $\mathbf{L}$ is the linking displacement between both constraints. On the other hand, $G_{P_0}$ and $X_{\mathbf{u}_0}$ can be decomposed into composition of subgroups as follows:

$$\left.\begin{array}{l} G_{P_0} = T_{P_0} \cdot R_{\mathbf{u}_1} = R_{\mathbf{u}_1} \cdot T_{P_0} \\ X_{\mathbf{u}_0} = T \cdot R_{\mathbf{u}_2} = R_{\mathbf{u}_2} \cdot T \end{array}\right\}$$

with $\mathbf{u}_1 \perp P_0$ and $\mathbf{u}_2 \| \mathbf{u}_0$.

If $\mathbf{u}_0 \not\perp P_0$, then $l_{11} \neq \pm 1$ (see table 4.3) and the only possible simplification for $R_c$ is:

$$R_c = \mathbf{A}\ \mathbf{Trans}(x'', y'', z'')\ \mathbf{Twix}(\theta)\ \mathbf{L}\ \mathbf{Twix}(\psi)\ \mathbf{B}$$

The simplified term, $\mathbf{Trans}(0, y, z)$, corresponds to the intersection of $G_{P_0}$ and $X_{\mathbf{u}_0}$. In terms of subgroups (table 4.3), we have

$$G_{P_0} \cdot X_{\mathbf{u}_0} = X_{\mathbf{u}_1} \cdot R_{\mathbf{u}_2} = R_{\mathbf{u}_1} \cdot T \cdot R_{\mathbf{u}_2} = R_{\mathbf{u}_1} \cdot X_{\mathbf{u}_2}$$

$$G_{P_0} \cap X_{\mathbf{u}_0} = T_{P_0}$$

with $\mathbf{u}_1 \perp P_0$ and $\mathbf{u}_2 \parallel \mathbf{u}_0$.

If $\mathbf{u}_0 \perp P_0$, then $\mathbf{u}_1 \parallel \mathbf{u}_0$, $l_{11} = \pm 1$, and $G_{P_0}$ becomes a subgroup of $X_{\mathbf{u}_0}$ (table 4.2). Consequently, $R_c$ can be expressed as

$$\begin{aligned} R_c &= \mathbf{A}\ \mathbf{Trans}(x''', y''', z''')\ \mathbf{Twix}(\theta)\mathbf{L}'\ \mathbf{Twix}(\psi)\ \mathbf{B} \\ &= \mathbf{A}\ \mathbf{Trans}(x''', y''', z''')\ \mathbf{Twix}(\theta + l'_{11}\psi)\ \mathbf{L}'\ \mathbf{B} \end{aligned}$$

where $\mathbf{L} = \mathbf{L}'\ \mathbf{Trans}(0, l_{24}, l_{34})$.

Notice that the necessary and sufficient condition for the equality

$$\mathbf{Twix}(\theta_1)\ \mathbf{L}\ \mathbf{Twix}(\theta_2) = \mathbf{Twix}(\psi)\ \mathbf{L}$$

to hold is that $l_{11} = \pm 1$, $l_{24} = 0$ and $l_{34} = 0$. In this case $\psi = \theta_1 + l_{11}\theta_2$.

Let us see another example. Imposing that the axes of the cylinders be aligned with the axes of their corresponding holes for the workpieces in fig. 4.3, the following expressions for both constraints will be obtained:

$$R_{12} = \mathbf{A}_{11}\mathbf{Trans}(x_1, 0, 0)\mathbf{Twix}(\theta_1)\mathbf{A}_{12}, \quad R_{12}^A = C_{\mathbf{u}_0}$$

$$R_{23} = \mathbf{A}_{21}\mathbf{Trans}(x_2, 0, 0)\mathbf{Twix}(\theta_2)\mathbf{A}_{22}, \quad R_{23}^A = C_{\mathbf{u}_1}$$

The composition of both constraints yields:

$$R_{12}R_{23} = \mathbf{A}_{11}\ \mathbf{Trans}(x_1, 0, 0)\ \mathbf{Twix}(\theta_1)\ \mathbf{L}\ \mathbf{Trans}(x_2, 0, 0)\ \mathbf{Twix}(\theta_2)\ \mathbf{A}_{22}$$

where the linking displacement is:

$$\mathbf{L} = \mathbf{A}_{12}\mathbf{A}_{21}.$$

Since $\mathbf{u}_0$ and $\mathbf{u}_1$ are parallel, and according to table 4.3, $l_{11} = \pm 1$; therefore, the composition of both constraints can be simplified leading to:

$$R_{12}R_{23} = \mathbf{A}_{11}\ \mathbf{Trans}(x_1 + l_{11}x_2, 0, 0)\ \mathbf{Twix}(\theta_1)\ \mathbf{L}\ \mathbf{Twix}(\theta_2)\ \mathbf{A}_{22} \qquad (4.6)$$

or, in other words,

$$R_{12}R_{23} = \mathbf{A}_{11}\ \hat{S}_1\ \mathbf{L}\ \hat{S}_2\ \mathbf{A}_{22}$$

where $\hat{S}_1 \in C_{\mathbf{u}}$ and $\hat{S}_2 \in R_{\mathbf{u}}$. Expression (4.6) is a regular representation for the composition of both constraints.

If, in addition to $l_{11} = \pm 1$, $l_{24} = 0$ and $l_{34} = 0$ ($\mathbf{u}_0 \bowtie \mathbf{u}_1$), a further simplification could be carried out and $R_{12}R_{23}$ becomes a trivial constraint. In this case $(R_{12}R_{23})^G \in C_{\mathbf{u}}$.

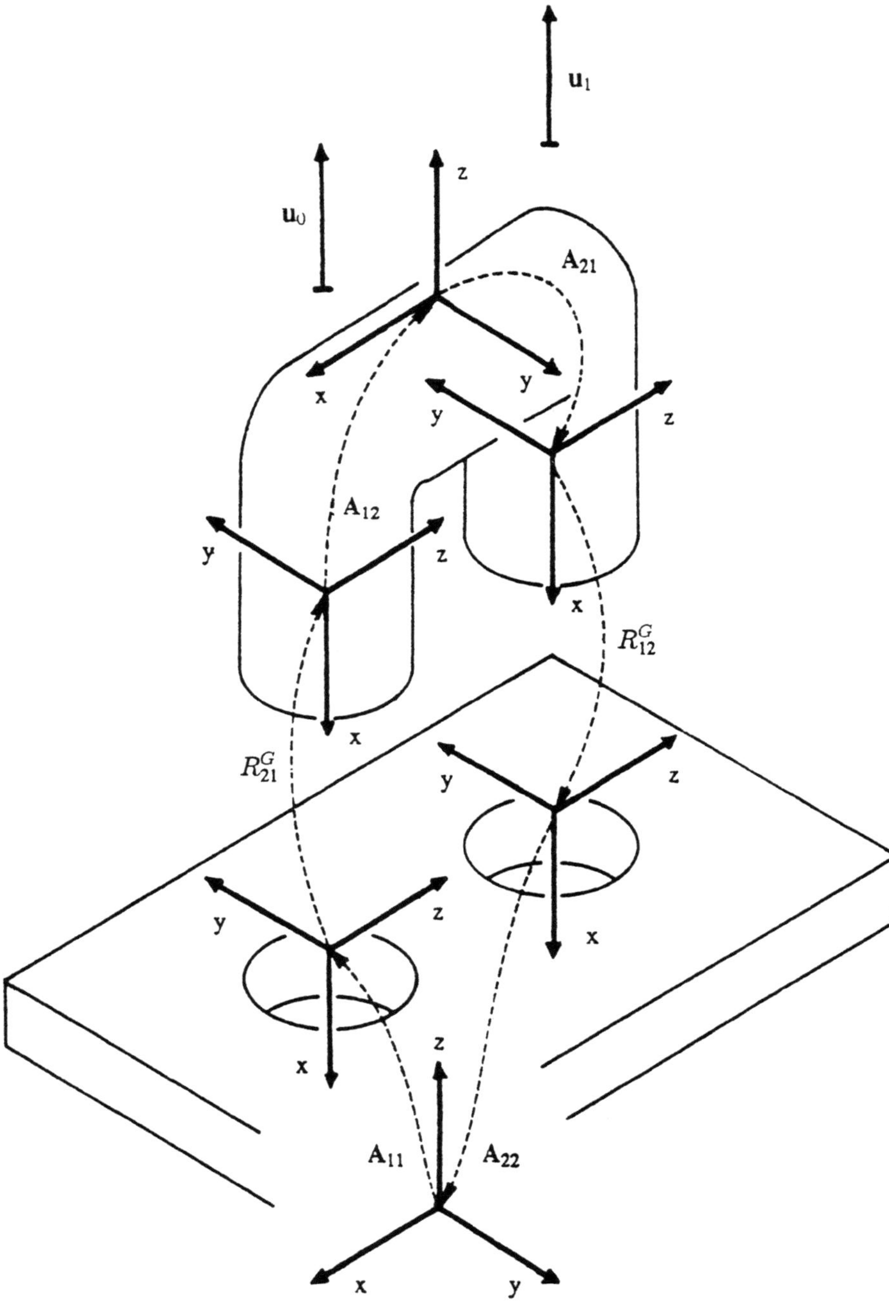


Figure 4.3: Insertion of a clamp. Geometric elements of definition, kinematic constraints and canonical subgroups involved.

Let us suppose that now we want to obtain $R_{12} \cap R_{23}^{-1}$, the equivalent constraint between $\mathcal{B}_1$ and $\mathcal{B}_2$. Then, $(R_{12} \cap R_{23}^{-1})^I$ can be easily obtained using table 4.3. Observe that the simplified term in 4.6, $\mathbf{Trans}(x_2, 0, 0)$, is $(R_{12} \cap R_{21}^{-1})^{IG}$. This term encompasses the remaining d.o.f. of body $\mathcal{B}_2$ with respect to $\mathcal{B}_1$, when $\mathcal{B}_3$ is kept rigidly linked, as in this case, to $\mathcal{B}_1$.

We have proved that the above constraints are not independent, but we have not checked their consistency. Depending on the relative dimensions of the involved workpieces, they may be inconsistent. The only thing we can say using this kind of symbolic manipulation is that, if $(R_{12} \cap R_{21}^{-1}) \neq \emptyset$, then $\mathcal{B}_2$ only have one translational degree of freedom with reference to $\mathcal{B}_1$. Checking consistency requires a geometrical analysis which requires, in turn, solving a kinematic equation. In our example, we would have to decide whether $R_{12}R_{23} = \mathbf{I}$ has a solution. Thus, although the previous ideas provide a theoretical framework within which it is easy to justify, for instance, when the composition of two constraints can be simplified, they must be complemented with an algorithm to obtain numerical values for the constrained d.o.f. if we care about dimensions. See [4] for new developments in this area.

### 4.3.4 Star-polygon transform

The above two basic operations are not enough for obtaining the equivalent constraint between any two bodies in an arbitrary graph of kinematic constraints. This fact can be easily proved by drawing a fully connected GR graph with four nodes and trying to obtain the equivalent constraint between any two of them through the iterative application of compositions and intersections of constraints.

The star-polygon transform is included here to provide a complete set of operations which make possible to obtain the equivalent constraint between any two bodies in an arbitrary GR graph.

The star-polygon transform consists in removing one node of the GR graph by fully connecting all the nodes connected to it with the equivalent constraint between them (fig. 4.2c). This operation can be seen as a generalized composition. Actually, when this transform is applied to a node of degree two, the result is the composition of two constraints.

The problem with this operations is that, once it has been applied, the involved constraints *share* variables. Thus, when a variable is assigned somewhere in the graph, it is necessary to take into account that it may be shared by another constraint. In the next section, an algorithm, which represents a way around this difficulty, is introduced. This algorithm is able to find the equivalent constraint between any two bodies without resorting to this operation.

## 4.4 Propagation of constraints

If, as the result of intersecting two constraints between the same two workpieces, the empty set is obtained, we say that they are inconsistent. The goal is now to verify the consistency of entire GR graphs. This can be stated as a problem of consistency in networks of relations. As it is pointed out in [10], any representation of the constraints that allow composition and intersection is sufficient for this purpose.

Informally, a GR graph is consistent if there exist configurations between workpieces whose defining coordinate transformations belong to the corresponding constraints. Obviously, a GR graph without cycles is always consistent; thus, it is easy to realize the important role of cycles in GR graphs.

Next, before introducing a general algorithm for propagating kinematic constraints, some few concepts on cycles in graphs are reminded.

### 4.4.1 Preliminaries on cycles

Two basic operations with cycles are the ***union*** and the ***ring sum***. The union of two cycles $C_1 = (V_1, E_1)$ and $C_2 = (V_2, E_2)$ is a graph $G = C_1 + C_2$ with node set $V_3 = V_1 \cup V_2$ and arc set $E_3 = E_1 \cup E_2$. The ring sum of two cycles $C_1$ and $C_2$ (written $C_1 \bigoplus C_2$) is another cycle or a set of disjoint cycles consisting of the node set $V_1 \cup V_2$ and of arcs that are either in $C_1$ or $C_2$, but not in both.

A set of cycles $\mathcal{H}$ in a graph $G = (V, E)$ is said to be a complete set of basic cycles if (i) every cycle in the graph can be expressed as a ring sum of some or all cycles in $\mathcal{H}$, and (ii) no cycle in $\mathcal{H}$ can be expressed as a ring sum of others in $\mathcal{H}$. The cardinality of a complete set of basic cycles is $\mu = | E | - | V | + 1$, which is called the cyclomatic number. Hence the maximum number of cycles is $2^\mu - 1$.

### 4.4.2 Isolation of blocks

When a kinematic constraint is posted, it can affect other workpieces different from those it is incident to, but, in general, a constraint is limited in its *scope*. In order to isolate subgraphs within which the effect of a constraint is limited, the following operations are applied:

1 Elimination of cutlines or bridges. This includes the elimination of pendant constraints (fig. 4.4a).

2 Split cutpoints or articulation nodes into two nodes to produce two disjoint subgraphs (fig. 4.4b).

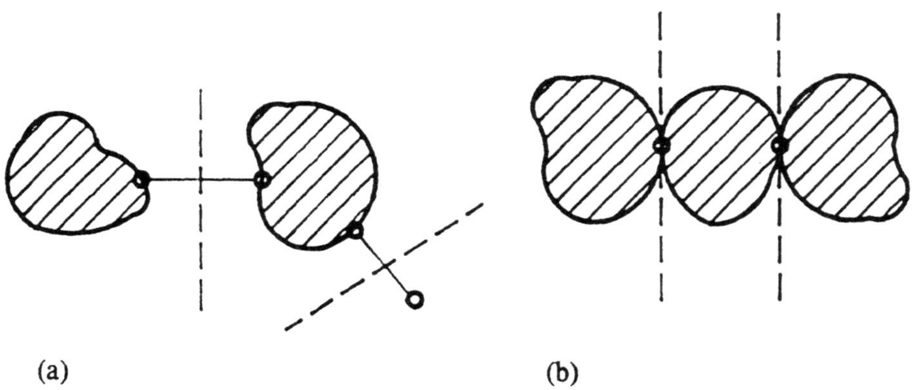


Figure 4.4: Operations applied for the isolation of blocks: (a) elimination of cutlines; and (b) splitting cutpoints.

As a result of these operations a set of subgraphs, or simply *blocks*, are obtained. A GR graph is consistent if each of its blocks is consistent.

Now, we can introduce a definition for an important subclass of graphs of kinematic constraints.

**Definition 5 (Trivial GR graph)** *A GR graph is said to be trivial iff the equivalent constraint between any two nodes in any of its blocks can be expressed a trivial constraint.*

It is obvious that a GR graph without cycles is always trivial.

Let us assume that the obtained blocks are planar graphs. This assumption, while not very restrictive, simplifies the treatment given below. Anyway, the provided results can be extended to non-planar graphs.

A plane representation of a graph divides the plane into *regions*. A region is characterized by the set of arcs forming its boundary. In a plane representation of a planar connected graph the set of cycles forming the interior regions, or *region cycles*, constitutes a complete set of basic cycles. The set of region cycles is not unique. Actually, there are $\binom{\mu+1}{\mu}$ different sets of region cycles. This can be easily seen by noting that a planar graph can be embedded on the surface of a sphere. The number of region cycles in the surface of a sphere would be $\mu + 1$, which are also the shortest cycles for a planar graph.

Let $X$ be the set of region cycles in a planar block $G$. The *cycle graph* of $X$ is the graph with vertex set $X$ and arcs joining two distinct nodes if and only if the corresponding cycles have an arc in common. This graph is denoted by $\mathcal{D}(G)$ and it can be easily proved that $\mathcal{D}(G)$ is a subgraph of the dual graph of $G$ (see [6, page 106]). Nodes in a $\mathcal{D}(G)$ graph stand for cycles and arcs in $\mathcal{D}(G)$, for *shared arcs* in $G$. For extension, constraints labeling a shared arc are called *shared constraints.* Note that an arc can only be shared by two region cycles.

Let $C_i$ be a region cycle whose arc set is labeled with the constraints

$$\{R_1, R_2, \dots, R_j, \dots, R_n\}$$

according to fig. 4.5. Then, the constraint $R_j$ can be substituted by

$$R_j^i = R_j \cap (R_{j-1}^{-1} \cdots R_1^{-1} \cdot R_n^{-1} \cdots R_{j+1}^{-1})$$

without modifying the consistency of the corresponding GR graph (fig. 4.5b). In order to simplify the notation, we will write

$$R_j^i = \cap^{C_i} R_j.$$

This is the basic mechanism for constraint propagation as it is shown below.

### 4.4.3 A filtering algorithm for propagating kinematic constraints

A general procedure to *propagate* the effect of constraints in GR graphs has been devised, either to characterize the set of configurations that satisfy all the constraints or to find out that there exist no such configurations.

The propagation process consists in *filtering* all constraints, that is eliminating from the constraints those displacements which cannot appear in any solution. Eventually, if all constraints are reduced to only one element, a single solution is obtained.

Global consistency in a block $G$ is checked by eliminating local inconsistencies; that is, by eliminating inconsistencies in region cycles – which is equivalent to ensure *node inconsistency* in $\mathcal{D}(G)$ –, and by eliminating inconsistencies between adjacent region cycles – which is equivalent to ensure *arc consistency* in $\mathcal{D}(G)$ (see [10] or [11]). The following procedure implements this idea.

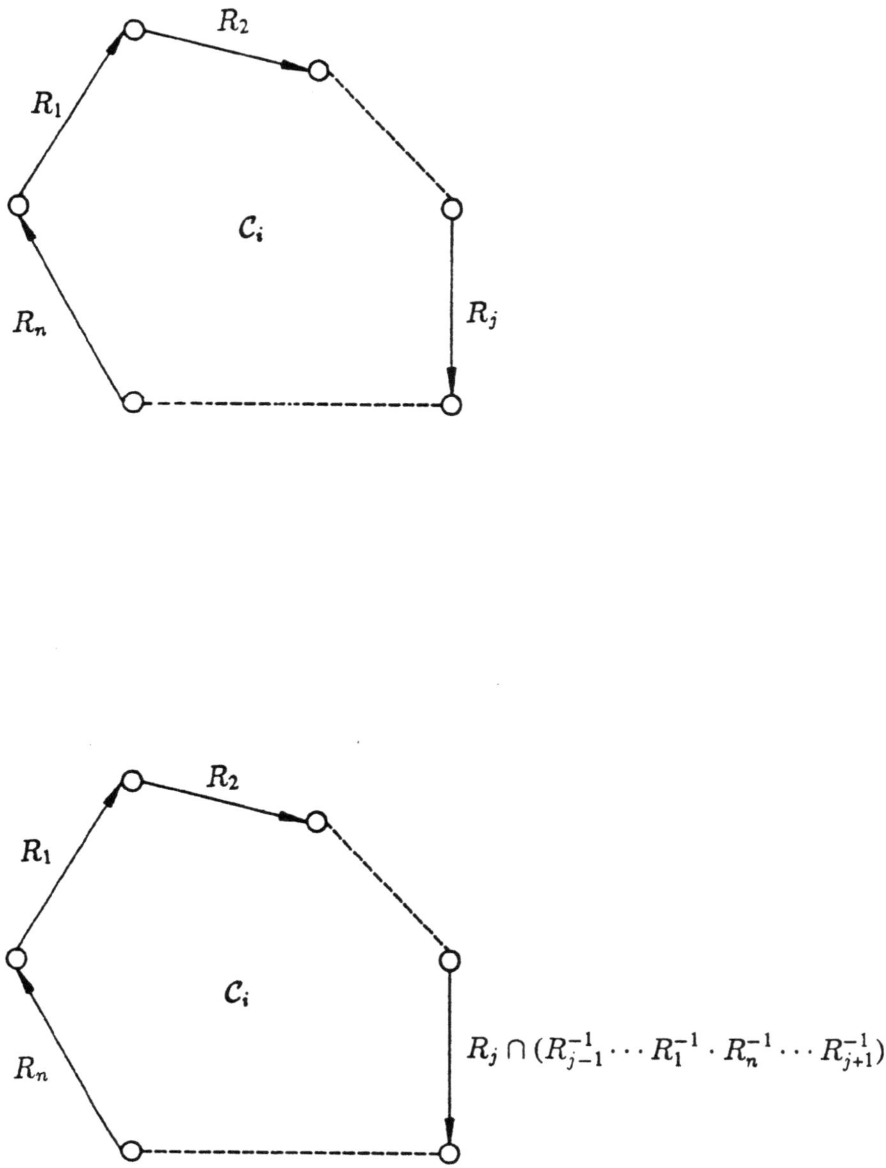


Figure 4.5: The basic mechanism for constraint propagation. A constraint $R_j$, labeling an arc in a cycle $C_i$, can be substituted by $\cap^{\mathcal{C}_i} R_j$.

**procedure** filter_constraints;<br>
**input:** G; /* a block of a GR graph*/<br>
**output:** G;<br>
**repeat**

stop:= true;

/* check node consistency */

**forall** region cycles $C_j$ **do**<br>
**forall** constraints $R_i$ in $C_j$ **do**<br>
/*1*/ $R_i^j := \cap^{C_j} R_i$;<br>
/*2*/ **if** $R_i^j == \emptyset$ **then exit();**<br>
**enddo;**<br>
**enddo;**

/* check arc consistency */

**forall** shared constraints $R_i$ **do**<br>
/*3*/ $R_T := \cap_k R_i^k$;<br>
/*4*/ **if** $R_T == \emptyset$ **then exit();**<br>
**if** $R_T \neq R_i$ **then**<br>
stop:= false;<br>
$R_i := R_T$;<br>
**endif;**<br>
**enddo;**<br>
**until** stop;<br>
**end.**

Geometric inconsistencies can be found either when $\cap^{C_j} R_i$ or when $\cap_k R_i^k$ become the empty set. In the first case, the cycle $C_j$ becomes inconsistent; in the second one, all the cycles sharing the constraint $R_i$ do. The problem of obtaining the minimum set of constraints that made a given GR graph become inconsistent is addressed in [17].

The above algorithm can be easily modified for its application for a topological analysis, stating the problem in terms of composition and intersections of associated subgroups instead of constraints. Then, sentences /*2*/ and /*4*/ can be removed and, if the outcome /*1*/ is not a subgroup for a particular $C_j$, it is not taken into account when computing /*3*/. In this case, the algorithm will halt when no progress is made, either because the graph is not trivial, or because all possible filterings have already been carried out. An example illustrating this idea is shown in the next section.

If the corresponding GR graph is planar, it is presumable that the complexity of the above algorithm is polynomial in the number of constraints [11].

The significance of the described algorithm is that it only needs to repeatedly handle a set of short cycles. Because of domain specific attributes, node and arc consistency provide a sufficient guarantee that there is a complete solution, in the same way the celebrated Waltz's filtering algorithm provides a complete solution for polyhedral scene labeling looking only for arc consistency [19].

## 4.5 Example

Let the workpieces in fig. 4.6a be elements of an assembly. The matings between complementary features of workpieces $\mathcal{B}_1$, $\mathcal{B}_2$ and $\mathcal{B}_3$ lead to the GR graph in fig. 4.6b, which only contains one block with one cycle. Thus, the equivalent constraint between any two workpieces can be obtained by simply reduction of the graph to a single edge linking them. For example, the equivalent constraint between $\mathcal{B}_2$ and $\mathcal{B}_3$, if different from null, will be a coset of

$$(T_{\mathbf{u}_1} \cdot T_{\mathbf{u}_2}) \cap T_{\mathbf{u}_3} = T_{\mathbf{u}_1} \cap T_{\mathbf{u}_3} = I,$$

i.e. $\mathcal{B}_2$ will remain fixed with reference to $\mathcal{B}_3$. This suggest that $\mathcal{B}_2$ and $\mathcal{B}_3$ must be put together before $\mathcal{B}_1$ is assembled, providing valuable information about the assembly sequence.

Now, let us consider all the workpieces in fig. 4.6a. Given the matings between their complementary features, the problem consists in deciding whether these matings are enough for fixing the relative location of these four workpieces. The corresponding GR graph appears in fig. 4.6c. Neither composition nor intersection of constraints can be applied to reduce it.

Using the algorithm proposed in the last section, we can write the following table:

| | $R_i^A$ | $\mathcal{C}_1$ | $\mathcal{C}_2$ | $\mathcal{C}_3$ | results 1$^{st}$ iteration | $\mathcal{C}_1$ | $\mathcal{C}_2$ | $\mathcal{C}_3$ | results 2$^{nd}$ iteration |
|---|---|---|---|---|---|---|---|---|---|
| $R_1$ | $T_{\mathbf{u}_1}$ | $T_{\mathbf{u}_1}$ | | | $T_{\mathbf{u}_1}$ | $I$ | | | $I$ |
| $R_2$ | $T_{\mathbf{u}_2}$ | $T_{\mathbf{u}_2}$ | | $I$ | $I$ | $I$ | | $I$ | $I$ |
| $R_3$ | $T_{\mathbf{u}_3}$ | $I$ | $I$ | | $I$ | $I$ | $I$ | | $I$ |
| $R_4$ | $C_{\mathbf{u}_4}$ | | ? | ? | $C_{\mathbf{u}_4}$ | | $C_{\mathbf{u}_4}$ | $C_{\mathbf{u}_4}$ | $C_{\mathbf{u}_4}$ |
| $R_5$ | $C_{\mathbf{u}_5}$ | | ? | | $C_{\mathbf{u}_5}$ | | $C_{\mathbf{u}_4}$ | | $C_{\mathbf{u}_4}$ |
| $R_6$ | $C_{\mathbf{u}_6}$ | | | ? | $C_{\mathbf{u}_6}$ | | | $C_{\mathbf{u}_4}$ | $C_{\mathbf{u}_4}$ |

taking into account that:

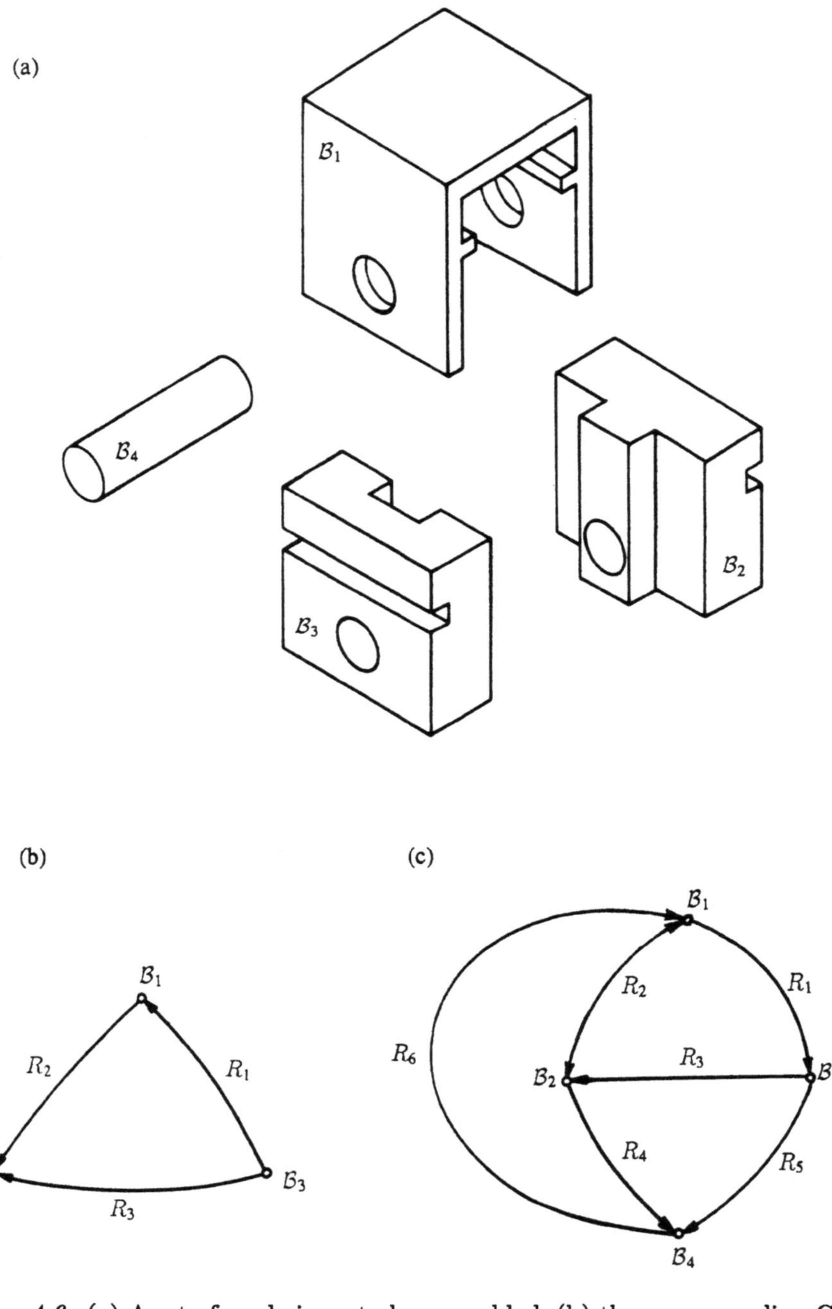


Figure 4.6: (a) A set of workpieces to be assembled; (b) the corresponding GR graph involving workpieces $\mathcal{B}_1$, $\mathcal{B}_2$ and $\mathcal{B}_3$; and (c) the GR graph involving all workpieces.

| | |
|---|---|
| $\mathbf{u}_1 \parallel \mathbf{u}_2$ | $\mathbf{u}_4 \bowtie \mathbf{u}_5$ |
| $\mathbf{u}_1 \perp \mathbf{u}_3$ | $\mathbf{u}_4 \bowtie \mathbf{u}_6$ |
| $\mathbf{u}_2 \perp \mathbf{u}_1$ | $\mathbf{u}_5 \bowtie \mathbf{u}_6$ |
| $\mathbf{u}_3 \perp \mathbf{u}_4$ | $\mathbf{u}_2 \perp \mathbf{u}_6$ |
| $\mathbf{u}_3 \perp \mathbf{u}_5$ | $\mathbf{u}_1 \perp \mathbf{u}_6$ |

which can be directly inferred from the linking displacements.

The outcomes of $(\cap^{\mathcal{C}_i} R_j)^A$ appear in row $R_j$ and column $\mathcal{C}_i$. In the first iteration, some of this subgroups cannot be obtained since the corresponding constraint is not trivial. In the third iteration no new progresses can be carried out and, since it was possible to compute $(\cap^{\mathcal{C}_i} R_j)^A$ for $i = 1..3$ and $j = 1..6$, the algorithm finishes after propagating all the constraints. Then, it can be said that, if all introduced constraints are geometrically consistent, then the bodies $\mathcal{B}_1$, $\mathcal{B}_2$ and $\mathcal{B}_3$ will remain fixed and, if they are considered as a subassembly, then the body $\mathcal{B}_4$ will have two d.o.f. with reference to it.

## 4.6 Summary

A kinematic constraint has been defined as a set of displacements which can be expressed as a composition of cosets of Euclidean subgroups. A constraint is said to be trivial when it can be reduced to a single coset. Trivial constraints include all kinematic lower pairs.

A characterization of the spatial relationships between bodies in assemblies as trivial kinematic constraints, as well as a tabulation of the outcomes of the composition and intersection of the corresponding subgroups has been given. The theoretical foundation for this systematization has been taken from [7].

A graph of kinematic constraints has been defined as a graph whose nodes correspond to workpieces and whose directed arcs are labeled with trivial kinematic constraints.

It has been shown that it is not always possible to obtain the equivalent constraint between any two bodies in a graph of kinematic constraints by simply composing and intersecting constraints, so that the graph is reduced to a single arc linking both bodies. An algorithm that provides a way around this difficulty has been proposed. This algorithm filters all the constraints in a graph of kinematic constraints. This process consists in eliminating from the constraints those displacements which cannot appear in any solution.

It has been shown how – relying on the composition and intersection of subgroups – it is possible to carry out a topological analysis of the motion possibility for a set of bodies linked by a set of trivial kinematic constraints. It has also been shown that it is not possible to derive geometric inconsistencies from this analysis.

# Acknowledgements

This research was funded by the *Fundación Areces*, under the project SEPETER, and the *Comisión Interministerial de Ciencia y Tecnología (CICYT),* under the project "Automatic spatial reasoning based on constraints." The author was also supported by a NATO fellowship.

The author thank Yanxi Liu and Robin Popplestone, from the University of Massachusetts at Amherst, for their useful comments during the preparation of this chapter.


# References


[1] J. Angeles, *Rational Kinematics.* Springer-Verlag, New York, 1988.

[2] O. Bottema and B. Roth, *Theoretical Kinematics.* Dover Publications, New York, 1979.

[3] J.E. Campbell. *Introductory Treatise on Lie's Theory of Finite Continuous Transformations Groups.* Clarendon Press, Oxford, 1903.

[4] E. Celaya and C. Torras. Finding Object Configurations that Satisfy Spatial Relationships. *ECAI*, Stockholm, 1990.

[5] N. Christofides. *Graph Theory. An Algorithmic Approach.* Academic Press, N.Y., 1975.

[6] N. Deo. *Graph Theory with Applications to Engineering and Computer Science.* Prentice-Hall, Englewood Cliffs, N.J.,1974.

[7] J.M. Hervé. Analyse Structurelle des Mécanismes par Groupe des Déplacements. *Mechanism and Machine Theory*, Vol. 13, pp. 437-450, 1978.

[8] N. Koutsou. *Planning Motion in Contact to Achieve Parts Mating.* Ph.D. dissertation, University of Edinburgh, 1986.

[9] Y. Liu. *Symmetry Groups in Robotic Assembly Planning.* Ph.D. dissertation, University of Massachusetts, 1990.

[10] A.K. Mackworth. Consistency in Networks of Relations. *Artificial Intelligence Journal*, Vol. 8, pp. 99-118, 1977.

[11] A.K. Mackworth and E.C. Freuder, The Complexity of Some Polynomial Network Consistency Algorithms for Constraint Satisfaction Problems. *Artificial Intelligence Journal*, Vol. 25, pp. 65-74, 1985.

[12] E. Mazer. LM-GEO Geometric Programming of Assembly Robots. In *Advanced Software in Robotics*, pp. 99-125, Elsevier Science Publishers B.V. (North-Holland), 1984.

[13] R.P. Paul. *Robot Manipulators: Mathematics, Programming, and Control.* Cambridge, MA. MIT Press, 1981.

[14] R.J. Popplestone, A.P. Ambler and I.M. Bellos. An interpreter for a language for describing assemblies. *Artificial Intelligence Journal*, Vol. 14, pp. 79-107, 1980.

[15] R.J. Popplestone, Group Theory and Robotics. In *Robotics Research. The First International Symposium*, edited by M. Brady and R. Paul. MIT Press, 1984.

[16] F. Thomas and C. Torras. A Group Theoretic Approach to the Computation of Symbolic Part Relations. *IEEE Journal of Robotics and Automation*, Vol.4, No.6, December 1988.

[17] F. Thomas and C. Torras. Inferring Feasible Assemblies from Spatial Constraints. *Technical Report of the Institute of Cybernetics*, IC-DT-1989.03,June 1989.

[18] J.M. Valade. Geometric Reasoning and Automatic Synthesis of Assembly Trajectory, *International Conference on Advanced Robots*, Tokyo, Japan, Sept. 9-10, 1985.

[19] D. Waltz. Generating Semantic Descriptions from Drawings of Scenes with Shadows. AI-TR-271, MIT, reprinted in *The Psychology of Computer Vision*, Winston (ed.), McGraw Hill, 1975.

[20] J.R. Woodwark. Some Speculations on Feature Recognition. In *Geometric Reasoning*, edited by D. Kapur and L. Mundy, MIT Press, 1989.

# Chapter 5

# Relative positioning of parts in assemblies using mathematical programming

Joshua U. Turner

In extending solid modeling technology to the modeling of assemblies of discrete parts, it is important to capture the position of each part in the assembly relative to the positions of its neighbors. By describing each part position in terms of relationships between various features of the part and mating features of its neighboring parts, it is possible for the solid modeling system to compute the modeling transformations needed to simulate the desired assembly configuration. In this chapter we formulate a mathematical programming approach to this problem. This approach is particularly useful in situations that arise in the solution of tolerancing problems.

It should be noted carefully that the relative positioning problem is distinct from the path planning problem or the assembly sequence problem. Here it is only the final position of each part that is of concern.

The AUTOPASS language of Lieberman and Wesley [1], [2], [3], first described a representation for feature-based assembly modeling, in which part positions are specified in terms of coplanarity and coaxiality relations between mating part features. Lee and Gossard [4] give a variant on this representation. Turner [5], [6], and Gossard et al. [7] give representations in which part positions are specified using relative positioning operators imbedded in a CSG-like tree.

Ambler and Popplestone [8], Lee and Andrews [9], and Rocheleau and Lee [10] give strategies for computing the modeling transformations implied by the feature relationships. Their methods attempt to satisfy all feature relationships among all parts simultaneously, and require that all relationships be satisfied exactly. Mullineux [11] extends the method of Rocheleau and Lee so that part relationships need not be exactly satisfied, but in so doing, creates a large, seemingly intractable unconstrained optimization problem. Rossignac [12] gives a strategy for computing the part positioning transformations through a concatenation of simple atomic operators. His method requires the user to specify the operator sequence, and does not allow for multiple mating feature relationships between a pair of parts.

In this chapter we extend previous work by formulating an efficient approach based on mathematical programming, that allows for part shapes in which it is not possible to satisfy some or all of the mating feature relationships exactly. This makes it possible to apply feature-based assembly modeling techniques to variational models of parts and assemblies which arise when tolerances are taken into account. The feature relationships are treated as inequalities. Mathematical programming is used to find the optimal configuration of the parts.

Although most problems can be solved in a sequential manner, positioning one part at a time, in some cases it is necessary to position several parts simultaneously. The mathematical programming formulation supports both sequential and simultaneous positioning operations.

If the nominal part positions are available, and if the actual part shapes are assumed to incorporate small variations to the nominal shapes, then the mathematical programming problem can be linearized, and solved quickly. This can be particularly important in the solution of problems involving tolerances, where the nominal part positions are known, and where we want to determine the effect of small variations applied to each of the parts upon the positions of the other parts, and ultimately, upon the overall functional requirements of the assembly. Further details as to the application of these methods to tolerancing problems may be found in Turner [5], [6], [13], [14], [15], [16], [17].

The next section gives a summary of the general mathematical programming schema. The rest of the chapter applies this schema to two common situations: 1) the assembly of planar polyhedra, and 2) the assembly of parts with mating patterns of holes and posts. By applying the assumption of small variations, both problems are linearized.

## 5.1 Relative Positioning by Mathematical Programming

Generally, regardless of whether a sequential or a simultaneous strategy is adopted, the simple relative positioning relationships presented in the preceding work suffer from a lack of generality. It is not clear how more complex assembly constraints would be modeled.

For instance, a coplanarity or coaxiality constraint may not be meaningful when two parts come in contact in several places. As an instance, suppose a flat plate is to be positioned so that it lies across the top of a U-shaped part. If the two ends of the U are not exactly coplanar with each other, then it will not be possible for coplanarity between the two parts to be maintained at both ends. Generally, in an actual part instance, the two ends of the U will not be exactly coplanar, even if they are nominally coplanar. As another example, if one part is a shaft to be inserted into a U-shape bracket so as to pass through a hole at each end of the U, and if the two holes are not coaxial, then it will not be possible for coaxiality between the two parts to be maintained at both ends.

Since such assembly situations are common, particularly when part variations are modeled, it is important to develop suitable relative positioning capabilities.

This section introduces a general approach to relative positioning based on mathematical programming. The general idea is as follows: Given a target object $B$ to be positioned relative to a given reference object $A$. $B$ is to be positioned such that certain of its features mate with certain features of $A$. The mating features should be aligned as closely as possible. Interference should be avoided.

In establishing a position for the target object $B$, there are six degrees of freedom (three translational, and three rotational). Lozano-Perez and Wesley [18] have observed that these six degrees of freedom may be conceived as six independent parameters, collectively spanning a six-dimensional *configuration space* that governs the position of $B$. The specification of a particular position for $B$ is equivalent to selecting a particular value for each of these six parameters (a particular point in the configuration space). The mathematical programming model may be formulated based on the following two observations:

1. The requirement that the various features of $B$ avoid interference with the corresponding features of $A$ is equivalent to a collection of **functional constraints** on the feasible points of the configuration space.

2. The requirement that mating pairs of features be aligned as closely as possible corresponds to one or more **objective functions**. on the space.

These non-interference constraints and alignment objectives can be based on any desired relationships between the part geometries, including, but not limited to, relationships between mating planar or axial features.

So the establishment of an optimum feasible position for $B$ can be formulated as a mathematical programming problem: The six dimensions of the configuration space of $B$ are the decision variables. The pairing of mating non-interfering features establishes one or more functional constraints on the feasible points of the configuration space. The goal of aligning one or more feature pairs as closely as possible determines one or more objective functions. These objective functions may be optimized sequentially, or simultaneously.

Where the relative positions of three or more parts are to be established simultaneously, the decision space will have six variables per part, for all the parts except one part selected as a fixed frame of reference.

The next two sections give two common examples of this schema, and show that in the case where small variations may be assumed, both can be solved as linear programming problems.

## 5.2 Mating Polyhedra

First, consider the case of assemblies composed of polyhedral parts. All mating pairs of features take the form of contacts between parallel planar faces.

The condition of non-interference between two parts could be evaluated by computing their volumetric intersection and then determining whether the result is a non-empty volume, but this is rather expensive: typical approaches grow quadratically with the number of faces of the two parts.

A less expensive non-interference test involves an examination of each pair of mating faces. If each face of a mating pair lies entirely outside the half-space bounded by the other face, then there is no interference. Since each of the two faces is planar, this condition can be evaluated by substituting the vertices of each face, in the plane equation of the other. Actually, we need consider only those vertices of each face which are part of its convex hull.

If the two mating faces only partially overlap, then this test is too restrictive. We need only be concerned with the region of overlap. The following procedure gives a simple approach to handling this most general case.

1. Assuming that the nominal position of part $B$ is already known, locate part $B$ so that it takes on its nominal position with respect to part $A$.
2. For each pair of mating faces $f_A$ and $f_B$, project $f_A$ onto the plane of $f_B$.
3. Compute the intersections of the boundaries of the two faces.

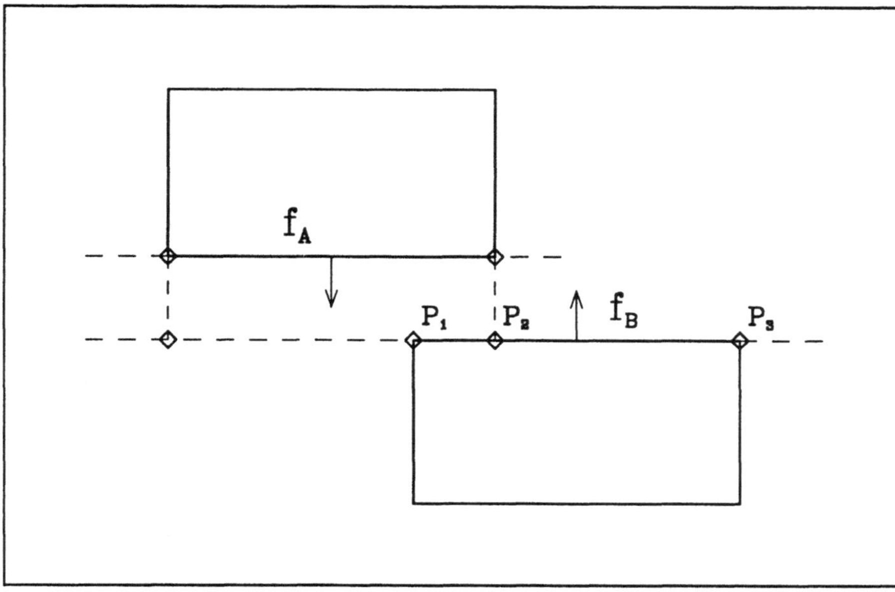


Figure 5.1: Procedure for Generating Non-interference Constraints

4. Compute the convex hull of: a) the vertices of intersection, b) the vertices of face $f_B$ that are interior to the projected face $f_A$, and c) the vertices of projected face $f_A$ that are interior to $f_B$.
5. Attach all of the vertices of this convex hull to face $f_B$, so that any transformation applied to part $B$ applies to these vertices as well.
6. Generate constraints to the effect that no vertex of this convex hull may appear on the material side of of the half-space bounded by $f_A$.

Providing that the nominal parts do not intersect, and that large deviations from nominal are not possible, this procedure will assure the non-intersection of the two parts in their final positions. Figure 5.1 illustrates the procedure.

Constraints are generated to the effect that $P_1$ and $P_2$ must remain below the plane of $f_A$. Note that $P_3$ is not constrained.

In addition to these non-interference constraints, one or more objective functions are required. The objective functions will specify that the two objects should be positioned so that for one or more pairs of mating faces, the contact between the two faces is maintained as much as possible. Maximum contact is maintained by minimizing the maximum distance between the two faces. For mating polyhedra it is sufficient to minimize the maximum distance between each vertex of the convex hull computed in the preceding procedure, and the plane of the corresponding face $f_A$. If contact is to be maintained between two or more pairs of mating faces, then a choice can be made whether these objectives are to be achieved simultaneously or sequentially. Sequential objectives give a result which is analogous to the fixturing of one part with respect to another using an ordered sequence of datum surfaces. Simultaneous objectives give an overall best fit without favoring one pair of mating features over another. In either case, well established techniques from mathematical programming can be applied.

The preceding formulation yields a linear problem in the vertices of the parts to be positioned. The vertex coordinates, in turn, are functions of the six positioning degrees of freedom. If small variations can be assumed, then these functions can be linearized, and the relative positions of the parts can be computed rapidly, using linear programming.

To make this discussion more concrete, let us look at a simple two-dimensional example. Suppose that an object $B$ is to be positioned relative to another object $A$. $B$ will be positioned by applying a rigid transformation to the coordinates of $B$. In two dimensions a rigid transformation may be expressed in terms of three parameters (two translational, and one rotational). These will be denoted $t_x$, $t_y$, and $\theta$. If the initial coordinates of any given point of $B$ are given by $p = (x, y)$, then after transformation, the same coordinates will be given by

$$p' = (x \cos\theta - y \sin\theta + t_x,\; x \sin\theta + y \cos\theta + t_y)$$

Let us assume that the nominal position of $B$ is already known. Under the assumption of small variations, we may assume that only small rotations will be necessary to accommodate any variations in $B$. This allows us to use the following approximations:

$$\sin\theta \approx \theta$$
$$\cos\theta \approx 1$$

Then the transformed coordinates of the point are given by

$$p' \approx (x - y\theta + t_x,\; x\theta + y + t_y)$$

Figure 5.2 gives the geometries of the two objects that will be used.

For convenience, the geometry of $A$ is given by identifying the equations of its bounding edges. The geometry of $B$ is given by identifying the coordinates of its vertices. The desired position of $B$ with respect to $A$ is shown with dashed lines. (The role of the parameter $b$ Is to allow for some variability in the shape of $A$. For now, assume that $b = 0$.)

The geometry of $B$ after transformation is shown in Figure 5.3.

Now non-interference constraints will be derived on the position of $B$, treating each of its four edges in turn. First, $B$ must be positioned so that its bottom edge falls above the line $y = 0$. This effectively constrains the y-coordinates of the two endpoints:

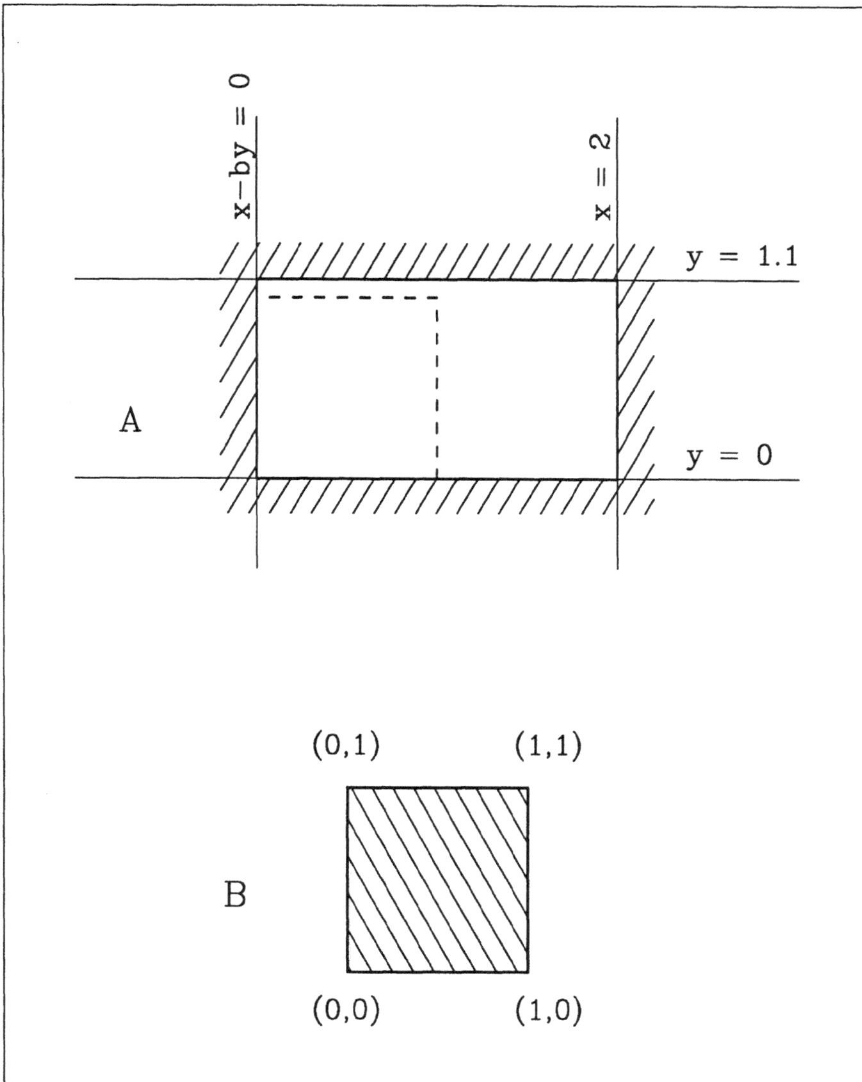


Figure 5.2: Reference Object (A) and Target Object (B)

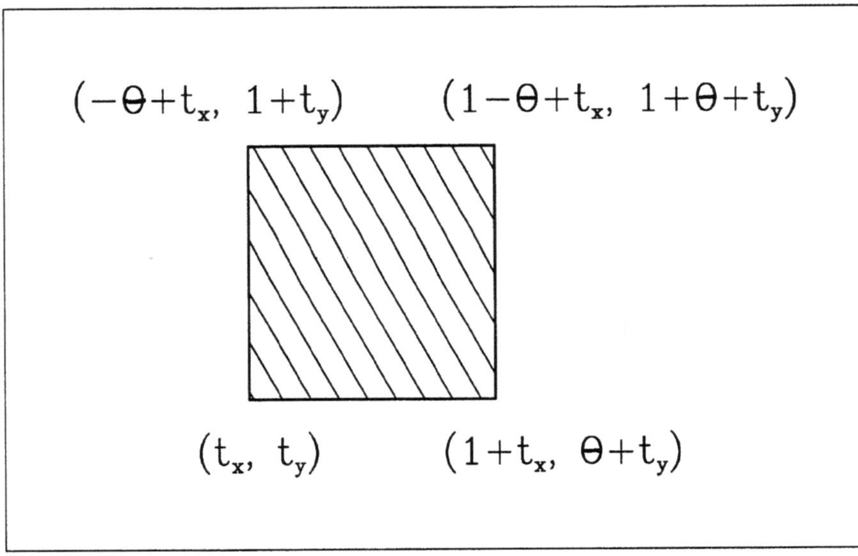


Figure 5.3: Target Object (B) after Transformation

$$t_y \geq 0 \tag{5.1}$$

$$\theta + t_y \geq 0 \tag{5.2}$$

Next, the left edge of $B$ must be to the right of the line $x - by = 0$. The following constraints are obtained by substituting the endpoints of the edge in the line equation:

$$t_x - b\,t_y \geq 0 \tag{5.3}$$

$$(-\theta + t_x) - b\,(1 + t_y) \geq 0 \tag{5.4}$$

Similarly, the non-interference constraints applied to the top edge, and the right edge give:

$$1 + t_y \leq 1.1 \tag{5.5}$$

$$1 + \theta + t_y \leq 1.1 \tag{5.6}$$

$$1 - \theta + t_x \leq 2 \tag{5.7}$$

$$1 + t_x \leq 2 \tag{5.8}$$

These eight functional constraints define a feasible region of the configuration space spanned by $t_x$, $t_y$, and $\theta$.

To determine an optimum position within the configuration space, the following goals are added: first, that along its left edge, $B$ should come as close as possible to touching $A$; second, that along its bottom edge, $B$ should come as close as possible to touching $A$ These two goals will be satisfied in sequence, resulting in a series of transformations in which $B$ is first moved into an optimum position with respect to its left edge, and then, holding that relationship fixed, moved into an optimum position with respect to its bottom edge.

The first goal may effectively be achieved by minimizing the maximum distance of the left edge of $B$ from the corresponding edge of $A$ Thus, the goal is to minimize the maximum of the left hand sides of equations 5.3 and 5.4. (There are well-known techniques in linear programming practice [19] for minimizing the maximum of several goals.)

Once this first goal has been achieved, the second goal may attempted. In order to fix the position of $B$ with respect to the first goal, equations 5.3 and 5.4 are replaced by equality constraints. The computed values of the left hand sides at the point at which the first goal is optimized, are used as right hand sides for these equality constraints. Now the second goal may be achieved in the same manner as the first.

These equations were set up and solved using linear programming. The parameter $b$ was introduced to allow for some variability in the shape of part $A$. The problem was solved for different values of $b$. The results are shown in Figure 5.4. Note that $B$ is always positioned to fall inside $A$, and that the two goals are achieved to the extent possible, with preference given to the first goal.

## 5.3 Mating Holes and Posts

Now consider an assembly consisting of two flat parts, one of which has a pattern of holes, and the other of which has a mating pattern of posts or pegs.

Let the hole radii be denoted $R_i$, and the hole positions be denoted $P_i$. Let the radii and positions of the posts be denoted $r_i$ and $p_i$ where $i = 1, ..., M$.

The minimum clearance between a mating hole and post is given by

$$R_i - r_i - ||P_i - p_i||$$

The non-interference constraints will specify that each clearance be non-negative.

Optimum alignment is achieved if the two parts are positioned so that the minimum clearance will be maximized over all the mating sites. So the objective will be to maximize minimum clearance. One way to achieve this objective is

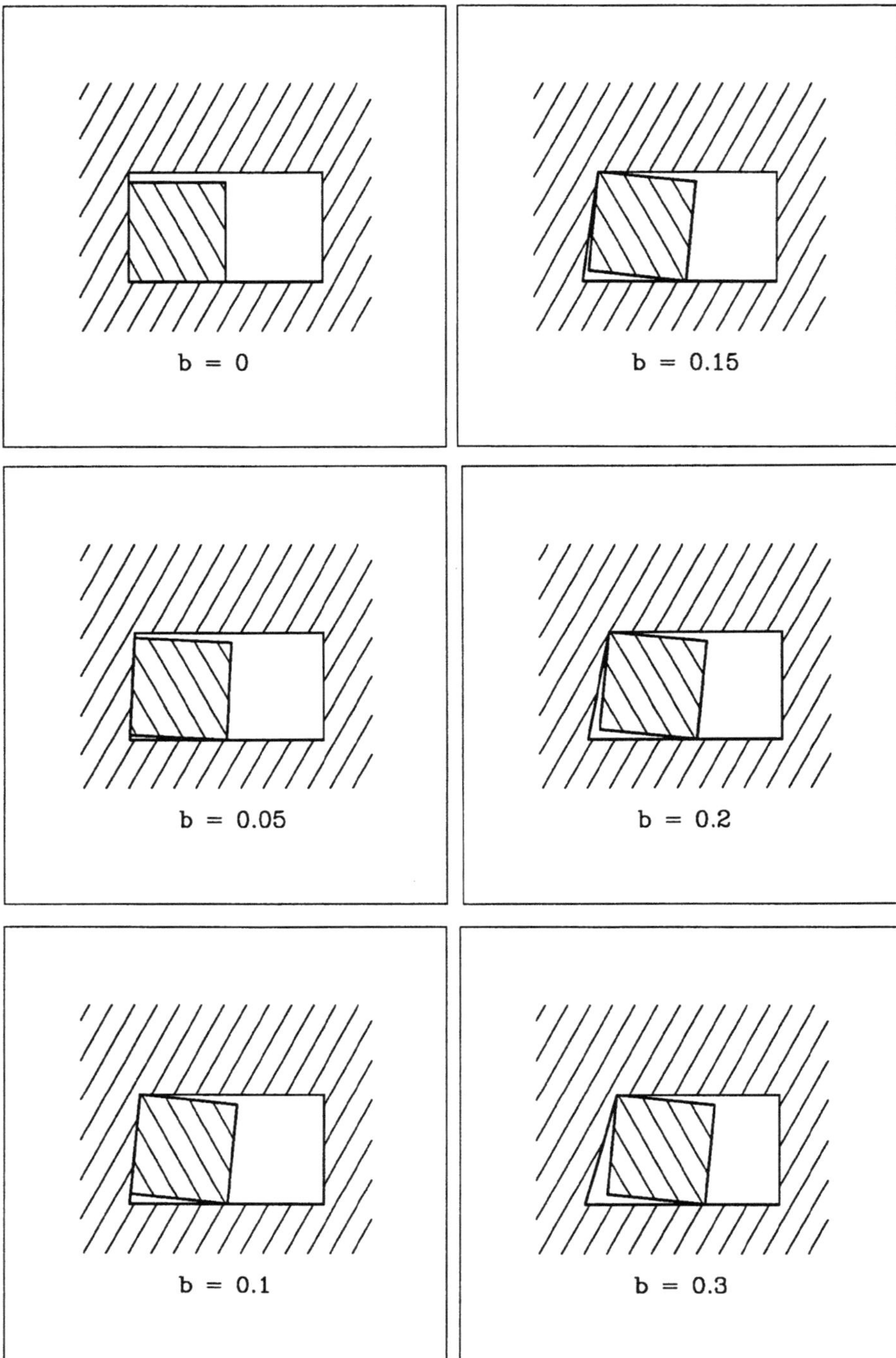


Figure 5.4: Solution of the LP Problem for Different Values of b

to introduce a new decision variable, $z$, and to maximize $z$ subject to the constraints:

$$||P_i - p_i|| + z \leq R_i - r_i \tag{5.9}$$

$$z \geq 0$$

Taking the position of the hole part as fixed, and positioning the post part relative to the hole part, the coordinates of the $p_i$ in equation 5.9 will be functions of the decision variables $t_x$, $t_y$, and $\theta$. Under the assumption of small variations, these coordinates may be given by

$$p_i' \approx (x_i - y_i\theta + t_x,\ x_i\theta + y_i + t_y)$$

as in the previous problem.

So far, this is a nonlinear problem, because $||P_i - p_i'||$ is a nonlinear function of the decision variables $t_x$, $t_y$, and $\theta$. The problem can be linearized at the expense of an approximation. The form of this approximation is as follows — rather than measure the exact distance between $P_i$ and $p_i'$, we can measure the *directed distance* between the two points. This distance is given by:

$$D_{ij} = (\cos\phi_j,\ \sin\phi_j) \cdot (P_i - p_i')$$

where "." indicates a dot product, and the left-hand term of the dot product gives the measurement direction. In other words, $D_{ij}$ is a signed value which gives the length of the projection of $(P_i - p_i')$ onto the direction vector $(\cos\phi_j,\ \sin\phi_j)$.

In general, this distance measure will understate the distance between the two points. However, by taking a number of different measurement directions, the largest of these distance measures will approximate the true distance. In particular, we will take

$$\phi_j = j\left(\frac{360}{n}\right) \quad j = 0, ..., n-1$$

where a larger value of $n$ will give a more accurate approximation.

Then the final form of equation 5.9 is:

$$D_{ij} + z \leq R_i - r_i$$

where $i = 1, ..., M$ and $j = 0, ..., n-1$. Thus there will be one constraint per value of $j$ for each hole-post pair.

These equations were set up and solved using linear programming. Figure 5.5 and Figure 5.6 give an example of two parts before and after positioning.

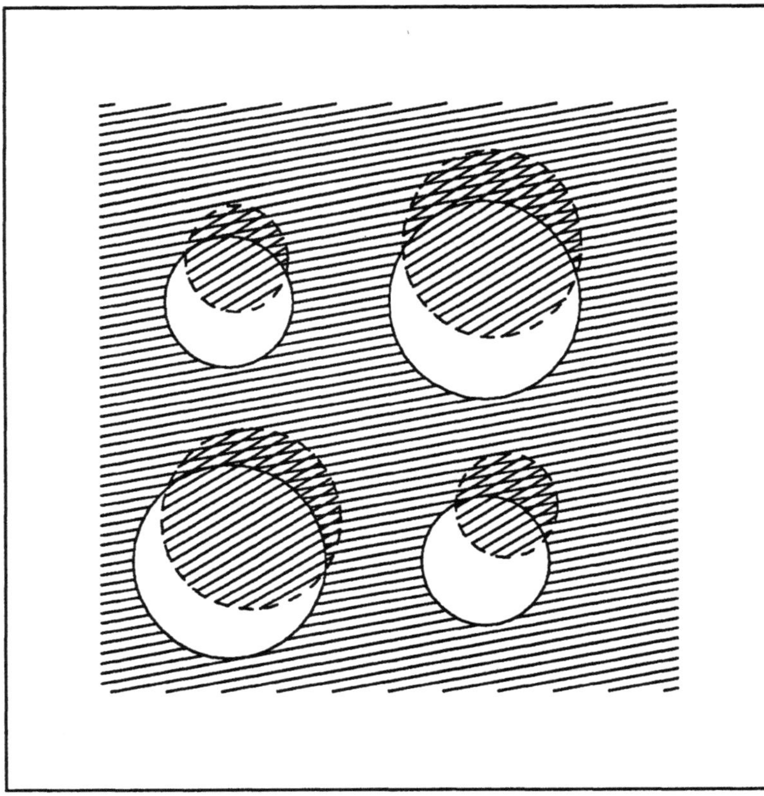

Figure 5.5: Hole Part and Post Part Before Positioning

## 5.4 Computer Time

It is well-known [19] that linear programming problems can be solved with computer time that is proportional to the cube of the number of decision variables. For the simple problems presented here, each having six decision variables, the computer time is negligible. For a large assembly of parts, good performance will be dependent on performing as many assembly operations as possible in a sequential fashion. If each part is assembled sequentially relative to an established frame of reference, then the total computer time will grow in a linear fashion with the number of parts in the assembly. On the other hand, if all assembly operations are performed simultaneously, then total computer time will grow as the cube of the number of parts. Fortunately, most assemblies are constructed in a sequential fashion.

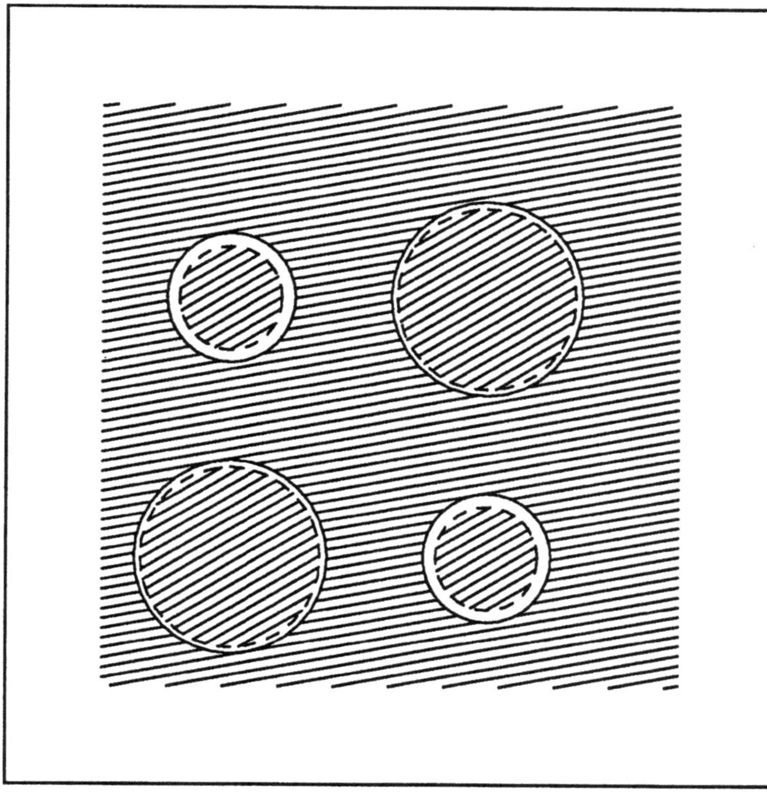

Figure 5.6: Hole Part and Post Part After Positioning

If nonlinear methods are used, then the computer times required will be significantly higher, but again, costs can be minimized by performing the assembly operations in a sequential fashion as much as possible.

## 5.5 Limitations

In both examples, it was necessary to invoke the assumption of small variations in order to linearize the equations. In many applications we would like to allow for situations in which large variations are required to position one part relative to another. However, the preceding strategies are not reliable if more than 10° of rotation is necessary.

But if we are willing to make relative positioning a two step process, this need not be a problem. Suppose we develop a gross positioning strategy intended to place the parts in approximate position to one another. This gross positioning strategy could be simply a matter of associating a local coordinate system with each of a pair of mating parts, where these local coordinate systems are derived from the part geometries, and then generating a transformation matrix that lines up the two coordinate systems.

Given such a gross positioning strategy, the linearized situations illustrated in this chapter can be used to provide a fine positioning strategy.

If a single application of the fine positioning strategy has unacceptable error due to linearization, then several successive linearizations may be necessary.

The approach was discussed in terms of establishing the position of a single target object $B$ relative to an existing frame of reference $A$. However, the approach can also be applied to determine the positions of several objects simultaneously. For instance, given three objects, $A$, $B$, and $C$, to be mutually positioned so that certain constraints apply. Taking one of the three objects as fixed, the constraints may be interpreted as conditions on a cross-product space formed from the configuration spaces of the other two objects.

## 5.6 Summary

This chapter has introduced a general approach to relative positioning based on mathematical programming. Non-interference conditions determine functional constraints. Alignment conditions determine one or more objective functions. These non-interference constraints and alignment objectives can be completely general. Mathematical programming is used to establish an optimum feasible position.

The treatment of alignment conditions as objective functions allows for the formulation of relationships in which conditions such as coplanarity, and coaxiality cannot be satisfied exactly. Where multiple alignment objectives are specified, these objectives can be satisfied in a given sequence, as illustrated in the first example, or can be satisfied simultaneously. (In this latter case, an overall objective is formulated — namely, minimize the maximum of any of the individual objectives.)

Two common situations were used to illustrate these ideas. It was shown that if small variations can be assumed, then the original nonlinear programs can be linearized. Because the linearizations are only accurate within a small range, several successive linearizations may sometimes be necessary to obtain acceptable answers. Code is currently being written to handle three-dimensional versions of these situations, using an existing solid modeling system.

## Acknowledgement

A version of this work first appeared in *Computer-Aided Design*, Volume 22, Number 7, September 1990, pp. 394-400.

This material is based upon work supported by the National Science Foundation under Grant Nos. DMC-8820733, and DDM-8908160. The United States Government has certain rights in this material. Any opinions, findings, and conclusions or recommendations expressed in this material are those of the author and do not necessarily reflect the views of the National Science Foundation.


## References


[1] L. I. Lieberman and M. A. Wesley, "Autopass: An automatic programming system for computer controlled mechanical assembly," *IBM Journal of Research and Development*, vol. 21, pp. 321–333, July 1977.

[2] M. A. Wesley, "Construction and use of geometric models," in *Computer Aided Design: Modelling, Systems Engineering, CAD-Systems* (J. Encarnacao, ed.), pp. 79–136, New York: Springer-Verlag, 1980.

[3] M. A. Wesley, T. Lozano-Perez, L. I. Lieberman, M. A. Lavin, and D. D. Grossman, "A geometric modeling system for automated mechanical assembly," *IBM Journal of Research and Development*, vol. 24, pp. 64–74, January 1980.

[4] K. Lee and D. C. Gossard, "A hierarchical data structure for representing assemblies: Part 1," *Computer-Aided Design*, vol. 17, pp. 15–19, January/February 1985.

[5] J. U. Turner, *Tolerances in Computer-Aided Geometric Design*. PhD thesis, Rensselaer Polytechnic Institute, May 1987.

[6] J. U. Turner, "Exploiting solid models for tolerance computations," in *Geometric Modeling for Product Engineering* (M. Wozny, J. Turner, and K. Preiss, eds.), North-Holland, 1990.

[7] D. C. Gossard, R. P. Zuffante, and H. Sakurai, "Representing dimensions, tolerances, and features in mcae systems," *IEEE Computer Graphics and Applications*, vol. 8, pp. 51–59, March 1988.

[8] A. P. Ambler and R. J. Popplestone, "Inferring the positions of bodies from specified spatial relationships," *Artificial Intelligence*, vol. 6, pp. 157–174, 1975.

[9] K. Lee and G. Andrews, "Inference of the positions of components in an assembly: Part 2," *Computer-Aided Design*, vol. 17, pp. 20–24, January/February 1985.

[10] D. N. Rocheleau and K. Lee, "System for interactive assembly modeling," *Computer-Aided Design*, vol. 19, pp. 65–72, March 1987.

[11] G. Mullineux, "Optimization scheme for assembling components," *Computer-Aided Design*, vol. 19, pp. 35–40, January/February 1987.

[12] J. R. Rossignac, "Constraints in constructive solid geometry," in *Proceedings 1986 Workshop on Interactive 3D Graphics*, (Chapel Hill, North Carolina), pp. 93–110, October 23-24 1986.

[13] J. U. Turner and M. J. Wozny, "Tolerances in computer-aided geometric design," *Visual Computer*, vol. 3, no. 4, pp. 214–226, 1987.

[14] J. U. Turner, M. J. Wozny, and D. D. Hoh, "Tolerance analysis in a solid modeling environment," in *Proceedings of the 1987 ASME Computers in Engineering Conference*, (New York, N. Y.), August 9-13 1987.

[15] J. U. Turner, "New methods for tolerance analysis in solid modeling," in *Proceedings 1988 International Conference on CIM*, (Troy, New York), pp. 306–314, IEEE Computer Society Press, May 23-25 1988.

[16] J. U. Turner and M. J. Wozny, "A framework for tolerances utilizing solid models," in *Proceedings Third International Conference on Computer-Aided Production Engineering*, (Ann Arbor, Michigan), June 1-3 1988.

[17] J. U. Turner, "Automated tolerancing using solid modeling technology," in *Proceedings AUTOFACT '88*, (Chicago, Illinois), Oct. 30 – Nov. 2 1988.

[18] T. Lozano-Perez and M. Wesley, "An algorithm for planning collision-free paths among polyhedral obstacles," *Communications of the ACM*, vol. 22, no. 10, 1979.

[19] F. S. Hillier and G. J. Lieberman, *Introduction to Operations Research*. Oakland, California: Holden-Day, 1980.

# Part II

# Assembly Planning

# Chapter 6

# Representations for assembly sequences

**Luiz S. Homem de Mello and Arthur C. Sanderson**

Several methodologies for representing assembly sequences have been utilized. These include representations based on directed graphs, on AND/OR graphs, on establishment conditions, and on precedence relationships. The latter includes two types: precedence relationships between the establishment of one connection between parts and the establishment of another connection, and precedence relatioships between the establishment of one connection and states of the assembly process. Those based on directed graphs and on AND/OR graphs are explicit representations since there is a mapping from the assembly tasks into the elements of the representations. Those based on establishment conditions and on precedence relationships are implicit representations because they consist of conditions that must be satisfied by the assembly sequences.

This chapter analyzes these five representations and shows how they are interrelated and how one can be derived from the others. The correctness and completeness of these representations are also addressed.

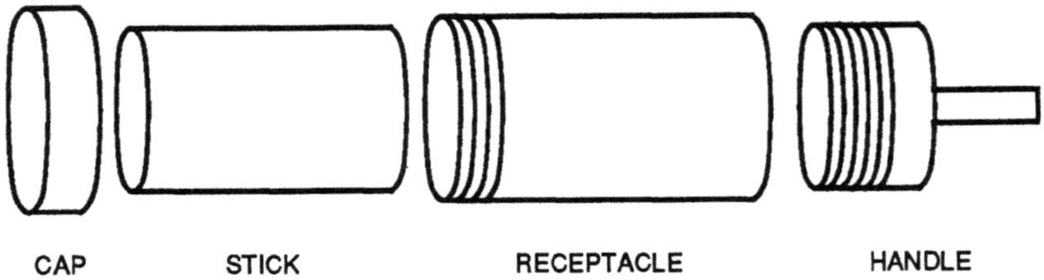


Figure 6.1: A four-part assembly in exploded view

## 6.1 Terminology and notation

A mechanical assembly is a composition of interconnected parts forming a stable unit. Each part is a solid rigid object that is its shape remains unchanged. Parts are interconnected whenever they have one or more compatible surfaces in contact. Surface contacts between parts reduce the degrees of freedom for relative motion. A cylindrical contact, for example, prevents any relative motion that is not a translation along the axis or a rotation around the axis. Attachments may act on surface contacts and eliminate all degrees of freedom for relative motion. For example, if a cylindrical contact has a pressure-fit attachment, then no relative motion between the parts is possible.

A subassembly is a nonempty subset of parts that either has only one element (i.e. only one part), or is such that every part has at least one surface contact with another part in the subset. Although there are cases in which it is possible to join the same pair of parts in more than one way, a unique assembly geometry will be assumed for each pair of parts. This geometry corresponds to their relative location in the whole assembly. A subassembly is said to be stable if its parts maintain their relative position and do not break contact spontaneously. All one-part subassemblies are stable.

The assembly process consists of a succession of tasks, each of which consists of joining subassemblies to form a larger subassembly. The process starts with all parts separated and ends with all parts properly joined to form the whole assembly. For the current analysis, it is assumed that exactly two subassemblies are joined at each assembly task, and that after parts have been put together, they remain together until the end of the assembly process.

It is also assumed that whenever two parts are joined all contacts between them are established. Due to this assumption, an assembly can be represented by a simple undirected graph $< P, C >$ in which $P = \{p_1, p_2, \cdots, p_N\}$ is the set of nodes, and $C = \{c_1, c_2, \cdots, c_L\}$ is the set of edges. Each node in $P$ corresponds to a part in the assembly, and there is one edge in $C$ connecting every pair of nodes whose corresponding parts have at least one surface contact. The elements of $C$ are referred to as *connections*, and the graph $< P, C >$ is referred to as the *assembly's graph of connections*. A connection encompasses

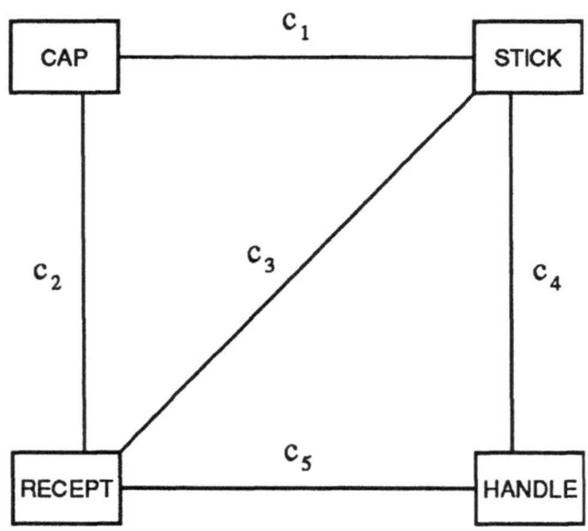


Figure 6.2: The graph of connections for the four-part assembly

all contacts between two parts. Figure 6.1 shows an assembly in exploded view, and figure 6.2 shows its corresponding graph of connections.

### 6.1.1 Assembly states

The state of the assembly process is the configuration of the parts at the begining (or at the end) of an assembly task. The configuration of parts is given by the contacts that have been established. Since whenever two parts are joined all contacts between them are established, the configuration of parts is given by the connections that have been established. Therefore, a state of the assembly process can be represented by an $L$-dimensional binary vector $\underline{x} = [x_1\, x_2 \cdots x_L]$ in which the $i^{th}$ component $x_i$ is true (T) or false (F) respectively if the $i^{th}$ connection is established in that state or not.

For example, the initial state of the assembly process for the product shown in figure 6.1 can be represented by the 5-dimensional binary vector [F F F F F] whereas the final state can be represented by [T T T T T]. If the first task of the assembly process is the joining of the cap to the receptacle, the second state of the assembly process can be represented by [F T F F F]

As mentioned above, it is assumed that whenever a subassembly is formed all connections between its parts are established. Therefore, any subassembly can be characterized by its set of parts. In the rest of this paper, references to subsets of parts should be understood as references to the subassemblies made up of those parts. It will always be clear from context what the whole assembly is. Because of this assumption, any state of the assembly process can also be represented by a partition of the set of parts of the whole assembly. For example, the initial state of the assembly process of the assembly shown in figure

6.1 can be represented by { {CAP}, {RECEPTACLE}, {STICK}, {HANDLE} } whereas the final state can be represented by { {CAP, RECEPTACLE, STICK, HANDLE} }. If the first task of the assembly process is the joining of the cap to the receptacle, the second state of the assembly process can be represented by { {CAP, RECEPTACLE}, {STICK}, {HANDLE} }.

Given an assembly's graph of connections and one of the two representations of assembly states described above (binary vector or partition), it is straightforward to obtain the other representation.

There are partitions of the set of parts of the whole assembly that cannot characterize a state of the assembly process. For example, the partition { {CAP, HANDLE}, {RECEPTACLE}, {STICK} } cannot characterize a state of the assembly process for the assembly shown in figure 6.1 because the subset {CAP, HANDLE} does not characterize a subassembly. Partitions that can characterize a state of the assembly process will be referred to as *state partitions*, and partitions that cannot characterize a state will be referred to as *nonstate partitions*.

Similarly, not all $L$-dimensional binary vectors can characterize a state. For example, for the assembly shown in figure 6.1 the 5-dimensional binary vector [T T F F F] does not correspond to a state because if connections $c_1$ and $c_2$ are established then connection $c_3$ should also be established. $L$-dimensional binary vectors that can characterize a state will be referred to as *state vectors* whereas $L$-dimensional binary vectors that cannot characterize a state will be referred to as *nonstate vectors*.

Any state of the assembly process can be associated to a simple undirected graph $< P, C_k >$ in which $P$ is the set of nodes of the assembly's graph of connections, and $C_k$ is the subset of connections ($C_k \subseteq C$) that is established in that state. This graph is referred to as the *state's graph of connections*. Except for the final state of the assembly process, a state's graph of connections has more than one component.

The subassembly predicate *sa* will be used to determine whether or not a subset of parts makes up a subassembly. The argument to this predicate is a subset of parts, and its value is either true or false depending on whether or not that subset of parts corresponds to a subassembly. For example, for the assembly shown in figure 6.1, $sa(\{$ RECEPTACLE, HANDLE $\}) =$ T whereas $sa(\{$ CAP, HANDLE $\}) =$ F. From the assembly's graph of connections it is straightforward to compute *sa* for any given subset of parts.

For the analysis in this chapter, a partition of the set of parts whose elements all satisfy the subassembly predicate is an assembly state representation, regardless of whether that state actually occurs in any of the different ways the assembly can be assembled. The corresponding $L$-dimensional binary vector is also an assembly state representation. And the corresponding configuration of parts is an assembly state. For example, for the assembly shown in figure

6.1, the partition { {CAP, RECEPTACLE, HANDLE}, {STICK} } as well as the corresponding 5-dimensional binary vector [F T F F T] are assembly state representations. Yet, since it was assumed that once parts are put together they remain together, the configuration of parts (i.e. the state) corresponding to these representations cannot occur in any assembly process. Once the cap and the handle are joined to the receptacle, it is no longer possible to join the stick.

We will use the subassembly-stability predicate *st* to determine whether or not a subassembly described by its set of parts is stable. The computation of *st* has been addressed elsewhere (e.g. the work of Boneschanscher[1]).

An assembly state representation for which all subassemblies satisfy the stability predicate is said to be an *stable assembly state representation.* For example, for the assembly shown in figure 6.1, the partition { {CAP, RECEPTACLE, HANDLE}, {STICK} } as well as the corresponding binary vector [F T F F T] are stable assembly state representations.

### 6.1.2 Assembly tasks

Given two subassemblies characterized by their sets of parts $\theta_i$ and $\theta_j$, we say that joining $\theta_i$ and $\theta_j$ is an assembly task if the set $\theta_k = \theta_i \cup \theta_j$ characterizes a subassembly. For example, for the assembly shown in figure 6.1, if $\theta_i =$ {RECEPTACLE} and $\theta_j =$ {HANDLE} then joining $\theta_i$ and $\theta_j$ is an assembly task, whereas if $\theta_i =$ {CAP} and $\theta_j =$ {HANDLE} then joining $\theta_i$ and $\theta_j$ is not an assembly task. The subassemblies $\theta_i$ and $\theta_j$ are the *input* subassemblies of the assembly task, and $\theta_k$ is the *output* subassembly of the assembly task. Due to the assumption of unique geometry, an assembly task can be characterized by its input subassemblies only, and it can be represented by a set of two subsets of parts. For example, for the assembly shown in figure 6.1, the joining of the cap to the receptacle is represented by { {CAP}, {RECEPTACLE} }.

Alternatively, a task can be seen as a decomposition of the output subassembly into the two input subassemblies. Therefore, an assembly task can be characterized by the output subassembly and the set of its connections that are in neither of the input subassemblies. In this view, an assembly task is represented by a set that contains a subset of parts and a subset of connections. For example, for the assembly shown in figure 6.1, the joining of the cap to the receptacle is represented by { {CAP, RECEPTACLE}, $\{c_2\}$ }.

The set of connections in the representation of an assembly task corresponds to a cut-set of the graph of connections of the task's output subassembly. Conversely, each cut-set of a subassembly's graph of connections corresponds to an assembly task[7]. Given the set of all cut-sets of a subassembly's graph of connections, the set of their corresponding assembly tasks is referred to as the *tasks of the subassembly.*

An assembly task is said to be *geometrically* feasible if there is a collision-free path to bring the two subassemblies into contact from a situation in which they are far apart. And an assembly task is said to be *mechanically* feasible if it is feasible to establish the attachments that act on the contacts between the two subassemblies. We will use the geometric-feasibility predicate *gf* and the mechanical-feasibility predicate *mf* to determine whether or not two subsets of parts characterize, respectively, a geometrically feasible and a mechanically feasible assembly task. These predicates take as argument a set of two subassemblies, each characterized by its set of parts. The computation of these predicates is discussed elsewhere[7].

### 6.1.3 Assembly sequences

Given an assembly that has $N$ parts, an ordered set of $N-1$ assembly tasks $\tau_1 \tau_2 \cdots \tau_{N-1}$ is an assembly sequence if there are no two tasks that have a common input subassembly, the output subassembly of the last task is the whole assembly, and the input subassemblies to any task $\tau_i$ is either a one-part subassembly or the output subassembly of a task that precedes $\tau_i$. To any assembly sequence $\tau_1 \tau_2 \cdots \tau_{N-1}$ there corresponds an ordered sequence $s_1 s_2 \cdots s_N$ of $N$ assembly states of the assembly process. The state $s_1$ is the state in which all parts are separated. The state $s_N$ is the state in which all parts are joined forming the whole assembly. And any two consecutive states $s_i$ and $s_{i+1}$ are such that only the two input subassemblies of task $\tau_i$ are in $s_i$ and not in $s_{i+1}$, and only the output subassembly of task $\tau_i$ is in $s_{i+1}$ and not in $s_i$. Therefore, an assembly sequence can also be characterized by an ordered sequence of states.

An example of an assembly sequence for the assembly shown in figure 6.1 is:

1. The first task ($\tau_1$) consists of joining the cap to the receptacle.
2. The second task ($\tau_2$) consists of joining the stick to the subassembly made up of the cap and the receptacle.
3. The third task ($\tau_3$) consists of joining the handle to the subassembly made up of the cap, the stick, and the receptacle.

An assembly sequence is said to be feasible if all its assembly tasks are geometrically and mechanically feasible, and the input subassemblies of all tasks are stable. The assembly sequence described above is feasible. An example of an unfeasible assembly sequence for the assembly shown in figure 6.1 is:

1. The first task ($\tau_1$) consists of joining the cap to the receptacle.
2. The second task ($\tau_2$) consists of joining the handle to the subassembly made up of the cap and the receptacle.

3. The third task ($\tau_3$) consists of joining the stick to the subassembly made up of the cap, the stick, and the receptacle.

This assembly sequence is infeasible because the third task ($\tau_3$) is not geometrically feasible since there is no collision free path to bring the stick into the receptacle, once both the cap and the handle have been joined to the receptacle.

An assembly sequence (not necessarily feasible) can be represented in different ways. We will use the following representations:

- An ordered list of task representations. The number of elements in this list is equal to the number of parts minus one.
- An ordered list of binary vectors. Each vector must correspond to a state (not necessarily stable). The number of elements in this list is the equal to the number of parts.
- An ordered list of partitions of the set of parts. Each partition must correspond to a state (not necessarily stable). The number of elements in this list is equal to the number of parts.
- An ordered list of subsets of connections. The number of elements in this list is equal to the number of parts minus one.

For example, the feasible assembly sequence for the product shown in figure 6.1 that was described above can be represented as follows:

- The three-element list of task representations

  ( { {CAP} , {RECEPTACLE} }
  { {CAP, RECEPTACLE} , {STICK} }
  { {CAP, RECEPTACLE, STICK} , {HANDLE} } )

- The four-element list of 5-dimensional binary vectors

  ( [F F F F F] [F T F F F] [T T T F F] [T T T T T] )

- The four-element list of partitions of the set of parts

  ( { {CAP} , {RECEPTACLE} , {STICK} , {HANDLE} }
  { {CAP, RECEPTACLE} , {STICK} , {HANDLE} }
  { {CAP, RECEPTACLE , STICK} , {HANDLE} }
  { {CAP, RECEPTACLE , STICK, HANDLE} } )

- The three-element list of sets of connections ( { $c_2$ } {$c_1$, $c_3$ } {$c_4$, $c_5$} ).

Given the assembly's graph of connections and an assembly sequence in any of these four representations, it is straightforward to obtain the other three representations.

Furthermore, these assembly sequence representations have the following properties:

- Any ordered list of binary vectors $(\underline{x}_1\ \underline{x}_2 \cdots \underline{x}_N)$, in which $\underline{x}_i = [x_{i1}\ x_{i2}\ x_{i3}\ \cdots\ x_{iL}]$, that represents one assembly sequence is such that

  $[(j > i) \wedge (x_{ik} = \mathrm{T})] \Rightarrow (x_{jk} = \mathrm{T})$.

  This corresponds to the fact that once a connection is established, it remains established until the end of the assembly process.

- Any ordered list of partitions of the set of parts $(\Theta_1\, \Theta_2 \cdots \Theta_N)$ that represents one assembly sequence is such that

  $[(j > i) \wedge (\theta_a \in \Theta_i)] \Rightarrow \exists \theta_b[(\theta_b \in \Theta_j) \wedge (\theta_a \subseteq \theta_b)]$.

  This corresponds to the fact that once parts are put together they remain together until the end of the assembly process.

- Any ordered list of sets of connections $(\gamma_1\, \gamma_2 \cdots \gamma_{N-1})$ that represents one assembly sequence is such that $\gamma_1 \cup \gamma_2 \cup \cdots \cup \gamma_{N-1} = C$ and $\gamma_i$ is a cut-set of the state's graph of connections associated to the $i^{th}$ state of the assembly process.

Since each assembly sequence can be represented by ordered lists, it is possible to represent the set of all assembly sequences by a set of lists, each corresponding to a different assembly sequence. Since many assembly sequences share common subsequences, attempts have been made to create more compact representations that can encompass all assembly sequences. The next five sections discuss different approaches towards representing all assembly sequences of a mechanical assembly.

## 6.2 Directed Graph Representation of Assembly Sequences

Given an assembly whose graph of connections is $< P , C >$, a directed graph can be used to represent the set of all assembly sequences. The nodes in this directed graph correspond to stable state partitions of the set $P$. These are the partitions $\Theta$ of $P$ such that if $\theta \in \Theta$ then $\theta$ is a stable subassembly of $P$. The edges in this directed graph are ordered pairs of nodes. For any edge, there are only two subsets $\theta_i$ and $\theta_j$ in the state partition corresponding to the first node that are not in the state partition corresponding to the second node. Also, there is only one subset $\theta_k$ in the state partition corresponding to the second node that are not in the state partition corresponding to the first node, and $\theta_k = \theta_i \cup \theta_j$. Furthermore, the assembly task that joins $\theta_i$ and $\theta_j$ is feasible. Therefore, each edge corresponds to an assembly task. This graph is referred to as *directed graph of feasible assembly sequences*, and it can be formally defined as follows:

**Definition 1** *The* **directed graph of feasible assembly sequences** *of an assembly whose set of parts is $P$ is the directed graph $< X_P\,, T_P >$ in which*

$$X_P = \left\{\Theta \mid \left[\Theta \in \Delta(P)\right] \wedge \left[\forall\theta(\theta \in \Theta) \Rightarrow \left(sa(\theta) \wedge st(\theta)\right)\right]\right\}$$

*is the assembly's set of stable states, and*

$$\begin{aligned} T_P = \Big\{(\Theta_i\,, \Theta_j) \mid \; & \left[(\Theta_i\,, \Theta_j) \in X_P \times X_P\right] \wedge \\ & \left[\mathcal{U}\left(\Theta_i - (\Theta_i \cap \Theta_j)\right) \in \Theta_j - (\Theta_i \cap \Theta_j)\right] \wedge \\ & \left[\left|\Theta_j - (\Theta_i \cap \Theta_j)\right| = 1\right] \wedge \left[\left|\Theta_i - (\Theta_i \cap \Theta_j)\right| = 2\right] \wedge \\ & \left[mf\left(\Theta_i - (\Theta_i \cap \Theta_j)\right)\right] \wedge \left[gf\left(\Theta_i - (\Theta_i \cap \Theta_j)\right)\right]\Big\} \end{aligned}$$

*is the assembly's set of feasible state transitions.*

The notation $\Delta(P)$ has been used to represent the set of all partitions of $P$, and the notation $\mathcal{U}(\{A, B, \cdots, Z\})$ has been used to represent $A \cup B \cup \cdots \cup Z$.

Figure 6.3 shows the directed graph of feasible assembly sequences for the assembly shown in figure 6.1. Each node of the graph in figure 6.3 is labeled by a partition of the set of parts that represents a stable assembly state. To facilitate the exposition, the nodes in figure 6.3 also have identification numbers placed at their upper left corners.

A path in the directed graph of feasible assembly sequences $< X_P\,, T_P >$ whose initial node is $\Theta_I = \{\,\{\,p_1\,\}\{\,p_2\,\}\cdots\{\,p_N\,\}\,\}$ and whose terminal node is $\Theta_F = \{\,\{\,p_1, p_2, \cdots, p_N\,\}\,\}$ corresponds to a feasible assembly sequence for the assembly $P$, and conversely. In such a path, the ordered sequence of edges corresponds to the ordered sequence of tasks, while the ordered sequence of nodes corresponds to the ordered sequence of states of the assembly process. For example, the feasible assembly sequence described in the previous section corresponds to nodes 1, 2, 7, and 13 of the graph shown in figure 6.3.

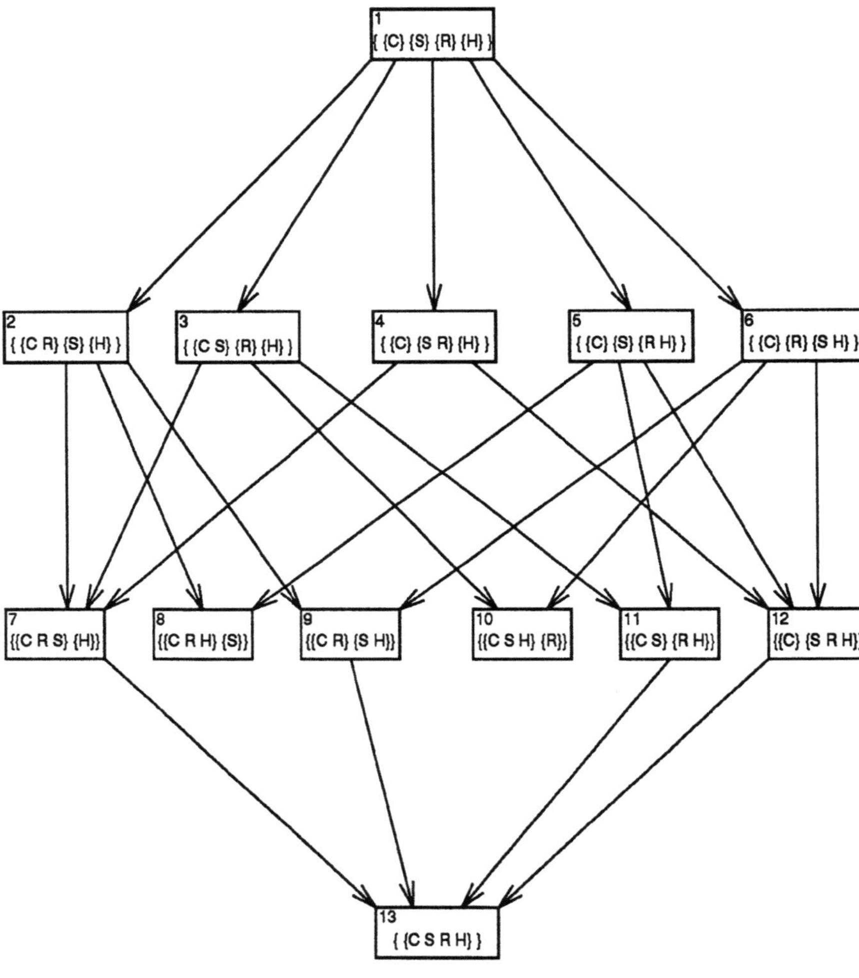


Figure 6.3: Directed graph of feasible assembly sequences for the assembly shown in figure 1

## 6.3 AND/OR Graph Representation of Assembly Sequences

An AND/OR graph can also be used to represent the set of all assembly sequences. The nodes in this AND/OR graph are the subsets of $P$ that characterize stable subassemblies. The hyperarcs correspond to the geometrically and mechanically feasible assembly tasks. Each hyperarc is an ordered pair in which the first element is a node that corresponds to a stable subassembly $\theta_k$, the second element is a set of two nodes $\{ \theta_i, \theta_j \}$ such that $\theta_i \cup \theta_j = \theta_k$ and the assembly task characterized by $\theta_i$ and $\theta_j$ is feasible. Each hyperarc is associated to a decomposition of the subassembly that corresponds to its first element and can also be characterized by this subassembly and the subset of all its connections that are not in the graphs of connections of the subsubassemblies in the hyperarc's second element. This subset of connections associated to a hyperarc corresponds to a cut-set in the graph of connections of the subassembly in the hyperarc's first element. This AND/OR graph can be formally defined as follows:

**Definition 2** *The* **AND/OR graph of feasible assembly sequences** *of an assembly whose set of parts is* $P = \{ p_1, p_2, \cdots p_N \}$ *is the* AND/OR *graph* $< S_P, D_P >$ *in which*

$$S_P = \left\{ \theta \in \Pi(P) \mid sa(\theta) \wedge st(\theta) \right\}$$

*is the set of stable subassemblies, and*

$$D_P = \Big\{ (\theta_k, \{\theta_i, \theta_j\}) \mid \quad [\theta_i, \theta_j, \theta_k \in S_P] \wedge [\mathcal{U}(\{\theta_i, \theta_j\}) = \theta_k] \wedge \\ [mf(\{\theta_i, \theta_j\})] \wedge [gf(\{\theta_i, \theta_j\})] \Big\}$$

*is the set of feasible assembly tasks.*

The notation $\Pi(P)$ has been used to represent the set of all subsets of $P$.

As an example, figure 6.4 shows part of the AND/OR graph for the assembly shown in figure 6.1. Each node of the graph in figure 6.4 is associated with a subset of parts that corresponds to a subassembly. To facilitate the exposition, both the nodes and the hyperarcs in figure 6.4 have identification numbers. There are only two stable subassemblies of the product shown in figure 6.1 whose corresponding nodes were not included in figure 6.4; they are the subassembly made up of the cap, the receptacle, and the handle, and the subassembly made up of the cap, the stick, and the handle. They were not included to avoid cluttering the figure. Since these subassemblies cannot be reached from the top node, they do not occur in any feasible assembly sequence.

From the AND/OR graph of feasible assembly sequences one can define feasible assembly trees as follows:

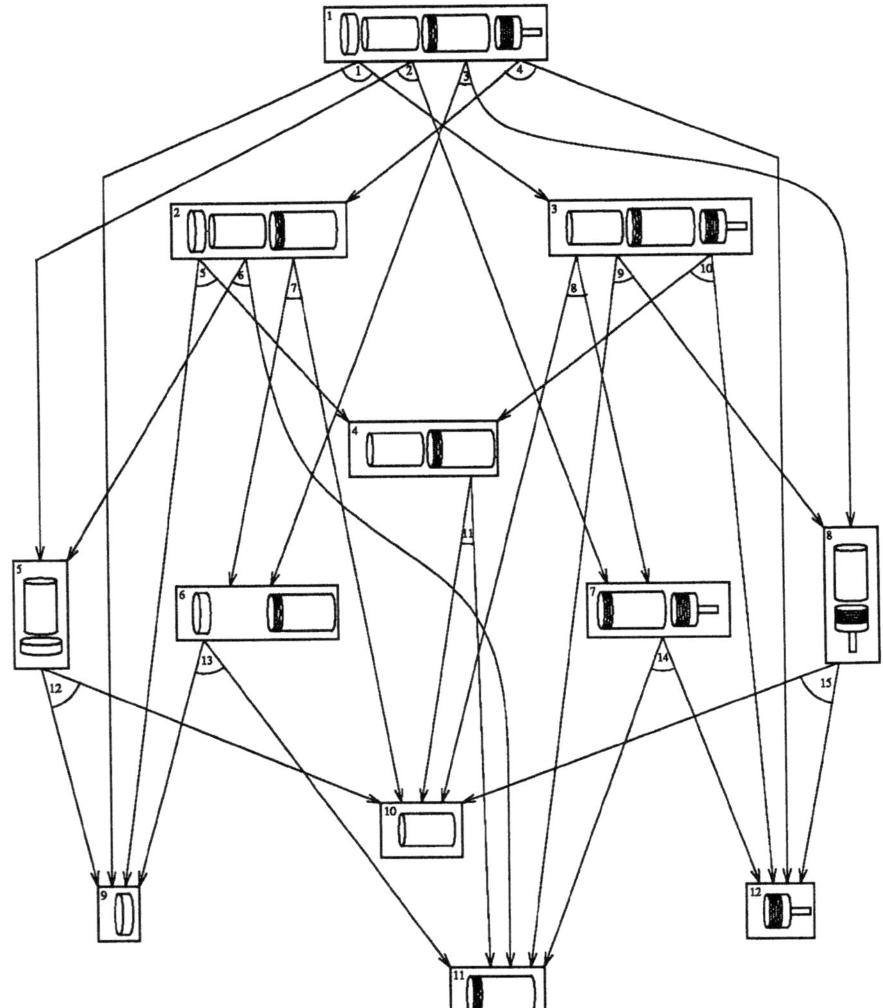


Figure 6.4: AND/OR graph of feasible assembly sequences for the assembly shown in figure 1

**Definition 3** *Given the* AND/OR *graph of feasible assembly sequences of an assembly whose set of parts is* $P = \{p_1, p_2, \cdots, p_N\}$ *any* AND/OR *path having* $\{p_1, p_2, \cdots, p_N\}$ *as its initial node, and having* $\{p_1\}, \{p_2\}, \cdots, \{p_N\}$ *as its terminal nodes is a* **feasible assembly tree** *of that assembly.*

An assembly tree induces a partial order among its hyperarcs: hyperarc $h_i$ is said to precede hyperarc $h_j$ if there is a node $n_k$ in the assembly tree such that $h_i$ is incident from $n_k$ and $h_j$ is incident to $n_k$. At least one sequence of the hyperarcs of an assembly tree is consistent with this partial order. Furthermore, every sequence of the hypearcs that is consistent with the partial order corresponds to a feasible assembly sequence.

### The Correspondence between the Directed Graph and the AND/OR Graph

Every feasible assembly sequence in the directed graph of feasible assembly sequences corresponds to a feasible assembly tree in the AND/OR graph of feasible assembly sequences. And every feasible assembly tree in the AND/OR graph of feasible assembly sequences corresponds to one or more feasible assembly sequences in the directed graph of feasible assembly sequences. The two theorems below establish the correspondence between assembly trees and assembly sequences. Proofs of these theorems are presented elsewhere[5].

**Theorem 1** *Given an assembly tree of an assembly, if* $h_1\, h_2 \cdots h_i = (\sigma_i, \Theta_i) \cdots \cdots h_{N-1}$ *is a sequence of all the hyperarcs of that assembly tree that is consistent with the partial order induced by the tree, then the sequence* $\Omega_1, \Omega_2 \cdots \Omega_N$ *in which* $\Omega_1 = \{\{p_1\}, \{p_2\}, \cdots \{p_N\}\}$ *and* $\Omega_{i+1} = (\Omega_i - \Theta_i) \cup \{\sigma_i\}$ *is a feasible assembly sequence of the assembly.*

**Theorem 2** *If* $\Omega_1, \Omega_2 \cdots \Omega_N$ *is an assembly sequence of an assembly whose set of parts is* $P = \{p_1, p_2, \cdots p_N\}$ *and*

$$
\begin{aligned}
&S_P^a = \Omega_1 \cup \Omega_2 \cup \cdots \cup \Omega_N \\
&\{\sigma_i\} = \Omega_{i+1} - (\Omega_i \cap \Omega_{i+1}) \text{ for } i = 1, 2, \cdots, N-1 \\
&\Theta_{\sigma_i} = \Omega_i - (\Omega_i \cap \Omega_{i+1}) \text{ for } i = 1, 2, \cdots, N-1 \\
&h_i = (\sigma_i, \Theta_{\sigma_i}) \text{ for } i = 1, 2, \cdots, N-1 \\
&H_P^a = \{h_1, h_2, \cdots, h_{N-1}\}
\end{aligned}
$$

*then* $< S_P^a, H_S^a >$ *is an assembly tree of that assembly.*

The useful feature of the AND/OR graph representation is that it encompasses all possible assembly sequences. One advantage of the AND/OR graph is that for most assemblies that have more than 5 parts it has fewer nodes than the directed graph of assembly states[6]. Furthermore, it explicitly shows the possibility of simultaneous execution of assembly tasks.

## 6.4 Establishment Condition Representation of Assembly Sequences

If we represent the states of the assembly process by $L$-dimensional binary vectors, then a set of logical expressions can be used to encode the directed graph of feasible assembly sequences. Let $\Xi_i = \{\underline{x}_1, \underline{x}_2, \cdots, \underline{x}_{K_i}\}$ be the set of states from which the $i^{th}$ connection can be established without precluding the completion of the assembly. The *establishment condition* for the $i^{th}$ connection is the logical function

$$F_i(\underline{x}) = F_i(x_1, x_2, \cdots, x_L) = \sum_{k=1}^{K} \prod_{i=1}^{L} \gamma_{k\,l}$$

where the sum and the product are the logical operations OR and AND respectively[1], and $\gamma_{k\,l}$ is either the symbol $x_l$ if the $l^{th}$ component of $\underline{x}_k$ is true (T), or the symbol $\overline{x}_l$ if the $l^{th}$ component of $\underline{x}_k$ is false (F). Clearly, $F_i(\underline{x}_k) = \text{T}$ if and only if $\underline{x}_k$ is an element of $\Xi_i$. It is often possible to simplify the expression of $F_i(\underline{x})$ using the rules of boolean algebra.

Any assembly sequence whose representation as an ordered sequence of states is $(\underline{x}_1\,\underline{x}_2\,\cdots\,\underline{x}_N)$ and whose representation as an ordered sequence of subsets of connections is $(\gamma_1\,\gamma_2\,\cdots\,\gamma_{N-1})$ is feasible if and only if it is such that if the $i^{th}$ connection is established in the $k^{th}$ task (i.e. $c_i \in \gamma_k$) then $F_i(\underline{x}_k) =$ T. Therefore, the set of establishment conditions is a correct and complete representation of assembly sequences.

Knowing $F_1(\underline{x})$, $F_2(\underline{x})$, $\cdots$, $F_L(\underline{x})$, and the assembly's graph of connections, it is straightforward to construct the assembly's directed graph of assembly states. This representation was first introduced by Bourjault[2].

### Obtaining the establishment conditions from the directed graph

The establishment conditions can be obtained from the directed graph of feasible assembly sequences by systematically looking at the edges that are incident from and to nodes that correspond to states from which the assembly can be completed.

As an example, the establishment conditions for the assembly shown in figure 6.1 that are obtained from its directed graph of feasilbe assembly sequences, which is shown in figure 6.3, are:

[1] The logical operation AND will be denoted either by the symbol "$\wedge$" or by the product of the two logical variables. Similarly, the logical operation OR will be denoted either by the symbol "$\vee$" or by the sum of the two logical variables.

$$
\begin{aligned}
F_1(x_1, x_2, x_3, x_4, x_5) \;=\;& \overline{x_1}\cdot\overline{x_2}\cdot\overline{x_3}\cdot\overline{x_4}\cdot\overline{x_5}+\overline{x_1}\cdot x_2\cdot\overline{x_3}\cdot\overline{x_4}\cdot\overline{x_5}+\\
& \overline{x_1}\cdot\overline{x_2}\cdot x_3\cdot\overline{x_4}\cdot\overline{x_5}+\overline{x_1}\cdot\overline{x_2}\cdot\overline{x_3}\cdot\overline{x_4}\cdot x_5+\\
& \overline{x_1}\cdot x_2\cdot\overline{x_3}\cdot x_4\cdot\overline{x_5}+\overline{x_1}\cdot\overline{x_2}\cdot x_3\cdot x_4\cdot x_5
\end{aligned}
$$

$$
\begin{aligned}
F_2(x_1, x_2, x_3, x_4, x_5) \;=\;& \overline{x_1}\cdot\overline{x_2}\cdot\overline{x_3}\cdot\overline{x_4}\cdot\overline{x_5}+x_1\cdot\overline{x_2}\cdot\overline{x_3}\cdot\overline{x_4}\cdot\overline{x_5}+\\
& \overline{x_1}\cdot\overline{x_2}\cdot x_3\cdot\overline{x_4}\cdot\overline{x_5}+\overline{x_1}\cdot\overline{x_2}\cdot\overline{x_3}\cdot x_4\cdot\overline{x_5}+\\
& x_1\cdot\overline{x_2}\cdot\overline{x_3}\cdot\overline{x_4}\cdot x_5+\overline{x_1}\cdot\overline{x_2}\cdot x_3\cdot x_4\cdot x_5
\end{aligned}
$$

$$
\begin{aligned}
F_3(x_1, x_2, x_3, x_4, x_5) \;=\;& \overline{x_1}\cdot\overline{x_2}\cdot\overline{x_3}\cdot\overline{x_4}\cdot\overline{x_5}+x_1\cdot\overline{x_2}\cdot\overline{x_3}\cdot\overline{x_4}\cdot\overline{x_5}+\\
& \overline{x_1}\cdot x_2\cdot\overline{x_3}\cdot x_4\cdot\overline{x_5}+\overline{x_1}\cdot\overline{x_2}\cdot\overline{x_3}\cdot x_4\cdot\overline{x_5}+\\
& \overline{x_1}\cdot\overline{x_2}\cdot\overline{x_3}\cdot\overline{x_4}\cdot x_5+\overline{x_1}\cdot x_2\cdot\overline{x_3}\cdot x_4\cdot\overline{x_5}+\\
& x_1\cdot\overline{x_2}\cdot\overline{x_3}\cdot\overline{x_4}\cdot x_5
\end{aligned}
$$

$$
\begin{aligned}
F_4(x_1, x_2, x_3, x_4, x_5) \;=\;& \overline{x_1}\cdot\overline{x_2}\cdot\overline{x_3}\cdot\overline{x_4}\cdot\overline{x_5}+\overline{x_1}\cdot x_2\cdot\overline{x_3}\cdot\overline{x_4}\cdot\overline{x_5}+\\
& \overline{x_1}\cdot\overline{x_2}\cdot x_3\cdot\overline{x_4}\cdot\overline{x_5}+\overline{x_1}\cdot\overline{x_2}\cdot\overline{x_3}\cdot\overline{x_4}\cdot x_5+\\
& x_1\cdot\overline{x_2}\cdot\overline{x_3}\cdot\overline{x_4}\cdot x_5+x_1\cdot x_2\cdot x_3\cdot\overline{x_4}\cdot\overline{x_5}
\end{aligned}
$$

$$
\begin{aligned}
F_5(x_1, x_2, x_3, x_4, x_5) \;=\;& \overline{x_1}\cdot\overline{x_2}\cdot\overline{x_3}\cdot\overline{x_4}\cdot\overline{x_5}+x_1\cdot\overline{x_2}\cdot\overline{x_3}\cdot\overline{x_4}\cdot\overline{x_5}+\\
& \overline{x_1}\cdot\overline{x_2}\cdot x_3\cdot\overline{x_4}\cdot\overline{x_5}+\overline{x_1}\cdot\overline{x_2}\cdot\overline{x_3}\cdot x_4\cdot\overline{x_5}+\\
& \overline{x_1}\cdot x_2\cdot\overline{x_3}\cdot x_4\cdot\overline{x_5}+x_1\cdot x_2\cdot x_3\cdot\overline{x_4}\cdot\overline{x_5}
\end{aligned}
$$

The first establishment condition ($F_1(x_1, x_2, x_3, x_4, x_5)$) corresponds to the fact that the only states in which connection $c_1$ (i.e. the connection between the cap and the stick) can be established without precluding the completion of the assembly are either the state in which no connection has been established (node 1 in figure 6.3), or the state in which only connection $c_2$ is established (node 2), or the state in which only connection $c_3$ is established (node 4), or the state in which only connection $c_5$ is established (node 5), or the state in which only connections $c_2$ and $c_4$ are established (node 9), or the state in which only connection $c_1$ and $c_2$ are not established (node 12). It should be noticed that there is no term corresponding to the state in which only connection $c_4$ is established (node 6); although it is feasible to establish connection $c_1$ the resulting state (node 10) is a dead-end from which the assembly cannot be completed.

The establishment conditions defined in this chapter can only discriminate between feasible and nonfeasible assembly sequences. There are sequences of states that "satisfy" the establishment conditions but are not assembly sequences and therefore cannot be feasible assembly sequences. For example, the sequence of states [F F F F F] [T T T F F] [T T T T T] and its corresponding sequence of subsets of connections ($\{c_1, c_2, c_3\}$ $\{c_4, c_5\}$) "satisfy" the above set of establishment conditions. Yet, they do not not correspond to an assembly sequence since it does not encompass exactly 3 assembly tasks.

It is possible to simplify the expressions of the establishment conditions using the rules of boolean algebra. The expressions above can be rewritten in disjunctive normal form[4], as:

$$
\begin{aligned}
F_1(x_1, x_2, x_3, x_4, x_5) &= \overline{x_1}\cdot\overline{x_2}\cdot\overline{x_3}\cdot\overline{x_4}+\overline{x_1}\cdot\overline{x_2}\cdot\overline{x_4}\cdot\overline{x_5}+\\
&\quad \overline{x_1}\cdot\overline{x_2}\cdot x_3\cdot x_4\cdot x_5+\overline{x_1}\cdot x_2\cdot\overline{x_3}\cdot\overline{x_5}\\
F_2(x_1, x_2, x_3, x_4, x_5) &= \overline{x_1}\cdot\overline{x_2}\cdot\overline{x_4}\cdot\overline{x_5}+x_1\cdot\overline{x_2}\cdot\overline{x_3}\cdot\overline{x_4}+\\
&\quad \overline{x_1}\cdot\overline{x_2}\cdot\overline{x_3}\cdot\overline{x_5}+\overline{x_1}\cdot\overline{x_2}\cdot x_3\cdot x_4\cdot x_5\\
F_3(x_1, x_2, x_3, x_4, x_5) &= \overline{x_1}\cdot\overline{x_3}\cdot\overline{x_5}+\overline{x_2}\cdot\overline{x_3}\cdot\overline{x_4}\\
F_4(x_1, x_2, x_3, x_4, x_5) &= \overline{x_1}\cdot\overline{x_3}\cdot\overline{x_4}\cdot\overline{x_5}+\overline{x_1}\cdot\overline{x_2}\cdot\overline{x_4}\cdot\overline{x_5}+\\
&\quad \overline{x_2}\cdot\overline{x_3}\cdot\overline{x_4}\cdot x_5+x_1\cdot x_2\cdot x_3\cdot\overline{x_4}\cdot\overline{x_5}\\
F_5(x_1, x_2, x_3, x_4, x_5) &= \overline{x_2}\cdot\overline{x_3}\cdot\overline{x_4}\cdot\overline{x_5}+\overline{x_1}\cdot\overline{x_2}\cdot\overline{x_4}\cdot\overline{x_5}+\\
&\quad \overline{x_1}\cdot\overline{x_3}\cdot x_4\cdot\overline{x_5}+x_1\cdot x_2\cdot x_3\cdot\overline{x_4}\cdot\overline{x_5}
\end{aligned}
$$

A second type of simplification is possible if we consider the nonstate vectors as DON'T CARE conditions. For the assembly shown in figure 6.1, there are 19 nonstate vectors:

| | | | | |
|---|---|---|---|---|
| [F F F T T] | [F F T F T] | [F F T T F] | [F T F T T] | [F T T F F] |
| [F T T F T] | [F T T T F] | [F T T T T] | [T F F T T] | [T F T F F] |
| [T F T F T] | [T F T T F] | [T F T T T] | [T T F F F] | [T T F F T] |
| [T T F T F] | [T T F T T] | [T T T F T] | [T T T T F] | |

If we consider these vectors as DON'T CARE combinations in the simplification process, the resulting expressions for the establishment conditions are:

$$F_1(x_1, x_2, x_3, x_4, x_5)=\overline{x_1}\cdot\overline{x_2}\cdot\overline{x_4}+\overline{x_1}\cdot x_2\cdot\overline{x_5}+\overline{x_1}\cdot\overline{x_2}\cdot x_5$$

$$F_2(x_1, x_2, x_3, x_4, x_5)=\overline{x_2}\cdot\overline{x_4}\cdot\overline{x_5}+x_1\cdot\overline{x_2}\cdot\overline{x_4}+\overline{x_1}\cdot\overline{x_2}\cdot x_4$$

$$F_3(x_1, x_2, x_3, x_4, x_5)=\overline{x_1}\cdot\overline{x_3}\cdot\overline{x_5}+\overline{x_2}\cdot\overline{x_3}\cdot\overline{x_4}$$

$$F_4(x_1, x_2, x_3, x_4, x_5)=\overline{x_1}\cdot\overline{x_4}\cdot\overline{x_5}+\overline{x_2}\cdot\overline{x_4}\cdot x_5+x_1\cdot x_2\cdot\overline{x_4}$$

$$F_5(x_1, x_2, x_3, x_4, x_5)=\overline{x_1}\cdot\overline{x_2}\cdot\overline{x_5}+\overline{x_1}\cdot x_4\cdot\overline{x_5}+x_1\cdot\overline{x_4}\cdot\overline{x_5}$$

Still a third type of simplification is possible if we consider DON'T CARE conditions the states that do not occur in any feasible assembly sequence. For the assembly shown in figure 6.1, there are two states that do not occur in any feasible assembly sequence; they correspond to nodes 8 and 10 in figure 6.3 and their corresponding binary vectors are: [F T F F T] and [T F F T F] If we also consider these vectors as DON'T CARE combinations in the simplification process, the resulting establishment conditions for the assembly shown in figure 6.1 are:

$$F_1(x_1, x_2, x_3, x_4, x_5) = \overline{x_1} \cdot \overline{x_4} + \overline{x_1} \cdot x_2 + \overline{x_1} \cdot x_5$$

$$F_2(x_1, x_2, x_3, x_4, x_5) = \overline{x_2} \cdot \overline{x_5} + x_1 \cdot \overline{x_2} + \overline{x_2} \cdot x_4$$

$$F_3(x_1, x_2, x_3, x_4, x_5) = \overline{x_3}$$

$$F_4(x_1, x_2, x_3, x_4, x_5) = \overline{x_1} \cdot \overline{x_4} + \overline{x_4} \cdot x_5 + x_2 \cdot \overline{x_4}$$

$$F_5(x_1, x_2, x_3, x_4, x_5) = \overline{x_2} \cdot \overline{x_5} + x_4 \cdot \overline{x_5} + x_1 \cdot \overline{x_5}$$

The expressions for the establishment conditions above are simpler than those listed previously. Although they can correctly discriminate between feasible and unfeasible assembly sequences, they are not as safe to be used in the real time control of the assembly process. For example, these expressions indicate, correctly, that the assembly sequence whose representation as an ordered sequence of states is ([F F F F F] [F T F F F] [F T F F T] [T T T T T]) and whose representation as an ordered sequence of subsets of connections is ($\{c_2\}$ $\{c_5\}$ $\{c_1\, c_3\, c_4\}$) is not feasible because $F_5(\text{F},\text{T},\text{F},\text{F},\text{F}) = \text{F}$, and therefore the second assembly task, in which the $5^{th}$ connection is established, is not feasible. But should the assembly process accidentally reach the state whose binary vector representation is [F T F F T] these expressions for the establishment conditions would indicate, incorrectly, that it is feasible to establish connections $c_1$, $c_3$, and $c_4$ and therefore to complete the assembly. This happens because this state ([F T F F T]) was considered a DON'T CARE condition.

## 6.5 Precedence relationships between the establishment of one connection and states of the assembly process

Two types of precedence relationships can be used to represent assembly sequences: precedence relationships between the establishment of one connection and states of the assembly process, and precedence relationships between the establishment of one connection between two parts and the establishment of another connection. This section addresses the former type and the next section addresses the latter.

We will use the notation $c_i \rightarrow S(\underline{x})$ to indicate that the establishment of the $i^{th}$ connection must precede any state $s$ of the assembly process for which the value of the logical function $S(\underline{x})$ is true. The argument of $S(\underline{x})$ is the $L$-dimensional binary vector representation of the state $s$. We will use a compact notation for logical combinations of precedence relationships. For example, we will write $c_i + c_j \rightarrow S(\underline{x})$ when we mean $[c_i \rightarrow S(\underline{x})] \vee [c_j \rightarrow S(\underline{x})]$.

An assembly sequence whose representation as an ordered sequence of binary vectors is $(\underline{x}_1\, \underline{x}_2 \cdots \underline{x}_N)$ and whose representation as an ordered sequence of subsets of connections is $(\gamma_1\, \gamma_2 \cdots \gamma_{N-1})$ satisfies the precedence relationship $c_i \rightarrow S(\underline{x})$ if

$$S(\underline{x}_k) \Rightarrow \exists l[(l < k) \wedge (c_i \in \gamma_l)] \text{ for } k = 1, 2, \cdots N$$

For example, for the assembly shown in figure 6.1, the assembly sequence whose representation as an ordered sequence of binary vectors is ( [F F F F F] [T F F F F] [T T T F F] [T T T T T] ) and whose representation as an ordered sequence of subsets of connections is $(\{c_1\}\, \{c_2\,,\, c_3\}\, \{c_4\,,\, c_5\})$ satisfies the precedence relationship $c_1 \rightarrow x_2 \cdot x_3$ because the only states for which $S(\underline{x}) = x_2 \cdot x_3$ is true are the third and the fourth, and the establishment of connection $c_1$ occurs on the first assembly task. This sequence does not satisfy the precedence relationship $c_4 \rightarrow x_1 \cdot x_2 \cdot x_3$ because for the third state the value of $S(\underline{x}) = x_1 \cdot x_2 \cdot x_3$ is true but the establishment of connection $c_4$ occurs on the third assembly task, which occurs after the third state.

Let $\Psi_S$ be the set of assembly states that never occur in any feasible assembly sequence. These include the unstable assembly states plus the stable states from which the final state cannot be reached plus the states that cannot be reached from the initial state. Let $\Psi_X = \{\underline{x}_1\,,\, \underline{x}_2\,,\, \cdots\,,\, \underline{x}_J\}$ be the set of all $L$-dimensional binary vectors that represent the assembly states in $\Psi_S$ Every element $\underline{x}_j$ of $\Psi_X$ is such that the value of the logical function $G(\underline{x}_j)$ is true, where

$$G(\underline{x}) = G(x_1\,,\, x_2\,,\, \cdots\,,\, x_L) = \sum_{k=1}^{K} \prod_{l=1}^{L} \lambda_{k\,l}. \tag{6.1}$$

The sum and the product in equation 6.1 are the logical operations OR and AND respectively, and $\lambda_{k\,l}$ is either the symbol $x_l$ if the $l^{th}$ component of $\underline{x}_k$ is true, or the symbol $\overline{x_l}$ if the $l^{th}$ component of $\underline{x}_k$ is false. In many cases the expression of $G(\underline{x})$ can be simplified using the rules of boolean algebra. Allowing for simplifications, but keeping the logical function as a sum of products (*disjunctive form*[4]), equation 6.1 can be rewritten as

$$G(\underline{x}) = \sum_{j=1}^{J'} g_j(\underline{x}) \tag{6.2}$$

where each term $g_j(\underline{x})$ is the product of a subset of $\{ x_1, x_2, \cdots, x_L, \overline{x_1}, \overline{x_2}, \cdots \cdots, \overline{x_L} \}$ that does not include both $x_i$ and $\overline{x_i}$ for any $i$. Each term $g_j(\underline{x})$ can be rewritten grouping all the nonnegated variables first and all the negated variables last, that is $g_j(\underline{x}) = x_a \cdot x_b \cdots x_h \cdot \overline{x_p} \cdot \overline{x_q} \cdots \overline{x_z}$.

Any assembly sequence that includes a state that is in $\Psi_S$ is an unfeasible assembly sequence. Therefore, a necessary condition for the feasibility of an assembly sequence whose representation as an ordered list of binary vectors is $(\underline{x}_1\,\underline{x}_2 \cdots \underline{x}_N)$ is that $G(\underline{x}_1) = G(\underline{x}_2) = \cdots = G(\underline{x}_N) = \text{F}$. This condition is equivalent to $g_j(\underline{x}_i) = \text{F}$ for $i = 1, 2, \cdots, N$ and for $j = 1, 2, \cdots, J'$. This necessary condition is also sufficient if the assembly has the following property:

**Property 1** *Given any two states $s_i$ and $s_j$ not necessarily in the same assembly sequence, let $\gamma_i$ and $\gamma_j$ be the sets of connections that are established in assembly tasks $\tau_i$ and $\tau_j$ from $s_i$ and $s_j$ respectively. If*

*$< P, C_i >$ is the state's graph of connections associated to $s_i$*

*$< P, C_j >$ is the state's graph of connections associated to $s_j$*

$\gamma_i \subset \gamma_j$

$C_i \subset C_j$ *and*

*$\tau_i$ is geometrically and mechanically feasible,*

*then*

*$\tau_j$ is geometrically and mechanically feasible,*

This property corresponds to the fact that if it is geometrically and mechanically feasible to establish a set of connections ($\gamma_j$) when many other connections $C_j$ have already been established, then it is also geometrically and mechanically feasible to establish fewer connections ($\gamma_i \subset \gamma_j$) when fewer other connections ($C_i \subset C_j$) have been established. Although many common assemblies have this property, there are assemblies that do not have it. An example of an assembly that does not have this property is presented elsewhere[5].

If the assembly has property 1, the following lemma establishes a necessary and sufficient condition for the feasibility of an assembly sequence.

**Lemma 3** *Given an assembly whose graph of connections is $< P, C >$ (with $C = \{c_1, c_2, \cdots, c_L\}$) let $\Psi_S$ be the set of states that do not occur in any feasible assembly sequence. If the assembly has property 1, then an assembly sequence is feasible if and only if it does not include any state in $\Psi_S$.*

If $(\underline{x}_1\,\underline{x}_2 \cdots \underline{x}_N)$ is an ordered list of binary vectors that represents an assembly sequence, the condition $g_j(\underline{x}_1) = g_j(\underline{x}_2) = \cdots = g_j(\underline{x}_N) = \text{F}$ corresponds to a precedence relationship. The following lemma establishes the correspondence.

**Lemma 4** *Given an assembly made up of N parts whose graph of connections is $< P, C >$ (with $C = \{c_1, c_2, \cdots, c_L\}$), let*

$$g(\underline{x}) = x_a \cdot x_b \cdots x_h \cdot \overline{x_p} \cdot \overline{x_q} \cdots \overline{x_z}$$

*with*

$$\{a, b, \cdots, h\} \cap \{p, q, \cdots, z\} = \emptyset$$

*and*

$$\{a, b, \cdots, h\} \cup \{p, q, \cdots, z\} \subset \{1, 2, \cdots, L\}$$

*If the assembly has the property 1 and if $(\underline{x}_1\, \underline{x}_2\, \cdots\, \underline{x}_N)$ is the representation of an assembly sequence as an ordered list of L-dimensional binary vectors, then the condition*

$$g(\underline{x}_1) = g(\underline{x}_2) = \cdots = g(\underline{x}_N) = \text{F}$$

*is equivalent to the condition*

$$c_p + c_q + \cdots + c_z \rightarrow S(\underline{x})$$

*where*

$$S(\underline{x}) = \prod_{i=1}^{L} \lambda_l \text{ and } \lambda_l = \begin{cases} x_l & \textit{if } l \in \{ab \cdots h\} \\ \text{true} & \textit{otherwise} \end{cases}$$

The product in this lemma is the logical operation AND. The logical function $S(\underline{x})$ is the product of the variables $x_k$ that are not negated in the expression of $g(\underline{x})$, that is $S(\underline{x}) = x_a \cdot x_b \cdots x_h$.

Applying lemma 4 to each of the $J'$ terms on the right side of equation 6.2 we obtain $J'$ precedence relationships. Given an assembly sequence, if it satisfies all $J'$ precedence relationships then it does not include any state in$\Psi_S$ and therefore is feasible. Conversely, if the assembly sequence does not include any state in $\Psi_S$ (and therefore it is a feasible assembly sequence) then it satisfies all precedence relationships. Therefore, the set of $J'$ precedence relationships is a correct and complete representation of the set of all feasible assembly sequences. This fact is established by the following theorem.

**Theorem 5** *Given an assembly made up of N parts whose graph of connections is $< P, C >$ (with $C = \{c_1, c_2, \cdots, c_L\}$), let*

$$G(\underline{x}) = \sum_{j=1}^{J'} g_j(\underline{x})$$

*be a disjunctive form of the logical function that is* true *if and only if* $\underline{x}$ *is a binary-vector representation of a state that does not occur in any feasible assembly sequence. Let* $A_j$ *be the set containing the indexes of the variables that are asserted in* $g_j(\underline{x})$. *Let* $N_j$ *be the set containing the indexes of the variables that are negated in* $g_j(\underline{x})$. *If the assembly has property 1, and if* $(\gamma_1, \gamma_2, \cdots, \gamma_{N-1})$ *is an ordered sequence of subsets of connections that represents an assembly sequence, then* $(\gamma_1, \gamma_2, \cdots, \gamma_{N-1})$ *satisfies the set of* $J'$ *precedence relationships*

$$\sum_{k \in N_j} c_k \rightarrow \prod_{i \in A_j} x_i \quad \text{for } j = 1, 2, \cdots, J'$$

*if and only if it corresponds to a feasible assembly sequence.*

An example will illustrate the use of theorem 5. For the assembly shown in figure 6.1, which has property 1, $\Psi_X = \{ [\text{F T F F T}] \; [\text{T F F T F}] \}$ (these binary vectors correspond to nodes 8 and 10 in the directed graph of assembly states shown in figure 6.3). Therefore,

$$\begin{aligned} G(\underline{x}) &= G(x_1, x_2, x_3, x_4, x_5) = \\ &\overline{x_1} \cdot x_2 \cdot \overline{x_3} \cdot \overline{x_4} \cdot x_5 + x_1 \cdot \overline{x_2} \cdot \overline{x_3} \cdot x_4 \cdot \overline{x_5} \end{aligned} \tag{6.3}$$

In this case the expression of $G(\underline{x})$ cannot be further simplified and we have

$$g_1(\underline{x}) = \overline{x_1} \cdot x_2 \cdot \overline{x_3} \cdot \overline{x_4} \cdot x_5 \quad A_1 = \{2, 5\} \quad N_1 = \{1, 3, 4\}$$

$$g_2(\underline{x}) = x_1 \cdot \overline{x_2} \cdot \overline{x_3} \cdot x_4 \cdot \overline{x_5} \quad A_2 = \{1, 4\} \quad N_2 = \{2, 3, 5\}$$

Therefore, the precedence relationships are:

$$c_1 + c_3 + c_4 \rightarrow x_2 \cdot x_5 \qquad c_2 + c_3 + c_5 \rightarrow x_1 \cdot x_4 \tag{Set 1}$$

A simpler set of precedence relationships can be obtained if in the simplification of $G(\underline{x})$ we set the nonstate vectors as DON'T CARE conditions. For the assembly shown in figure 6.1, there are 19 nonstate vectors:

| | | | | |
|---|---|---|---|---|
| [F F F T T] | [F F T F T] | [F F T T F] | [F T F T T] | [F T T F F] |
| [F T T F T] | [F T T T F] | [F T T T T] | [T F F T T] | [T F T F F] |
| [T F T F T] | [T F T T F] | [T F T T T] | [T T F F F] | [T T F F T] |
| [T T F T F] | [T T F T T] | [T T T F T] | [T T T T F] | |

Considering the above 19 nonstate vectors as DON'T CARE conditions in the simplification of $G(\underline{x})$ yields

$$G(\underline{x}) = G(x_1, x_2, x_3, x_4, x_5) = \overline{x_1} \cdot x_2 \cdot x_5 + x_1 \cdot \overline{x_2} \cdot x_4 \tag{6.4}$$

Therefore, the precedence relationships are

$$c_1 \rightarrow x_2 \cdot x_5 \qquad c_2 \rightarrow x_1 \cdot x_4 \tag{Set 2}$$

This set is simpler and yet equivalent to Set 1.

It should be noticed that an unfeasible assembly sequence, such as the assembly sequence whose representation as an ordered sequence of subsets of connections is $(\{c_2\}\,\{c_5\}\,\{c_1, c_3, c_4\})$, does not satisfy both sets of precedence relationships above (i.e. Sets 1 and 2). It should also be noticed that there are ordered sequences of $N-1$ subsets of connections and their corresponding ordered sequence of binary vectors, such as $(\{c_1\}\,\{c_2\}\,\{c_3, c_4, c_5\})$ and ( [F F F F F] [T F F F F] [T T F F F] [T T T T T] ) that do not represent an assembly sequence, but satisfy Sets 1 and 3 of precedence relationships. The precedence relationships obtained using the result of theorem 5 can only discriminate the feasible from the unfeasible assembly sequences. The information in the assembly's graph of connections allows the discrimination of assembly sequences from ordered sequences of subsets of connections that do not correspond to assembly sequences.

In order to be able to discriminate the representations of feasible assembly sequences from any sequence of $N-1$ subsets of connections, the set $\Psi_X$ must also include all nonstate vectors, and, of course, these combinations should not be considered DON'T CARE conditions.

**Corollary 6** *Given an assembly made up of $N$ parts whose graph of connections is $< P, C >$ (with $C = \{c_1, c_2, \cdots, c_L\}$), let*

$$G(\underline{x}) = \sum_{j=1}^{J'} g_j(\underline{x})$$

*be a disjunctive form of the logical function that is true if and only if $\underline{x}$ is either a non-state binary-vector or a binary-vector representation of a state that does not occur in any feasible assembly sequence. Let $A_j$ be the set containing the indexes of the variables that are asserted in $g_j(\underline{x})$. Let $N_j$ be the set containing the indexes of the variables that are negated in $g_j(\underline{x})$. If the assembly has property one, then an ordered sequence of $N-1$ subsets of connections satisfies the set of $J'$ precedence relationships*

$$\sum_{k \in N_j} c_k \rightarrow \prod_{i \in A_j} x_i \quad \text{for } j = 1, 2, \cdots, J'$$

*if and only if it represents a feasible assembly sequence.*

For the assembly shown in figure 6.1 there are two assembly states that do not occur in any feasible assembly sequence. And there are nineteen 5-dimensional nonstate vectors which were listed above. Let $G(\underline{x})$ be the logical function that is true if and only if $\underline{x}$ is one of these twenty-one 5-dimensional vectors, that is

$$\begin{aligned}
G(\underline{x}) \;=\; & \overline{x_1} \cdot x_2 \cdot \overline{x_3} \cdot \overline{x_4} \cdot x_5 + x_1 \cdot \overline{x_2} \cdot \overline{x_3} \cdot x_4 \cdot \overline{x_5} + \\
& \overline{x_1} \cdot \overline{x_2} \cdot \overline{x_3} \cdot x_4 \cdot x_5 + \overline{x_1} \cdot \overline{x_2} \cdot x_3 \cdot \overline{x_4} \cdot x_5 + \\
& \overline{x_1} \cdot \overline{x_2} \cdot x_3 \cdot x_4 \cdot \overline{x_5} + \overline{x_1} \cdot x_2 \cdot \overline{x_3} \cdot x_4 \cdot x_5 + \\
& \overline{x_1} \cdot x_2 \cdot x_3 \cdot \overline{x_4} \cdot \overline{x_5} + \overline{x_1} \cdot x_2 \cdot x_3 \cdot \overline{x_4} \cdot x_5 + \\
& \overline{x_1} \cdot x_2 \cdot x_3 \cdot x_4 \cdot \overline{x_5} + \overline{x_1} \cdot x_2 \cdot x_3 \cdot x_4 \cdot x_5 + \\
& x_1 \cdot \overline{x_2} \cdot \overline{x_3} \cdot x_4 \cdot x_5 + x_1 \cdot \overline{x_2} \cdot x_3 \cdot \overline{x_4} \cdot \overline{x_5} + \\
& x_1 \cdot \overline{x_2} \cdot x_3 \cdot \overline{x_4} \cdot x_5 + x_1 \cdot \overline{x_2} \cdot x_3 \cdot x_4 \cdot \overline{x_5} + \\
& x_1 \cdot \overline{x_2} \cdot x_3 \cdot x_4 \cdot x_5 + x_1 \cdot x_2 \cdot \overline{x_3} \cdot \overline{x_4} \cdot \overline{x_5} + \\
& x_1 \cdot x_2 \cdot \overline{x_3} \cdot \overline{x_4} \cdot x_5 + x_1 \cdot x_2 \cdot \overline{x_3} \cdot x_4 \cdot \overline{x_5} + \\
& x_1 \cdot x_2 \cdot \overline{x_3} \cdot x_4 \cdot x_5 + x_1 \cdot x_2 \cdot x_3 \cdot \overline{x_4} \cdot x_5 + \\
& x_1 \cdot x_2 \cdot x_3 \cdot x_4 \cdot \overline{x_5}. \qquad (6.5)
\end{aligned}$$

The first two terms in this function correspond to the two states that do not occur in any feasible assembly sequence and the other 19 terms correspond to the nonstate vectors. Using the rules of boolean algebra to simplify this function, we obtain

$$\begin{aligned}
G(\underline{x}) \;=\; & \overline{x_1} \cdot x_2 \cdot x_5 + \overline{x_1} \cdot x_2 \cdot x_3 + x_1 \cdot \overline{x_2} \cdot x_4 + x_1 \cdot \overline{x_2} \cdot x_3 + \\
& x_1 \cdot x_2 \cdot \overline{x_3} + \overline{x_3} \cdot x_4 \cdot x_5 + x_3 \cdot \overline{x_4} \cdot x_5 + x_3 \cdot x_4 \cdot \overline{x_5}.
\end{aligned}$$

Therefore, the precedence relationships are:

$$\begin{array}{llll}
c_1 \rightarrow x_2 \cdot x_5 & c_1 \rightarrow x_2 \cdot x_3 & c_2 \rightarrow x_1 \cdot x_4 & c_2 \rightarrow x_1 \cdot x_3 \\
c_3 \rightarrow x_1 \cdot x_2 & c_3 \rightarrow x_4 \cdot x_5 & c_4 \rightarrow x_3 \cdot x_5 & c_5 \rightarrow x_3 \cdot x_4.
\end{array} \qquad \text{(Set 3)}$$

The ordered sequences of subsets of connections $(\{c_1, c_2\}\{c_3, c_4, c_5\})$ which does not correspond to an assembly sequence but satisfies Sets 1 and 2 of precedence relationships does not satisfy Set 3. The third state satisfies $S(\underline{x}) = x_1 \cdot x_2$ but connection $c_3$ is established during the third assembly task, which occurs after the third assembly state; therefore, this sequence does not satisfy the precedence relationship $c_3 \rightarrow x_1 \cdot x_2$.

But it should be noticed that Set 3 of precedence relationships will be "satisfied" for ordered sequences of subsets of connections containing fewer than $N-1$ subsets. For example, the sequence $(\{c_1, c_2, c_3\}\{c_4, c_5\})$ "satisfies" Set 3 of precedence relationships. Yet, this sequence does not correspond to a feasible assembly sequence because it does not contain exactly $N-1$ subsets of connections.

## 6.6 Precedence relationships between the establishment of one connection and the establishment of another connection

We will use the notation $c_i < c_j$ to indicate the fact that the establishment of connection $c_i$ must precede the establishment of connection $c_j$. And we will use the notation $c_i \leq c_j$ to indicate the fact that the establishment of connection $c_i$ must precede or be simultaneous with the establishment of connection $c_j$. Furthermore, we will use a compact notation for logical combinations of precedence relationships; for example, we will write $c_i < c_j \cdot c_k$ when we mean $(c_i < c_j) \wedge (c_i < c_k)$, and we will write $c_i + c_j < c_k$ when we mean $(c_i < c_k) \vee (c_j < c_k)$.

An assembly sequence whose representation as an ordered sequence of binary vectors is $(\underline{x}_1 \underline{x}_2 \cdots \underline{x}_N)$ and whose representation as an ordered sequence of subsets of connections is $(\gamma_1 \gamma_2 \cdots \gamma_{N-1})$ satisfies the precedence relationship $c_i < c_j$ if $c_i \in \gamma_a$, $c_j \in \gamma_b$, and $a < b$. Similarly, the sequence satisfies $c_i \leq c_j$ if $c_i \in \gamma_a$, $c_j \in \gamma_b$, and $a \leq b$. For example, for the assembly shown in figure 6.1, the assembly sequence whose representation as an ordered sequence of binary vectors is ([F F F F F] [T F F F F] [T T T F F] [T T T T T]) and whose representation as an ordered sequence of subsets of connections is $(\{c_1\} \{c_2, c_3\} \{c_4, c_5\})$ satisfies the precedence relationships $c_2 < c_4$ and $c_2 \leq c_3$ but does not satisfy the precedence relationships $c_2 < c_3$ and $c_2 \leq c_1$.

Each feasible assembly sequence of a given assembly can be uniquely characterized by a logical expression consisting of the conjunction of precedence relationships between the establishment of one connection and the establishment of another connection. For example, for the assembly shown in figure 6.1, the assembly sequence ([F F F F F] [T F F F F] [T T T F F] [T T T T T]) can be uniquely characterized by the following conjunction of precedence relationships

$$(c_1 < c_2) \wedge (c_2 < c_4) \wedge (c_2 \leq c_3) \wedge (c_3 \leq c_2) \wedge (c_4 \leq c_5) \wedge (c_5 \leq c_4)$$

The set of all $M$ feasible assembly sequences can be uniquely characterized by a disjunction of $M$ conjunctions of precedence relationships in which each conjunction characterizes one assembly sequence. Clearly, this logical combination of precedence relationships constitutes a correct and complete representation for the set of all assembly sequences.

It is often possible to simplify this logical combination of precedence relationships using the rules of boolean algebra. Further simplification is possible if one notices that there are logical combinations of precedence relationships that cannot be satisfied by any assembly sequence. For the assembly shown in figure 6.1, for example, the combination $(c_1 < c_2) \wedge (c_2 < c_3) \wedge (c_3 < c_4) \wedge (c_4 < c_5)$ cannot be satisfied by any assembly sequence. These combinations can be set as don't care conditions in the simplification of the logical combination of precedence relationships.

It is possible to obtain simpler precedence relationships and the rest of this section describes two ways to do that.

### Precedence relationships derived from those obtained in the previous section

The precedence relationships obtained in the previous section have the form

$$c_p + c_q + \cdots + c_z \rightarrow x_a \cdot x_b \cdot \cdots \cdot x_h$$

which is equivalent to the disjunction of precedence relationships

$$(c_p \rightarrow x_a \cdot x_b \cdot \cdots \cdot x_h) \vee (c_q \rightarrow x_a \cdot x_b \cdot \cdots \cdot x_h) \vee \cdots$$
$$\cdots \vee (c_z \rightarrow x_a \cdot x_b \cdot \cdots \cdot x_h).$$

It is straigth forward to see that the precedence relationship

$$c_p \rightarrow x_a \cdot x_b \cdot \cdots \cdot x_h$$

is equivalent to

$$c_p \leq c_a + c_b + \cdots + c_h.$$

Therefore, the disjunction of precedence relationships above is equivalent to

$$(c_p \leq c_a + c_b + \cdots + c_h) \vee (c_q \leq c_a + c_b + \cdots + c_h) \vee \cdots$$
$$\cdots \vee (c_z \leq c_a + c_b + \cdots + c_h)$$

Each set of precedence relationships obtained in the previous section yields a conjunction of disjunctions of precedence relationships. For example, for the simple product shown in figure 6.1, we obtained, in the previous section, Set 1 of precedence relationships. That set is equivalent to the following conjunction of disjunctions:

$$[(c_1 \leq c_2 + c_5) \vee (c_3 \leq c_2 + c_5) \vee (c_4 \leq c_2 + c_5)] \wedge$$
$$[(c_2 \leq c_1 + c_4) \vee (c_3 \leq c_1 + c_4) \vee (c_5 \leq c_1 + c_4)]$$

### Precedence relationships derived from the set of feasible assembly sequences

Another simpler precedence relationship representation of all assembly sequences can be derived from the set of feasible assembly sequences if the assembly has the property 1 described in the previous section as well as the following property:

**Property 2** *If the subsets $\theta_1, \theta_2, \cdots, \theta_k$ of the set of parts $P$ characterize stable subassemblies, then the set $\theta = \theta_1 \cup \theta_2 \cup \cdots \cup \theta_k$ also characterizes a stable subassembly.*

Like in the case of property 1, many common assemblies have this second property. Yet, there are assemblies that do not have it. An example of an assembly that does not have this property is presented elsewhere[5].

The following theorem indicates how to obtain a simple precedence relationship representation of the assembly sequences for assemblies that have both properties 1 and 2.

**Theorem 7** *Given an assembly made up of $N$ parts whose graph of connections is $< P, C >$ (with $C = \{c_1, c_2, \cdots, c_L\}$), let*

$$\{\, (\gamma_{11}\gamma_{21}\cdots\gamma_{(N-1)1}), (\gamma_{12}\gamma_{22}\cdots\gamma_{(N-1)2}), \cdots, (\gamma_{1M}\gamma_{2M}\cdots\gamma_{(N-1)M}) \,\}$$

*be a set of $M$ ordered sequences of subsets of connections that represent feasible assembly sequences. If the assembly has properties 1 and 2, then any ordered sequence of $N-1$ subsets of connections that represents an assembly sequence corresponds to a feasible assembly sequence if it satisfies the set of $2L$ precedence relationships:*

$$c_i \leq \sum_{j=1}^{M} T_{ij} \quad i = 1, 2, \cdots, L \quad \text{and} \quad \sum_{j=1}^{M} H_{ij} \leq c_i \quad i = 1, 2, \cdots, L$$

*where*

$$T_{ij} = \prod_{k=1}^{L} \lambda_{ik} \text{ with } \lambda_{ik} = \begin{cases} c_k & \textit{if } c_k \in \gamma_{lj} \textit{ and } l \geq i \\ \text{true} & \textit{otherwise.} \end{cases}$$

$$H_{ij} = \prod_{k=1}^{L} \lambda_{ik} \text{ with } \lambda_{ik} = \begin{cases} c_k & \textit{if } c_k \in \gamma_{lj} \textit{ and } \leq i \\ \text{true} & \textit{otherwise.} \end{cases}$$

The sum and the product in this theorem are the logical operations OR and AND respectively. Each term $T_{ij}$ (for $i = 1, 2, \cdots, L$, and for $j = 1, 2, \cdots, M$) is the product of the variables corresponding to the connections that are established at the same time or after the establishment of connection $c_i$ in the $j^{th}$ sequence. Similarly, each term $H_{ij}$ (for $i = 1, 2, \cdots, L$, and for $j = 1, 2, \cdots, M$) is the product of the variables corresponding to the connections that are established at the same time or before the establishment of connection $c_i$ in the $j^{th}$ sequence. Precedence relationships that have T on either side are always satisfied.

An example will illustrate the use of theorem 7. The assembly shown in figure 6.1 has properties 1 and 2. For that assembly, the set of feasible sequences can be obtained from the directed graph shown in figure 6.3. There are ten feasible assembly sequences and they are:

$$(\{c_1\}\{c_2,c_3\}\{c_4,c_5\}) \quad (\{c_1\}\{c_5\}\{c_2,c_3,c_4\}) \quad (\{c_2\}\{c_1,c_3\}\{c_4,c_5\})$$
$$(\{c_2\}\{c_4\}\{c_1,c_3,c_5\}) \quad (\{c_3\}\{c_1,c_2\}\{c_4,c_5\}) \quad (\{c_3\}\{c_4,c_5\}\{c_1,c_2\})$$
$$(\{c_4\}\{c_2\}\{c_1,c_3,c_5\}) \quad (\{c_4\}\{c_3,c_5\}\{c_1,c_2\}) \quad (\{c_5\}\{c_1\}\{c_2,c_3,c_4\})$$
$$(\{c_5\}\{c_3,c_4\}\{c_1,c_2\})$$

Applying the result of theorem 7 to the above set of feasible sequences for the assembly shown in figure 6.1, the precedence relationships having connection $c_1$ alone on one side are:

$$\begin{aligned} c_1 \quad \leq \quad & c_2 \cdot c_3 \cdot c_4 \cdot c_5 + c_2 \cdot c_3 \cdot c_4 \cdot c_5 + c_3 \cdot c_4 \cdot c_5 + \\ & c_3 \cdot c_5 + c_2 \cdot c_4 \cdot c_5 + c_2 + c_3 \cdot c_5 + c_2 + c_2 \cdot c_3 \cdot c_4 + c_2 \end{aligned}$$

and

$$\begin{aligned} \mathrm{T} + \mathrm{T} + c_2 \cdot c_3 + c_2 \cdot c_3 \cdot c_4 \cdot c_5 + c_2 \cdot c_3 + c_2 \cdot c_3 \cdot c_4 \cdot c_5 + \\ c_2 \cdot c_3 \cdot c_4 \cdot c_5 + c_2 \cdot c_3 \cdot c_4 \cdot c_5 + c_5 + c_2 \cdot c_3 \cdot c_4 \cdot c_5 \quad \leq \quad c_1. \end{aligned}$$

Using the rules of boolean algebra, these two precedence relationships can be simplified yielding $c_1 \leq c_2 + c_3 \cdot c_5$ and $\mathrm{T} \leq c_1$ . The second precedence relationship is always satisfied and can be ignored. Similarly, applying the result of theorem 7, simplifying the logical expressions, and deleting those precedence relationships that have T on either side, we obtain four addional precedence relationships. The resulting set of precedence relationships is:

$$\begin{array}{lll} c_1 \leq c_2 + c_3 \cdot c_5 & c_2 \leq c_1 + c_3 \cdot c_4 & c_3 \leq c_1 \cdot c_5 + c_2 \cdot c_4 \\ c_4 \leq c_5 + c_2 \cdot c_3 & c_5 \leq c_4 + c_1 \cdot c_3. & \end{array} \qquad \text{(Set 4)}$$

Set 4 of precedence relationships still contains some redundancies and can be shown to be equivalent to:

$$c_3 \leq c_1 \cdot c_5 + c_2 \cdot c_4. \qquad \text{(Set 5)}$$

The simplest way to see the equivalence is to verify that Set 5 correctly discriminates the feasible assembly sequences from the unfeasible ones.

It should be noticed that an unfeasible assembly sequence, such as the assembly sequence whose representation as an ordered sequence of subsets of connections is $(\{c_2\}\,\{c_5\}\,\{c_1,c_3,c_4\})$, does not satisfy Set 5 of precedence relationships. It should also be noticed that there are ordered sequences of subsets of connections, such as $(\{c_3\}\,\{c_1,c_4\}\,\{c_2,c_5\})$, that do not represent an assembly sequence, but satisfy Set 5 of precedence relationships. The precedence relationships obtained using the result of theorem 7 can only discriminate the feasible from the unfeasible assembly sequences. The information in the assembly's graph of connections allows the discrimination of assembly sequences from ordered sequences of subsets of connections that do not correspond to assembly sequences.

Theorem 7 is a sufficient condition. The set of precedence relationships obtained using this theorem is correct but not necessarily complete. For example if the set of $M$ ordered sequences of subsets of connections is

$$\{\ (\{c_3\}\,\{c_1,c_2\}\,\{c_4,c_5\})\ (\{c_3\}\,\{c_4,c_5\}\,\{c_1,c_2\})\ (\{c_1\}\,\{c_5\}\,\{c_2,c_3,c_4\})\ \}$$

the resulting precedence relationships are:

$$\begin{array}{lll} c_1 \leq c_2 & c_2 \leq c_1 + c_3 \cdot c_4 & c_3 \leq c_2 \cdot c_4 \\ c_4 \leq c_2 \cdot c_3 + c_5 & c_5 \leq c_4 & c_1 \cdot c_3 \leq c_2 \\ c_3 \cdot c_5 \leq c_4 & c_1 + c_3 \cdot c_4 \leq c_5. & \end{array} \qquad \text{(Set 6)}$$

Some of the feasible sequences, such as $(\{c_1\}\,\{c_2,c_3\}\,\{c_4,c_5\})$, satisfy Set 6 of precedence relationships. Other feasible sequences, such as $(\{c_2\}\,\{c_1,c_3\}\,\{\ c_4, c_5\ \})$ and $(\{c_5\}\,\{c_1\}\,\{c_2,c_3,c_4\})$, do not satisfy Set 6. But if the set of $M$ ordered sequences of subsets of connections includes the representations of all feasible assembly sequences, then the resulting set of precedence relationships constitutes a correct and complete representation of the feasible assembly sequences. This leads to the following corollary:

**Corollary 8** *If the set of of $M$ ordered sequences of subsets of connections in theorem 7 includes the representations of all feasible assembly sequences, the resulting set of precedence relationships is a correct and complete representation of the feasible assembly sequences.*

Finally, it should be noticed that these precedence relationships can be obtained by answering the following two questions for each connection:

1. What connections must be undone when the $i^{th}$ connection is established?
2. What connections must not be left to be done after the $i^{th}$ connection is established?

De Fazio and Whitney[3] have proposed questions similar to these. The proof of theorem 7[5] shows that if the assembly has properties 1 and 2, then these questions lead to a correct and complete precedence relationship representation of assembly sequences. Furthermore, in order to answer these questions one must actually know what all the feasible assembly sequences are.

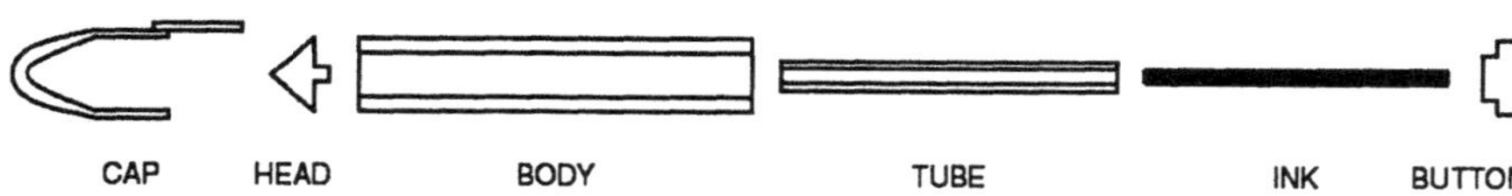


Figure 6.5: The ball-point pen

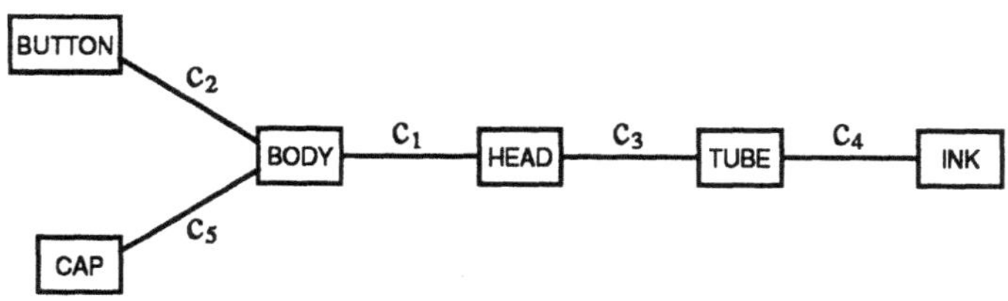


Figure 6.6: The graph of connections for the ball-point pen shown in figure 6.5

## 6.7 The Ball Point Pen Assembly

Both Bourjault[2] and De Fazio and Whitney[3] have used a ball point pen to illustrate their algorithms for the generation of mechanical assembly sequences. Figure 6.5 shows the ball point pen. It contains 6 parts, namely the cap, the body, the button, the head, the tube, and the ink. Although the ink is not actually rigid, in this analysis it can be considered rigid as long as the subassembly made up of the tube and the ink is considered unstable.

Figure 6.6 shows the ball point pen graph of connections. It has 6 nodes and 5 connections. In order to be consistent with previous work[2, 3], it is assumed that there is no contact between the ink and the head. Therefore, the graph of connections does not include an edge connecting the node corresponding to the ink to the node corresponding to the head.

### Directed graph of feasible assembly sequences

Figure 6.7 shows the directed graph of feasible assembly sequences. Each node of the graph in figure 6.7 is labeled by a 5-dimensional binary vector that represents a stable assembly state.

### AND/OR graph of feasible assembly sequences

Figure 6.8 shows the AND/OR graph of feasible assembly sequences for the ball point pen. Figure 6.8 does not include the nodes of the AND/OR graph that cannot be reached from the top node, since they do not occur in any feasible assembly sequence. These nodes correspond to the following subassemblies: { CAP, BODY }, { BODY, HEAD, BUTTON }, { CAP, BODY, BUTTON }, { BODY, HEAD, TUBE, BUTTON }, { CAP, BODY, HEAD, BUTTON }, and { CAP, BODY, HEAD, TUBE, BUTTON }.

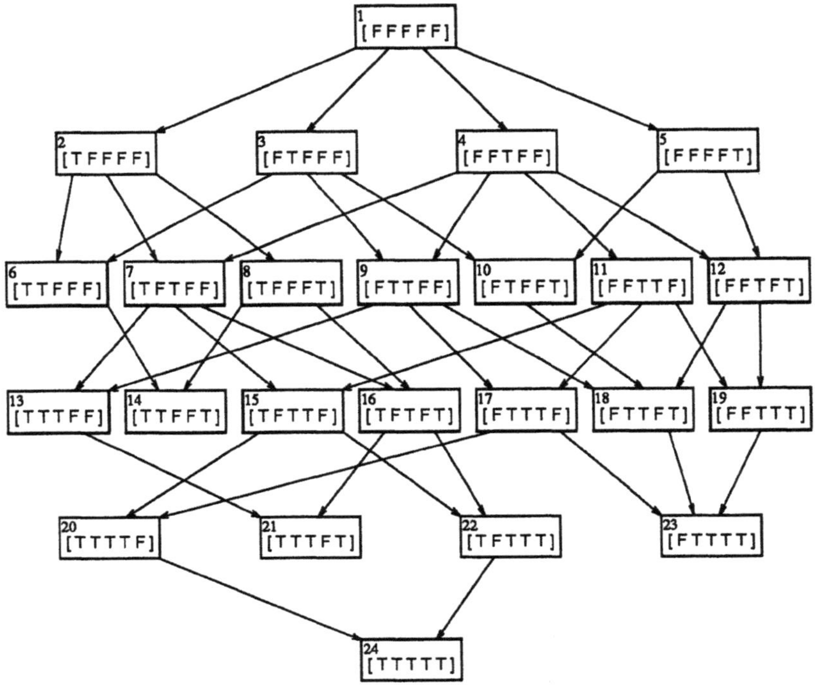


Figure 6.7: The directed graph of feasible assembly sequences of the ball-point pen shown in figure 6.5

**Establishment conditions**

The establishment conditions that are obtained from the directed graph of feasible assembly sequences are:

$$\begin{aligned}
F_1(\underline{x}) &= \overline{x_1}\cdot x_2\cdot x_3\cdot x_4\cdot\overline{x_5}+\overline{x_1}\cdot\overline{x_2}\cdot x_3\cdot x_4\cdot\overline{x_5}+\\
&\quad \overline{x_1}\cdot\overline{x_2}\cdot x_3\cdot\overline{x_4}\cdot\overline{x_5}+\overline{x_1}\cdot\overline{x_2}\cdot\overline{x_3}\cdot\overline{x_4}\cdot\overline{x_5}
\end{aligned}$$

$$\begin{aligned}
F_2(\underline{x}) &= x_1\cdot\overline{x_2}\cdot x_3\cdot x_4\cdot x_5+x_1\cdot\overline{x_2}\cdot x_3\cdot x_4\cdot\overline{x_5}+\overline{x_1}\cdot\overline{x_2}\cdot x_3\cdot x_4\cdot\overline{x_5}+\\
&\quad \overline{x_1}\cdot\overline{x_2}\cdot x_3\cdot\overline{x_4}\cdot\overline{x_5}+\overline{x_1}\cdot\overline{x_2}\cdot\overline{x_3}\cdot\overline{x_4}\cdot\overline{x_5}
\end{aligned}$$

$$\begin{aligned}
F_3(\underline{x}) &= x_1\cdot\overline{x_2}\cdot\overline{x_3}\cdot\overline{x_4}\cdot x_5+x_1\cdot\overline{x_2}\cdot\overline{x_3}\cdot\overline{x_4}\cdot\overline{x_5}+\\
&\quad \overline{x_1}\cdot x_2\cdot\overline{x_3}\cdot\overline{x_4}\cdot\overline{x_5}+\overline{x_1}\cdot\overline{x_2}\cdot\overline{x_3}\cdot\overline{x_4}\cdot\overline{x_5}
\end{aligned}$$

$$\begin{aligned}
F_4(\underline{x}) &= x_1\cdot\overline{x_2}\cdot x_3\cdot\overline{x_4}\cdot x_5+x_1\cdot\overline{x_2}\cdot x_3\cdot\overline{x_4}\cdot\overline{x_5}+\\
&\quad \overline{x_1}\cdot x_2\cdot x_3\cdot\overline{x_4}\cdot\overline{x_5}+\overline{x_1}\cdot\overline{x_2}\cdot x_3\cdot\overline{x_4}\cdot\overline{x_5}
\end{aligned}$$

$$\begin{aligned}
F_5(\underline{x}) &= x_1\cdot x_2\cdot x_3\cdot x_4\cdot\overline{x_5}+x_1\cdot\overline{x_2}\cdot x_3\cdot x_4\cdot\overline{x_5}+\\
&\quad x_1\cdot\overline{x_2}\cdot x_3\cdot\overline{x_4}\cdot\overline{x_5}+x_1\cdot\overline{x_2}\cdot\overline{x_3}\cdot\overline{x_4}\cdot\overline{x_5}
\end{aligned}$$

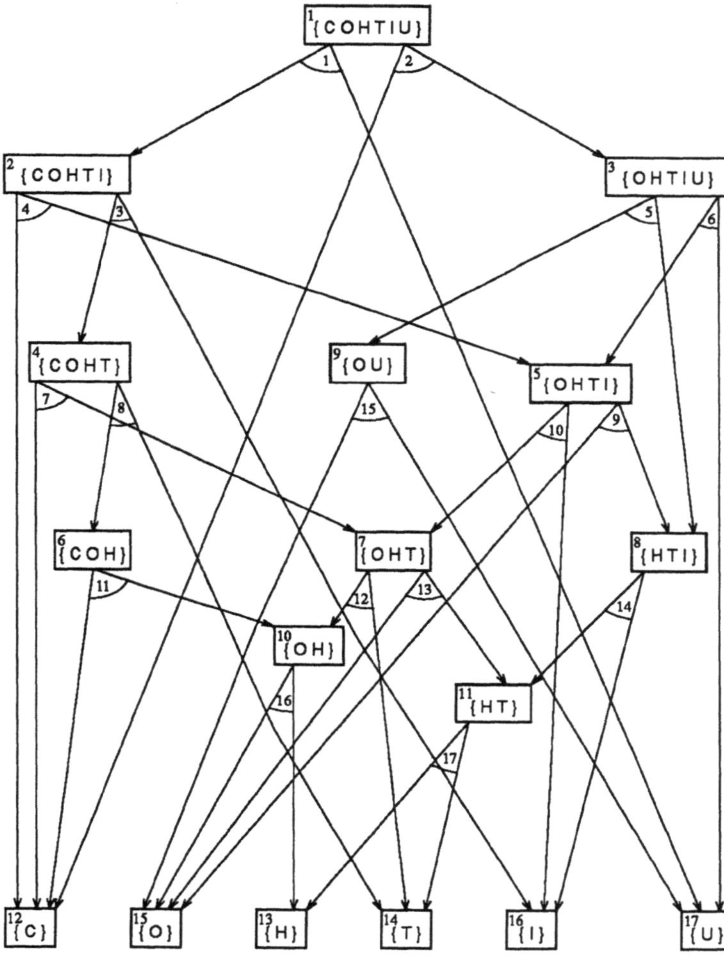


Figure 6.8: AND/OR graph of feasible assembly sequences for the ball-point pen
C = CAP O = BODY H = HEAD T = TUBE I = INK U = BUTTON

Using the rules of boolean algebra to simplify the logical expressions above, we obtain:

$$F_1(\underline{x}) = \overline{x_1} \cdot \overline{x_2} \cdot \overline{x_4} \cdot \overline{x_5} + \overline{x_1} \cdot x_3 \cdot x_4 \cdot \overline{x_5}$$

$$F_2(\underline{x}) = \overline{x_1} \cdot \overline{x_2} \cdot \overline{x_4} \cdot \overline{x_5} + \overline{x_2} \cdot x_3 \cdot x_4 \cdot \overline{x_5} + x_1 \cdot \overline{x_2} \cdot x_3 \cdot x_4$$

$$F_3(\underline{x}) = \overline{x_1} \cdot \overline{x_3} \cdot \overline{x_4} \cdot \overline{x_5} + x_1 \cdot \overline{x_2} \cdot \overline{x_3} \cdot \overline{x_4}$$

$$F_4(\underline{x}) = x_1 \cdot \overline{x_2} \cdot x_3 \cdot \overline{x_4} + \overline{x_1} \cdot x_3 \cdot \overline{x_4} \cdot \overline{x_5}$$

$$F_5(\underline{x}) = x_1 \cdot x_3 \cdot x_4 \cdot \overline{x_5} + x_1 \cdot \overline{x_2} \cdot \overline{x_4} \cdot \overline{x_5}$$

Further simplification is possible if the simplification process takes into account the fact that some states do not occur in any feasible assembly sequence. These states are:

[F F F F T] [F F F T F] [F F F T T] [F F T F T] [F F T T T]
[F T F F T] [F T F T F] [F T F T T] [F T T F T] [F T T T T]
[T F F T F] [T F F T T] [T T F F F] [T T F F T] [T T F T F]
[T T F T T] [T T T F F] [T T T F T]

and the resulting further simplified establishment conditions are:

$$F_1(\underline{x}) = \overline{x_1} \cdot x_4 + \overline{x_1} \cdot \overline{x_2} = \overline{x_1} \cdot (\overline{x_2} + x_4)$$

$$F_2(\underline{x}) = \overline{x_1} \cdot \overline{x_2} + \overline{x_2} \cdot x_4 = \overline{x_2} \cdot (\overline{x_1} + x_4)$$

$$F_3(\underline{x}) = \overline{x_3}$$

$$F_4(\underline{x}) = \overline{x_4} \cdot x_3$$

$$F_5(\underline{x}) = \overline{x_5} \cdot x_1$$

As pointed out in section 6.4, these last expressions can discriminate correctly between feasible and unfeasible assembly sequences but they are not safe for real time control. For example, these expressions indicate, correctly, that the assembly sequence whose representation as an ordered sequence of states is ([F F F F F] [T F F F F] [T T F F F] [T T T F F] [T T T T F] [T T T T T]) and whose representation as an ordered sequence of subsets of connections is ($\{c_1\}$ $\{c_2\}$ $\{c_3\}$ $\{c_4\}$ $\{c_5\}$) is not feasible because $F_2($[T F F F F]$) = $ F, and therefore the second assembly task, in which the $2^{nd}$ connection is established, is not feasible. But should the assembly process accidentally reach the state whose binary vector representation is [T T F F F] these expressions for the establishment conditions would indicate, incorrectly, that it is feasible to establish connections $c_3$, $c_4$, and $c_5$ and therefore to complete the assembly. This happens because this state ([T T F F F]) was considered a DON'T CARE condition.

### Precedence relationships between the establishment of one connection and states of the assembly process

There are 18 states that do not occur in any feasible assembly sequence, and they have been listed previously. Therefore, the logical function that is true if and only if is argument is a binary vector representation of a state that does not occur in any feasible assembly sequence is

$$\begin{aligned} G(\underline{x}) = {} & \overline{x_1} \cdot \overline{x_2} \cdot \overline{x_3} \cdot \overline{x_4} \cdot x_5 + \overline{x_1} \cdot \overline{x_2} \cdot \overline{x_3} \cdot x_4 \cdot \overline{x_5} + \overline{x_1} \cdot \overline{x_2} \cdot \overline{x_3} \cdot x_4 \cdot x_5 + \\ & \overline{x_1} \cdot \overline{x_2} \cdot x_3 \cdot \overline{x_4} \cdot x_5 + \overline{x_1} \cdot \overline{x_2} \cdot x_3 \cdot x_4 \cdot x_5 + \overline{x_1} \cdot x_2 \cdot \overline{x_3} \cdot \overline{x_4} \cdot x_5 + \\ & \overline{x_1} \cdot x_2 \cdot \overline{x_3} \cdot x_4 \cdot \overline{x_5} + \overline{x_1} \cdot x_2 \cdot \overline{x_3} \cdot x_4 \cdot x_5 + \overline{x_1} \cdot x_2 \cdot x_3 \cdot \overline{x_4} \cdot x_5 + \\ & \overline{x_1} \cdot x_2 \cdot x_3 \cdot x_4 \cdot x_5 + x_1 \cdot \overline{x_2} \cdot \overline{x_3} \cdot x_4 \cdot \overline{x_5} + x_1 \cdot \overline{x_2} \cdot \overline{x_3} \cdot x_4 \cdot x_5 + \\ & x_1 \cdot x_2 \cdot \overline{x_3} \cdot \overline{x_4} \cdot \overline{x_5} + x_1 \cdot x_2 \cdot \overline{x_3} \cdot \overline{x_4} \cdot x_5 + x_1 \cdot x_2 \cdot \overline{x_3} \cdot x_4 \cdot \overline{x_5} + \\ & x_1 \cdot x_2 \cdot \overline{x_3} \cdot x_4 \cdot x_5 + x_1 \cdot x_2 \cdot x_3 \cdot \overline{x_4} \cdot \overline{x_5} + x_1 \cdot x_2 \cdot x_3 \cdot \overline{x_4} \cdot x_5. \end{aligned}$$

The expression of $G(\underline{x})$ can be simplified using the rules of boolean algebra. A simpler disjunctive form of this function is

$$G(\underline{x}) \quad = \quad x_1 \cdot x_2 \cdot \overline{x_4} + \overline{x_3} \cdot x_4 + \overline{x_1} \cdot x_5$$

Therefore, the set of precedence relationships

$$c_4 \rightarrow x_1 \cdot x_2 \qquad c_3 \rightarrow x_4 \qquad c_1 \rightarrow x_5 \qquad \text{(Set 7)}$$

is a correct and complete representation of the assembly sequences. Set 7 of precedence relationships is the same set that was obtained by De Fazio and Whitney[3].

**Precedence relationships between the establishment of one connection and the establishment of another connection**

From the Set 7 of precedence relationships obtained in the previous section, the following conjunctions of precedence relationships can be derived:

$$(c_4 \leq c_1 + c_2) \wedge (c_3 \leq c_4) \wedge (c_1 \leq c_5) \qquad \text{(Set 8)}$$

Another precedence relationship representation can be obtained using theorem 7 presented in section 6.6. There are 12 feasible assembly sequences and they are:

({1}{3}{4}{2}{5}) ({1}{3}{4}{5}{2}) ({1}{3}{5}{4}{2})

({1}{5}{3}{4}{2}) ({2}{3}{4}{1}{5}) ({3}{1}{4}{2}{5})

({3}{1}{4}{5}{2}) ({3}{1}{5}{4}{2}) ({3}{2}{4}{1}{5})

({3}{4}{1}{2}{5}) ({3}{4}{1}{5}{2}) ({3}{4}{2}{1}{5}).

Applying the result of theorem 7 to the above set of feasible assembly sequences for the assembly shown in figure 6.5 and using the rules of boolean algebra, we obtain the following precedence relationship representation of assembly sequences:

$$\begin{array}{cc} c_1 \leq c_5 & c_2 \leq \mathrm{T} \\ c_3 \leq c_2 \cdot c_4 + c_1 \cdot c_4 \cdot c_5 & c_4 \leq c_2 + c_1 \cdot c_5 \\ c_5 \leq \mathrm{T} & \\ \mathrm{T} \leq c_1 & \mathrm{T} \leq c_2 \\ \mathrm{T} \leq c_3 & c_3 \leq c_4 \\ c_1 \leq c_5 & \end{array} \qquad \text{(Set 9)}$$

Set 9 can be further simplified. The precedence relationships that have T on either side are always satisfied and can be dropped. Furthermore, since $c_1 \leq c_5$ the second and third precedence relationships can be simplified to $c_3 \leq c_2 \cdot c_4 + c_1 \cdot c_4$ and $c_4 \leq c_2 + c_1$ respectively. Furthermore, $c_4 \leq c_2 + c_1$ requires that either $c_4 \leq c_2$ or $c_4 \leq c_1$; in both cases $c_3 \leq c_2 \cdot c_4 + c_1 \cdot c_4$

is also satisfied because $c_3 \leq c_4$. Therefore, the following set of precedence relationships

$$c_1 \leq c_5 \qquad c_3 \leq c_4 \qquad c_4 \leq c_2 + c_1 \qquad \text{(Set 10)}$$

constitutes a correct and complete representation of the the set of feasible assembly sequences. Set 10 is similar to set 7.

# References


[1] N. Boneschanscher et al. Subassembly stability. In *Proceedings of AAAI-88*, pages 780–785. Morgan Kaufman, August 1988.

[2] A. Bourjault. *Contribution a une Approche Méthodologique de L'Assemblage Automatisé: Elaboration Automatique des Séquences Opératoires.* Thèse d'état, Université de Franche-Comté, Besançon, France, November 1984.

[3] T. L. De Fazio and D. E. Whitney. Simplified Generation of All Mechanical Assembly Sequences. *IEEE J. Robotics Automat.*, RA-3(6):640–658, December 1987. Corrections *ibid* RA-4(6):705–708, December 1988.

[4] H. D. Ebbinghaus et al. *Mathematical Logic.* Springer Verlag, 1984.

[5] L. S. Homem de Mello. *Task Sequence Planning for Robotic Assembly.* PhD thesis, Carnegie Mellon University, May 1989.

[6] L. S. Homem de Mello and A. C. Sanderson. Task Sequence Planning for Assembly. In *12th World Congress on Scientific Computation*, volume 3, pages 390–392. IMACS – International Association for Mathematics and Computers in Simulation, July 1988.

[7] L. S. Homem de Mello and A. C. Sanderson. Automatic Generation of Mechanical Assembly Sequences. Technical Report CMU-RI-TR-88-19, The Robotics Institute – Carnegie Mellon University, December 1988.

[8] L. S. Homem de Mello and A. C. Sanderson. AND/OR Graph Representation of Assembly Plans. *IEEE Trans. Robotics Automat.*, 6(2):188–199, April 1990.

# Chapter 7

# A basic algorithm for the generation of mechanical assembly sequences

**Luiz S. Homem de Mello and Arthur C. Sanderson**

This chapter presents an algorithm for the generation of mechanical assembly sequences that is correct and complete. The algorithm takes a description of the assembly and returns the AND/OR graph representation of assembly sequences[10]. It is assumed that exactly two parts or subassemblies are joined at each time, and that after parts have been put together they remain together. It is also assumed that whenever parts are joined forming a subassembly, all contacts between the parts in that subassembly are established. Furthermore, it is assumed that the feasibility of joining two subassemblies is independent of how those subassemblies were built. These assumptions are consistent with the trend towards product designs that are suitable for automatic assembly[1, 3].

The correctness of the algorithm is based on the assumption that it is always possible to decide correctly whether two subassemblies can be joined, based on geometrical and physical criteria. This chapter presents an approach to compute this decision. An experimental implementation for the class of products made up of polyhedral and cylindrical parts having planar or cylindrical contacts among themselves is described.

The amount of computation involved in generating the AND/OR graph representation of assembly plans depends on the number of parts that make up the product, on how those parts are interconnected, and also on the resulting AND/OR graph. Bounds for the amount of computation involved are presented.

The algorithm described in this chapter operates on a relational model of an assembly. The relational model used in this work provides an efficient data structure which maintains contact geometry and connection information at one level of representation and complete part geometry at a second level. This hierarchy permits many of the planning decisions, such as local geometric feasibility, to be made by accessing only the highest level of the representation. In this sense, much of the planning process is carried out in terms of an abstraction of the actual assembly parts description. That is, the algorithmic structure which we describe here operates primarily in the relational graph domain, and not on the full part geometry unless needed for evaluation of a particular feasibility predicate. The relational graph structures the search for assembly plans by organizing the cut-sets of the graph. This organization of the algorithm and data structure are a principal contribution of this chapter.

Many different types of feasibility predicates could be incorporated into this overall structure. In this chapter, we discuss *internal* task predicates which are related to constraints imposed by other factors such as availability of resources, and stability predicates which assess the stability of the resulting subassemblies. We do not attempt to explore these criteria and constraints exhaustively here. We have focussed on the evaluation of local geometric feasibility as the principal criterion implemented for the studies described in this chapter. This local geometric criterion is a minimal constraint for all feasible assemblies and provides a good means to illustrate and evaluate the performance of the algorithm. This chapter should be viewed as a framework for assembly sequence planning which provides a basis for the incorporation of many different possible specific physical and geometric criteria. In many cases, the physical and geometric reasoning required for such criteria are active topics of research in themselves, and the development of extended feasibility criteria will be based on results of that research.

## 7.1 A Relational Model for Assemblies

A mechanical assembly is a composition of interconnected parts forming a stable unit. Each part is a solid rigid object, that is, its shape remains unchanded. Parts are interconnected whenever they have one or more surfaces in contact. Surface contacts between parts reduce the degrees of freedom for relative motion. A cylindrical contact, for example, prevents any relative motion that is not a translation along the axis or a rotation around the axis. Attachments may act on surface contacts and eliminate all degrees of freedom for relative motion. For example, if a cylindrical contact has a pressure-fit attachment, then no relative motion between the parts is possible.

The representations of products developed for high level robot programming languages (e.g. AUTOPASS[14]) emphasized the geometric aspects such as the shape of the parts and the contacts between parts. That emphasis is consistent with the goal of generating a sequence of robot actions that will join two subassemblies, given the sequence in which parts or subassemblies should be put together. However for the generation of the assembly sequences, a purely geometric description of the product is not sufficient. There are sequences that would be feasible from a geometric point of view, but are unfeasible in practice due to forces resulting from fasteners. Therefore, a model of assemblies to be used in generating assembly sequences must represent explicitly the fastenings that bind one part to another.

The representation of assemblies used by the algorithms described in sections 7.2 and 7.3 is a relational model that includes three types of entities: parts, contacts, and attachments. It also includes a set of relationships between entities. Both entities and relationships can have attributes. Formally, the relational model of an assembly is a 5-tuple $< P, C, A, R,$ *a-functions* $>$ in which

- $P$ is a set of symbols, each of which corresponds to one part in the assembly. No two elements of $P$ correspond to the same part.
- $C$ is a set of symbols, each of which corresponds to a contact between surfaces of two parts of the assembly. No two elements of $C$ correspond to the same contact. The two surfaces must be compatible. An example of a pair of compatible surfaces are a cylindrical shaft and a cylindrical hole. The same pair of parts may have more than one contact. And the same surface of one part may be in contact with surfaces of two or more other parts.
- $A$ is a set of symbols, each of which corresponds to an attachment that acts on a set of contacts. No two elements of $A$ correspond to the same attachment. An attachment always has an agent, which can be either the attached contact, or another contact, or a part. The access to an attachment may be blocked by one or more parts.
- $R$ is a set of symbols, each of which corresponds to a relationship between pairs of elements of $P \cup C \cup A$. No two elements of $R$ correspond to the same relationship.
- *a-functions* is a set of attribute functions[1] whose domains are subsets of $P \cup C \cup A \cup R$. These functions associate entities or relationships to their characteristics such as the type of attachment, the entities related by a relationship, and the shape of a part.

Examples of each type of entity, of relationships and of attribute functions will be discussed.

[1] A function is defined as a subset of the cartesian product of two sets (the domain and the range) that has no two pairs whose first elements are the same, and such that every element in the domain appears in one pair.

This definition of a relational model representation of assemblies is sufficiently general to encompass a large class of assemblies including those with rigid parts and no internal mechanisms. The set of functions can be enlarged to include all the information that might be necessary to generate assembly sequences. In practice, it may be convenient to restrict the class of assemblies represented. Our current experimental implementation has the following restrictions:

- The contacts between parts involve one of the following pairs of compatible surfaces:
  - planar surface and another planar surface,
  - cylindrical shaft and cylindrical hole,
  - polyhedral shaft and polyhedral hole,
  - threaded cylindrical shaft and threaded cylindrical hole.
- The types of attachments are:
  - glue attachment,
  - pressure fit attachment,
  - clip attachment,
  - screw attachment.

Examples of attribute functions are the following:

- The function that associates a part to a description of its shape:

  $$shape : P \rightarrow \mathcal{S}$$

  where $\mathcal{S}$ is the set of all shape descriptions.

- The function that associates a part to a description of its location:

  $$location : P \rightarrow \mathcal{T}$$

  where $\mathcal{T}$ is the set of all $4 \times 4$ homogeneous transformation matrices. The matrix $T_i$ associated to part $p_i$ corresponds to the position and orientation of a reference frame attached to part $p_i$ with respect to a global frame of reference for the whole assembly. The choice of this global frame of reference is arbitrary, but the same global reference must be used for all parts.

- The function that associates a contact to its type:

  $$\textit{type-of-contact}: C \rightarrow \text{contact-types}$$

  where contact-types = { planar, cylindrical, slot, threaded-cylindrical }.

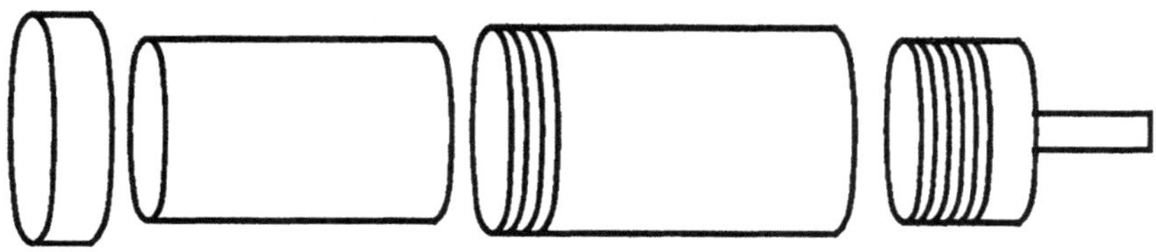


Figure 7.1: A four-part assembly in exploded view

- The function that associates a planar contact to the coordinates, with respect to the assembly's global frame of reference, of a vector normal to the contact plane:

  $normal : \{c \mid [c \in C] \wedge [\textit{type-of-contact}\ (c) = \text{planar}]\} \rightarrow \mathbf{R}^3$

- The function that associates a planar contact to the part-relationship that relates the contact to the part that is back of the contact, i.e. the part that is such that the normal to the contact plane points to the part's outside:

  $back : \{c \mid [c \in C] \wedge [\textit{type-of-contact}(c) = \text{planar}]\} \rightarrow R$

  This function must be consistent with the function *normal*.

- The function that associates a planar contact to the part-relationship that relates the contact to the part that is forward of the contact i.e. the part that is such that the normal to the contact plane points to the part's inside:

  $forward : \{c \mid [c \in C] \wedge [\textit{type-of-contact}(c) = \text{planar}]\} \rightarrow R$

  This function must be consistent with the function *normal*.

- The function that associates a part or a contact to its part-contact relationships:

  $\textit{part-contact-relationships}: P \cup C \rightarrow \Pi(R)$

  where $\Pi(R)$ is the set of all subsets of $R$.

- The function that associates a part-contact relationship to its part:

  $part : \{r \mid [r \in R] \wedge [\textit{type-of-relationship}(r) = \text{part-contact}]\} \rightarrow P$

- The function that associates a part-contact relationship to its contact:

  $contact : \{r \mid [r \in R] \wedge [\textit{type-of-relationship}(r) = \text{part-contact}]\} \rightarrow C$

The relational model of an assembly must be consistent. For example, if $part(r_1) = p_1$ and $contact(r_1) = c_1$ then $r_1 \in \textit{part-contact-relationships}\ (p_1)$ and $r_1 \in \textit{part-contact-relationships}\ (c_1)$ must hold. Furthermore, the relational

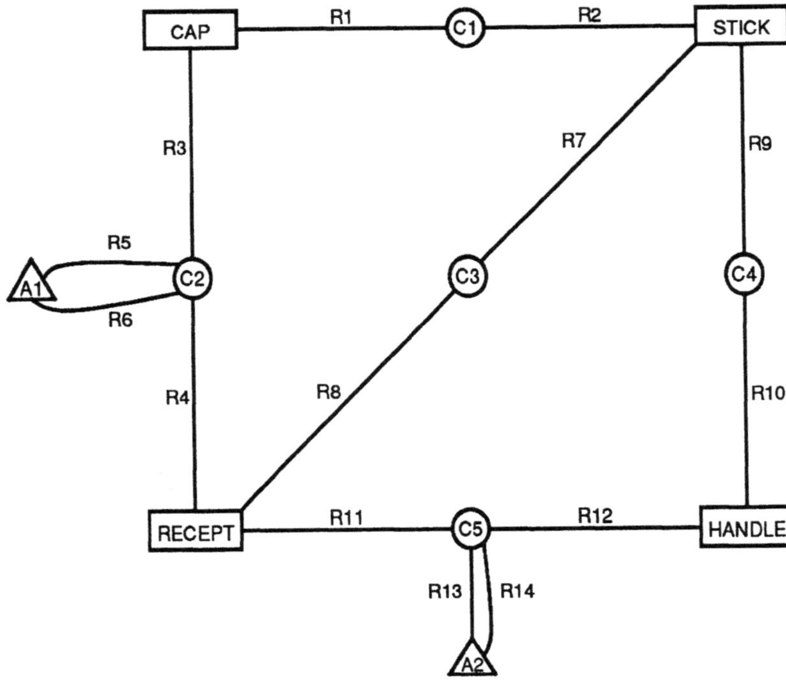


Figure 7.2: The relational model graph for the assembly shown in figure 3

model of an assembly must satisfy some syntactic constraints, the most important of which are:

- every contact must have exactly two part-contact relationships;
- every part must have at least one part-contact relationship, except in the case the assembly has only one part;
- every attachment must have at least one target-attachment relationship, and at least one agent-attachment relationship.

The relational model of an assembly can be represented by a graph plus the associated attribute functions. Figure 7.1 shows a simple assembly, and figure 7.2 shows its corresponding relational model graph.

The nodes in figure 7.2 correspond to the entities. Nodes corresponding to part entities are rectangles, nodes corresponding to contact entities are circles, and nodes corresponding to attachment entities are triangles. All nodes contain labels indicating their corresponding entities. The attribute functions associated with the contact entities C1, C2 and C3 are shown in Table 7.1.

The labeled lines connecting two nodes in figure 7.2 correspond to the relationships. Except for R5, R6, R13, and R14, all relationships are part-contact.

Table 7.1: Attribute Functions for the Contact Entities in Figure 4

| | C1 | C2 | C3 |
|---|---|---|---|
| *type-of-contact* | planar | threaded-cylindrical | cylindrical |
| *normal* | (0 1 0) | nil | nil |
| *back* | CAP | nil | nil |
| *forward* | STICK | nil | nil |
| *axis* | nil | ((0 0 0) (0 1 0)) | ((0 0 0) (0 1 0)) |
| *part-contact relationships* | (R1 R2) | (R3 R4) | (R7 R8) |
| *target-attachment relationships* | nil | (R5) | nil |
| *agent-attachment relationships* | nil | (R6) | nil |

Relationships R5 and R13 are target-attachment; they indicate that the contacts C2 and C5, respectively, are attached. Relationships R6 and R14 are agent-attachment; they indicate that the agents of the attachments are the target contacts themselves.

Given the relational model of an assembly $< P, C, A, R,$ *a-functions* $>$ a number of other useful representations may be generated. For example, the graph of connections of the assembly, as defined by Bourjault[4], is the simple graph $< V\ ,\ E >$ in which

$$V = P$$

$$E = \{(p_i, p_j) | [p_i \in P] \wedge [p_j \in P] \wedge \exists c \exists r_1 \exists r_2 [[c \in C] \wedge [\{r_1, r_2\} = \text{part-contact-relationships}(c)] \wedge [p_i = part(r_1)] \wedge [p_j = part(r_2)]]\}$$

Figure 7.3 shows the graph of connections for the simple assembly shown in figure 7.1.

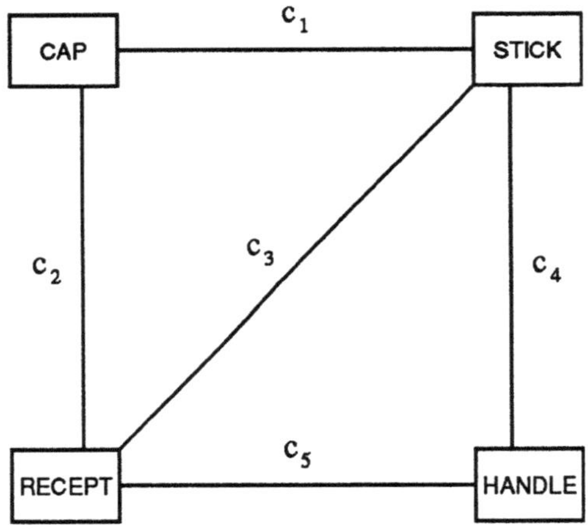


Figure 7.3: The graph of connections for the four-part assembly

### 7.1.1 Subassemblies

A subassembly is a nonempty subset of parts that either has only one element (i.e. only one part), or is such that every part has at least one surface contact with another part in the subset. Although there are cases in which it is possible to join the same pair of parts in more than one way, a unique assembly geometry will be assumed for each pair of parts. This geometry corresponds to their relative location in the whole assembly. A subassembly is said to be stable if its parts maintain their relative position and do not break contact spontaneously. All one-part subassemblies are stable.

Given the relational model of an assembly $< P, C, A, R,$ *a-functions* $>$ the relational model of a subassembly of that assembly is a relational model $< P_S, C_S, A_S, R_S,$ *a-functions$_S$* $>$ in which $P_S \in P$, $C_S \in C$, $A_S \in A$, $R_S \in R$, and every function in *a-functions$_S$*, is a subset of the corresponding function in *a-functions*. In addition to the syntactic constraints mentioned above that every relational model of an assembly must satisfy, the relational model $< P_S, C_S, A_S, R_S,$ *a-functions$_S$* $>$ of a subassembly of $< P, C, A, R,$ *a-functions* $>$ must also satisfy the constraint:

$$\forall c \forall r_1 \forall r_2 [[c \in C] \wedge [\{r_1, r_2\} = \textit{part-contact-relationships}(c)] \wedge [\text{part}(r_1) \in P_S] \wedge \text{part}(r_2) \in P_S] \rightarrow [c \in C_S]$$

This constraint corresponds to the assumption that whenever parts are joined forming a subassembly all contacts between the parts in that subassembly are established. It requires that those contacts in the model of the assembly whose two part-contact relationships involve parts in the subassembly must also be in the model of the subassembly. For example, for the assembly shown in figure 7.1, there is no subassembly relational model in which $P_S$ = {CAP, RECEPTACLE, STICK} and $C_S$ = {C2 C3}. If both the cap and the stick are in

$P_S$, then contact C1 must also be in $C_S$. This constraint allow the characterization of any subassembly $< P_S, C_S, A_S, R_S,$ *a-functions*$_S >$ of an assembly $< P, C, A, R,$ *a-functions* $>$ by its set of parts $P_S$ only. This feature will be used in the algorithm for the generation of mechanical assembly sequences described in the subsequent sections. In that algorithm, the intermediate subassemblies will be characterized by their sets of parts. Given a subset of parts $P_S$, there is a corresponding subgraph $< V_S , E_S >$ of the assembly's graph of connections $< V , E >$. In this subgraph, the set of nodes $V_S$ includes all the elements of $V$ that correspond to the parts in $P_S$. And the set of edges $E_S$ includes all the elements of $E$ that have both end points in $V_S$. A subset of parts $P_S$ characterize a subassembly if and only if the corresponding subgraph $< V_S , E_S >$ is connected (i.e. has only one component). A predicate that is satisfied only by the subsets of parts that correspond to subassemblies can be defined as follows:

**Definition 1** *The* subassembly *predicate associated to subassemblies of assembly* $\Psi =< P, C, A, R,$ *a-functions* $>$ *is the predicate*

$$sa_\Psi : \Pi(P) \rightarrow \{\text{true}, \text{false}\}$$

*with* $sa_\Psi(\theta) = \text{true}$ *if the subgraph* $< V_S , E_S >$ *in which*

$$V_S = \theta$$

$$\begin{aligned} E_S = \{ \ (p_i, p_j) | [p_i \in \theta] \wedge [p_j \in \theta] \wedge \\ \exists c \exists r_1 \exists r_2 [[c \in C] \wedge [\{r_1, r_2\} = \textit{part-contact-relationship}(c)] \\ \wedge [p_i = \textit{part}(r_1)] \wedge [p_j = \textit{part}(r_2)]] \ \} \end{aligned}$$

*is connected.*

## 7.2 Decompositions of an Assembly

The problem of generating the assembly sequences for a product can be transformed into the problem of generating the *disassembly* sequences for the same product. Since assembly tasks are not necessarily reversible, the equivalence of the two problems will hold only if each task used in *disassembly* is the reverse of a feasible assembly task, regardless of whether this reverse task itself is feasible or not. The expression *disassembly task*, therefore, refers to the reverse of a feasible assembly task.

As mentioned in the introduction, it was assumed that exactly two parts or subassemblies are joined at each time. It was also assumed that whenever parts are joined forming a subassembly, all contacts between the parts in that subassembly are established. In the *disassembly problem*, each task splits one subassembly into two smaller subassemblies, maintaining all contacts between the parts in either of the smaller subassemblies.

A decomposition approach can be used to solve the *disassembly problem.* In this approach the problem of *disassembling* one assembly is decomposed into two distinct subproblems, each being to *disassemble* one subassembly. Every decomposition must correspond to a *disassembly task.* If solutions to both subproblems that result from the decompositions are found, a solution to the original problem can then be obtained by combining the solutions to the two subproblems and the task corresponding to the decomposition. For subassemblies that contain one part only, a trivial solution containing no assembly task always exists. This decomposition approach lends itself to an AND/OR graph representation of assembly sequences[10]. The correspondence between the AND/OR graph and the directed graph representations of assembly sequences is discussed elsewhere[9].

From now on, references to products, to assemblies, or to subassemblies are references to their relational models, which are always assumed to be consistent and to satisfy the syntactic constraints of a relational model of an assembly. A real product will be referred to as a *physical product,* a real assembly as a *physical assembly,* and a real subassembly as a *physical subassembly.*

A decomposition of an assembly $< P, C, A, R,$ *a-functions* $>$ is a pair of its subassemblies $< P_{S1}, C_{S1}, A_{S1}, R_{S1},$ *a-functions*$_{S1} >$ and $< P_{S2}, C_{S2}, A_{S2}, R_{S2},$ *a-functions*$_{S2} >$ such that $P_{S1} \cup P_{S2} = P$ and $P_{S1} \cap P_{S2} = \emptyset$. The set $C_{S1-S2} = C - (C_{S1} \cup C_{S2})$ is referred to as the *contacts of the decomposition*; they are the contacts that belong to $C$ and do not belong to either $C_{S1}$ or $C_{S2}$. The contacts of a decomposition of an assembly define a cut-set in that assembly's graph of connections. Conversely, a cut-set in the graph of connections of an assembly define a decomposition of that assembly.

A decomposition of an assembly is said to be feasible if it satisfies two predicates: *TASK-FEASIBILITY*, and *SUBASSEMBLY-STABILITY*. These predicates reflect the feasibility of joining the physical subassemblies to produce the physical assembly.

The *TASK-FEASIBILITY* predicate is true if it is feasible to join the two subassemblies to form the assembly. It depends on a number of conditions such as the existence of a collision-free path to bring the two subassemblies into contact, the accessibility of fasteners, and the availability of force applying devices. These conditions can be subdivided into two categories: internal and external. The internal conditions depend exclusively on the assembly. The existence of a collision-free path and the accessibility of fasteners are internal conditions. The external conditions depend also on the devices available to execute the assembly. The availability of force applying devices is an external condition. There is some freedom in establishing the set of conditions as well as their precise definition. For example, it may be required that the (collision-free) path to bring the two subassemblies into contact be a combination of a straight-line translation with a rotation whose axis is parallel to the straight line of translation.

The *SUBASSEMBLY-STABILITY* predicate is true if in the physical subassembly there is a nonempty set of orientations for which there is a part $p$ such that if $p$ is fixed the other parts maintain their relative positions and do not break contact spontaneously. The stability of subassemblies depends on a number of conditions such as the gravity and the friction in the contacts.

As discussed in section 7.1, the subassemblies of a given assembly $\Psi = < P, C, A, R,$ *a-functions* $>$ can be characterized by their sets of parts. In our current implementation, the two predicates described above are defined as follows:

**Definition 2** *The* **task-feasibility** *predicate associated to subassemblies of assembly* $\Psi = < P, C, A, R,$ *a-functions* $>$ *in which* $P = \{p_1, p_2, \cdots, p_N\}$, *is the predicate*

$$gf_\Psi : \Pi(P) \times \Pi(P) \rightarrow \{\text{true}, \text{false}\}$$

*with* $gf_\Psi(\theta_1, \theta_2) = \text{true}$ *if and only if*

- $sa_\Psi(\theta_1) = \text{true}$, $sa_\Psi(\theta_2) = \text{true}$, $sa_\Psi(\theta_1 \cup \theta_2) = \text{true}$, *and* $\theta_1 \cap \theta_2 = \emptyset$;
- *there is a collision-free path that is a combination of a straight-line translation with a rotation whose axis is parallel to the straight line of translation, and that brings the two physical subassemblies of* $\Psi$ *characterized by* $\theta_1$ *and* $\theta_2$ *into contact from a situation in which they are sufficiently far apart; and*
- *it is feasible to establish the attachments that act on the set of contacts between parts in* $\theta_1$ *and parts in* $\theta_2$.

**Definition 3** *The* **subassembly-stability** *predicate associated to subassemblies of assembly* $\Psi = < P, C, A, R,$ *a-functions* $>$ *in which* $P = \{p_1, p_2, \cdots, p_N\}$, *is the predicate*

$$st_\Psi : \Pi(P) \rightarrow \{\text{true}, \text{false}\}$$

*with* $st_\Psi(\theta) = \text{true}$ *if and only if* $sa_\Psi(\theta) = \text{true}$.

The *TASK-FEASIBILITY* predicate is computed by first checking whether or not there is an incremental motion that separates the two subassemblies from their assembled relative position and that is a combination of a straight-line translation with a rotation whose axis is parallel to the straight line of translation. If no incremental motion exists, the decomposition is unfeasible. This predicated is called the *local geometric feasibility* predicate and is viewed here as one of a hierarchy of such predicates to be evaluated.

For many types of contacts there are very few feasible motions between the parts that are feasible. For example, the only direction along which a pin in a hole can translate is the direction of the axis. Similarly, the only feasible relative motion for two parts or subassemblies that have a threaded-cylindrical contact is a combination of a straight-line translation with a rotation whose axis is parallel to the straight line of translation.

Whenever the part or subassembly has such a constraining contact, the feasibility of local motion can be determined by checking the compatibility of the most restrictive contact with all other contacts. In the case of the pin in the hole, this consists of checking whether a translation of the pin along its axis is not blocked by any of the other contacts between the pin and the other part or subassembly.

This analysis is more difficult when the part or subassembly to be *disassembled* is constrained by planar contacts only. Each planar contact leaves an infinite number of unconstrained directions along which translation is possible. All these directions have positive (i.e. greater than or equal to zero) projection over the normal to the surface of the blocking part, pointing towards the outside of the blocking part.

In order to decide that a set of planar contacts does not completely constrain one part or subassembly, one must verify that there is a nonzero solution to the system of linear inequalities:

$$\sum_{j=1}^{3} n_{ij} x_j \geq 0 \qquad i = 1, 2, \cdots, N$$

where $\underline{n}_i = [n_{i1}\ n_{i2}\ n_{i3}]$ is the normal to the surface of the $i^{th}$ contact. This system of linear inequalities defines a polyhedral convex cone. It has been shown[7] that such a polyhedral convex cone can be *built up* from its (unique) $d$-dimensional face and its $(d+1)$-dimensional faces (if any), where $d = 3 - \text{rank}(M)$, and $M$ is the matrix of the coefficients $n_{ij}$. If $d$ is greater than zero, then the polyhedral convex cone has a face of dimension greater than zero and therefore the system of inequalities has a nonzero solution. If $d$ is equal to zero, then the system of inequalities has a nonzero solution only if the polyhedral convex cone has at least one one-dimensional face. The existence of a one-dimensional face can be determined by checking the $N \cdot (N-1)$ pairwise intersections of the planes corresponding to the inequalities. Each intersection of two distinct planes is a line. If one of the two unity vectors, $\underline{t}$ and $-\underline{t}$ along the intersection line of two planes has positive (i.e. greater than or equal to zero) projection over all the normal vectors $\underline{n}_1, \underline{n}_2, \cdots, \underline{n}_N$, then the half-line defined by that vector ($\underline{t}$ or $-\underline{t}$) is a one-dimensional face of the polyhedral convex cone.

If there is a nonzero solution to the system of inequalities, then the part or subassembly is not completely constrained. Otherwise the subassembly is completely constrained, and the decomposition is unfeasible.

Once the feasible incremental motions are determined, their corresponding global motions can be tested for collisions by finding the intersection of the

**procedure** *FEASIBILITY-TEST(decomposition, assembly)*
**begin**
**return** *AND(TASK-FEASIBILITY(decomposition, assembly),*
*SUBASSEMBLY-STABILITY(decomposition)* )
**end**

Figure 7.4: Procedure *FEASIBILITY-TEST*

volume swept by the motion of one part and the other part. This predicate is referred to as *global geometric feasibility* and provides a second level of test in the hierarchy. In our current system we have not implemented this test but have instead used *virtual* contacts to describe blocking relationships equivalent to contacts. In the product shown in figure 7.1, for example, if the stick did not touch the handle, the local analysis as described above would indicate that the stick can translate (incrementally) along its axis. In a case like this, a virtual planar contact, analogous to C4 on figure 7.2, would be added to the relational model indicating the blocking of the stick by the handle.

The feasibility of establishing the attachments that act on the set of contacts between parts in two subassemblies can be determined by inspection of the relational model of the assembly. Our current implementation includes routines that check whether the attachments acting on the contacts of the decomposition are not blocked in the resulting assembly, and are not present in either one of the subassemblies.

For the discussion in the next section, which presents the algorithm for generating the assembly sequences, it is assumed that there exist correct algorithms for computing the *TASK-FEASIBILITY* and the *SUBASSEMBLY-STABILITY* predicates discussed above, and that they are combined into the procedure *FEASIBILITY-TEST* shown in figure 7.4.

## 7.3 The Algorithm for Generating All Assembly Sequences

As discussed in the previous section, a decomposition approach has been used to generate all assembly sequences. The basic idea underlying the approach is to enumerate the decompositions of the assembly and to select those that are feasible. The decompositions are enumerated by enumerating the cut-sets of the assembly's graph of connections. Knowledge of the feasible decompositions allows the construction of the AND/OR graph representation of assembly plans. Each feasible decomposition corresponds to a hyperarc in the AND/OR graph connecting the node corresponding to the assembly to the two nodes corresponding to the two subassemblies. The same process is repeated for the subassemblies and subsubassemblies until only single parts are left.

```
procedure GET-FEASIBLE-DECOMPOSITIONS(assembly)
begin
fsbl-dec ← NIL
graph ← GET-GRAPH-OF-CONNECTIONS(assembly)
cut-sets ← GET-CUT-SETS(graph)
while cut-sets is not empty do
  begin loop1
  next-cut-set ← FIRST(cut-sets)
  cut-sets ← TAIL(cut-sets)
  next-dec ← GET-DECOMPOSITION(next-cut-set)
  if FEASIBILITY-TEST(next-dec)
    then fsbl-dec ← UNION(fsbl-dec, LIST(next-dec))
  end loop1
return fsbl-dec
end procedure
```

Figure 7.5: Procedure *GET-FEASIBLE-DECOMPOSITIONS*

It has been shown[6, 12] that the set of all cut-sets of a graph $< V, E >$ is a subspace of the vector space over the Galois field modulo 2 associated with the graph. The vectors in this vector space are the elements of $\Pi(E)$, the set of all subsets of $E$. It has also been shown that the fundamental system of cut-sets relative to a spanning tree is a basis of the cut-set subspace. Therefore, the cut-sets of a graph can be enumerated by constructing a spanning tree of the graph, finding the fundamental system of cut-sets relative to that spanning tree, and computing all the combinations of fundamental cut-sets. In our current implementation, the cut-sets are enumerated using a more efficient approach. We first look at all connected subgraphs having the cardinality of their set of nodes smaller than or equal to half of the cardinality of the set of nodes in the whole graph. For each of these subgraphs, the set of edges of the whole graph that have only one end in the subgraph defines a cut-set if their removal leaves the whole graph with exactly two components.

Figure 7.5 shows the procedure *GET-FEASIBLE-DECOMPOSITIONS* which takes as input the relational model of an assembly and returns all feasible decompositions of that assembly. The procedure first generates the graph of connections for the input assembly and computes the cut-sets of this graph. Each cut-set corresponds to a decomposition. The procedure *GET-DECOMPOSITIONS* is used to find the decomposition that corresponds to a cut-set, and the procedure *FEASIBILITY-TEST* discussed in the previous section is used to check whether or not that decomposition is feasible. The feasible decompositions are stored in the list *fsbl-dec* which was empty at the beginning. After all cut-sets have been processed, the procedure returns the list *fsbl-dec*.

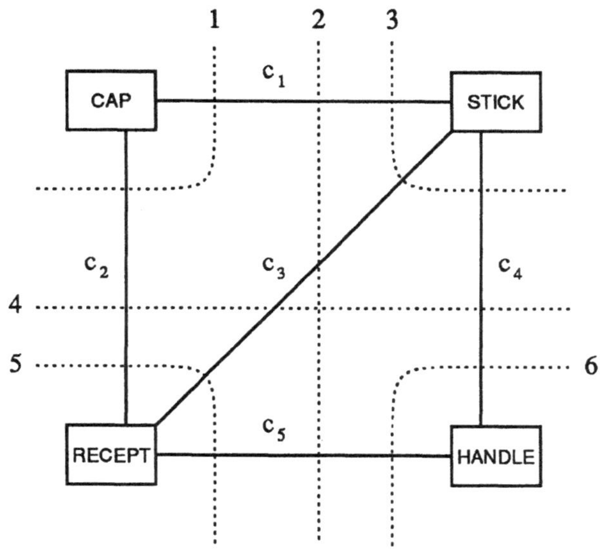


Figure 7.6: The cut sets of the graph of connections for the assembly shown in figure 7.1

An example will illustrate the computation of the feasible decompositions of an assembly. When passed the relational model of the assembly in figure 7.1, procedure *GET-FEASIBLE-DECOMPOSITIONS* will compute the graph of connections shown in figure 7.3, and all its cut-sets, which are indicated in figure 7.6. The analysis of those cut-sets will indicate the feasible decompositions. The first cut-set yields a feasible decomposition since it is feasible to join the cap and the subassembly made up of the three other parts. The second cut-set also yields a feasible decomposition because it is feasible to join the subassembly consisting of the cap plus the receptacle, and the subassembly consisting of the stick plus the handle. The third cut-set, however, does not yield a feasible decomposition, since it is not possible to join the stick and the subassembly made up of the three other parts. Similarly, the fourth and the sixth cut-sets yield feasible decompositions while the fifth cut-set does not. Therefore, procedure *GET-FEASIBLE-DECOMPOSITIONS* will return a list containing the four decompositions that correspond to the first, second, fourth, and sixth cut-sets.

Figure 7.7 shows the procedure *GENERATE-AND-OR-GRAPH* which takes the relational model of an assembly, and returns the AND/OR graph representation of all assembly sequences for that assembly. The nodes in the AND/OR graph returned are pointers to relational models of assemblies.

Procedure *GENERATE-AND-OR-GRAPH* uses the lists *closed* and *open* to store the pointers to the relational models of the subassemblies whose decompositions into smaller subassemblies respectively have and have not been generated.

```
procedure GENERATE-AND-OR-GRAPH(assembly)
begin
open ← LIST(GET-POINTERS(LIST(assembly)))
closed ← NIL
harcs ← NIL
while open is not empty do
  begin loop1
  next-sub ← FIRST(open)
  open ← TAIL(open)
  closed ← UNION(closed, LIST(next-sub) )
  dec-of-next-sub ← GET-FEASIBLE-DECOMPOSITIONS(next-sub)
  while dec-of-next-sub is not empty do
    begin loop2
    next-decomposition ←FIRST(dec-of-next-sub)
    dec-of-next-sub ← TAIL(dec-of-next-sub)
    subs ← GET-POINTERS(next-decomposition)
    harcs ← UNION(harcs, LIST(LIST(next-sub, subs)))
    while subs is not empty do
      begin loop3
      next-sub ← FIRST(subs)
      subs ← TAIL(subs)
      if next-sub is not in open or in closed, add it to open;
        otherwise ignore it
      end loop3
    end loop2
  end loop1
return LIST(closed, harcs)
end procedure
```

Figure 7.7: Procedure *GENERATE-AND-OR-GRAPH*

The procedure takes one element of *open* at a time, moves it to *closed*, and uses procedure *GET-FEASIBLE-DECOMPOSITIONS* to generate all decompositions of the relational model pointed by that element. For each decomposition, procedure *GENERATE-AND-OR-GRAPH* uses the procedure *GET-POINTERS* to get the pointers to the relational models of the subassemblies. Procedure *GET-POINTERS* checks whether each resulting subassembly has appeared before or not. If the subassembly has appeared before, its pointer is used, otherwise a new pointer is created. The new pointers are inserted into *open*. Each decomposition yields one hyperarc of the AND/OR graph.

Figure 7.8 shows the resulting AND/OR graph for the product shown in figure 7.1.

A more efficient implementation of the method for the generation of assembly sequences presented above will include additional tests aimed at avoiding unnecessary computation[2]. One such test is to check whether the feasibility of a decomposition follows from the feasibility of other decompositions. For example, the feasibility of the decomposition corresponding to hyperarc 10 in figure 7.8 follows from the feasibility of the decompositions corresponding to hyperarcs 4 and 5. If it was geometrically and mechanically feasible to *disassemble* the handle from the whole assembly (hyperarc 4), then it is geometrically and mechanically feasible to *disassemble* the handle from a subassembly. And since the subassembly made up of the stick and the receptacle is stable (hyperarc 5), it follows that the decomposition corresponding to hyperarc 10 is feasible. This test indicates that if the decompositions corresponding to hyperarcs 4 and 5 have already been analyzed and found to be feasible, then it is not necessary to perform the computation corresponding to procedure *FEASIBILITY-TEST* in the analysis of the decomposition that corresponds to hyperarc 10. Similarly, another additional test would check whether the unfeasibility of a decomposition follows from the unfeasibility of other decompositions already analyzed. More efficient implementations that include these types of tests have been reported recently[2, 8, 15].

## 7.4 Analysis of the Algorithm

### Correctness and completeness of *GET-FEASIBLE-DECOMPOSITIONS*

The partial correctness of the algorithm *GET-FEASIBLE-DECOMPOSITIONS* is immediate. The list *fsbl-dec* is initially empty. Only feasible decompositions are added to the list *fsbl-dec*. Therefore, the list returned by *GET-FEASIBLE-DECOMPOSITIONS* does not contain any element that is not a feasible decomposition of the assembly input. The total correctness follows from the fact that there is only a finite number of cut-sets in a graph. The list *cut-sets* contains initially all cut-sets of the graph of functional connection corresponding to the

[2]Our current implementation consists of the basic algorithms presented in the text and does not yet include these additional tests.

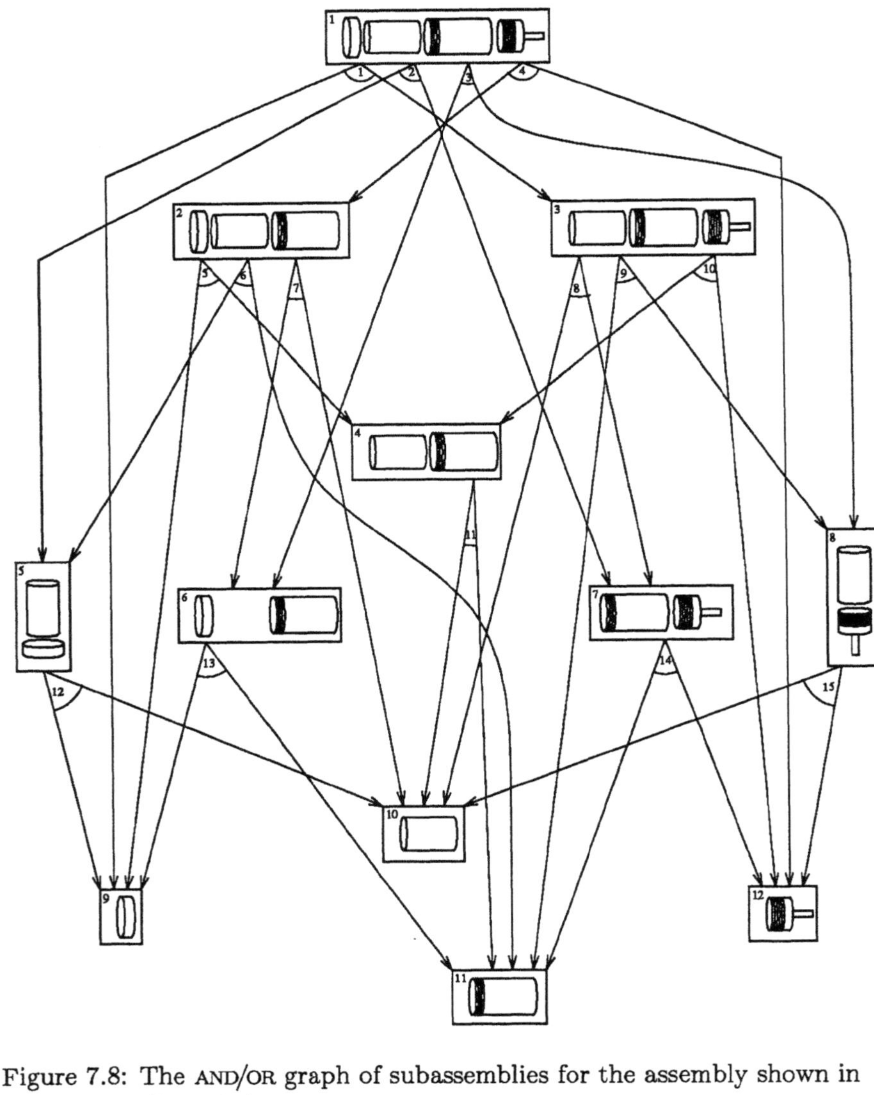

Figure 7.8: The AND/OR graph of subassemblies for the assembly shown in figure 7.1

assembly input. At each execution of *loop1*, one element is removed from the list *cut-sets*. Therefore, after a finite number of executions of *loop1* the list *cut-sets* becomes empty, and the algorithm terminates.

There is a one-to-one correspondence between cut-sets in the graph of connections of an assembly, and the decompositions of that assembly. Therefore, since algorithm *GET-FEASIBLE-DECOMPOSITIONS* goes over all cut-sets of the graph of connections, all feasible decompositions will be generated and the algorithm *GET-FEASIBLE-DECOMPOSITIONS* is complete.

It was assumed that the algorithm for generating the cut-sets of a graph is correct and complete. The enumeration of the cut-sets of a graph is studied in graph theory; for example, Deo[6] and Liu[12] discuss that problem. It was also assumed that it is possible to decide correctly whether a decomposition is feasible or not, based on geometrical and physical criteria, as discussed in the section 7.2.

**Correctness and completeness of *GENERATE-AND-OR-GRAPH***

List *closed* is updated at only one point, and it only gets elements that were previously in the *open* list. The *open* list contains initially a pointer to the relational model of the assembly input, which is a node of the AND/OR graph. List *open* is updated inside *loop3* where it gets pointers to the relational models of the subassemblies that are part of a feasible decomposition, and therefore, are nodes of the AND/OR graph. Therefore, the elements in the *open* list, and consequently the elements in the *closed* list, are always pointers to relational models either of the original assembly, or of subassemblies that take part of a feasible decomposition.

The *harcs* list is initially empty. It is updated only inside *loop2* where it gets the hyperarc corresponding to a feasible decomposition. Therefore, algorithm *GET-FEASIBLE-DECOMPOSITIONS* can only return a set of nodes and a set of hyperarcs of the AND/OR graph. This establishes the partial correctness of *GENERATE-AND-OR-GRAPH.*

List *open* gets only subassemblies and no subassembly is inserted more than once. Since there is a finite number of subassemblies, the algorithm terminates. This establishes the total correctness of *GENERATE-AND-OR-GRAPH.*

Since algorithm *GET-FEASIBLE-DECOMPOSITIONS* is complete, all possible decompositions of all subassemblies that are inserted into the list *open* yield a hyperarc. Furthermore, all subassemblies that result from a decomposition are inserted into list *open*, and later are moved to list *closed.* Therefore, the first list returned contains all subassemblies that resulted from some decomposition, and the second list returned contains one hyperarc for each decomposition of each subassembly. This establishes the completeness of *GENERATE-AND-OR-GRAPH.*

### Complexity

The amount of computation involved in the generation of the AND/OR graph for a given assembly depends on the number $N$ of parts that make up the assembly, on how interconnected those parts are, and also on the resulting AND/OR graph.

The number of prospective decompositions (i.e. cut-sets of the graphs of functional connections) that must be analyzed will be used in this section as a measure of the amount of computation involved in the generation of all assembly sequences[3]. Two models for how the parts in the assembly are interconnected are considered in order to provide bounds in the estimate of computational complexity:

1. a *weakly connected* assembly in which there are $N-1$ connections between the $N$ parts, with the $i^{th}$ connection being between the $i^{th}$, and the $(i+1)^{th}$ parts;
2. a *strongly connected* assembly in which every part is connected to every other part.

And three possibilities for the resulting AND/OR graph are considered:

1. a *balanced tree* AND/OR graph in which there is at most one hyperarc leaving each node and this hyperarc points to two nodes whose corresponding subassemblies either have the same number of parts, or their number of parts differ by one;
2. a *one-part-at-a-time tree* AND/OR graph in which there is at most one hyperarc leaving each node, and this hyperarc points to two nodes one of which corresponds to a one-part subassembly; and
3. a *network* AND/OR graph in which there are as many hyperarcs leaving each node as there are cut-sets in the graph of functional connections of the node's corresponding subassembly.

The resulting total number $D$ of decompositions that must be analyzed as a function of the number $N$ of parts that make up the assembly for each possible combination of how the parts are interconnected and the type of the resulting AND/OR graph is:

1. Weakly connected assemblies:
   (a) Balanced tree AND/OR graph: the number of prospective decompositions that must be analyzed is $N-1$ for the initial assembly, $N-2$ for all subassemblies, $N-4$ for all subsubassemblies, and so on.

[3]The overall complexity of algorithm GENERATE-AND-OR-GRAPH should take into account the computation involved in generating the cut-sets of the graph of functional connections.

Therefore[4],

$$\begin{aligned} D &= (N-1)+(N-2)+(N-4)+\cdots+(N-2^{\mathrm{int}(\log_2 N)}) \\ &= \sum_{i=0}^{\mathrm{int}(\log_2 N)} (N-2^i) = \\ &= N\cdot[\mathrm{int}(\log_2 N)+1] - 2^{[\mathrm{int}(\log_2 N)+1]} + 1 \end{aligned}$$

(b) One-part-at-a-time tree AND/OR graph: the number of prospective decompositions that must be analyzed is $N-1$ for the initial assembly, $N-2$ for the $(N-2)$ part subassembly, $N-3$ for the $(N-3)$-part subassembly, and so on. Therefore,

$$\begin{aligned} D &= (N-1)+(N-2)+(N-3)+\cdots+2+1 \\ &= \sum_{i=1}^{N-1}(N-i) = \frac{N\cdot(N-1)}{2} \end{aligned}$$

(c) Network AND/OR graph: the number of prospective decompositions that must be analyzed is $N-1$ for the $N$-part subassembly, $N-2$ for each of the two $(N-1)$-part subassemblies, $N-3$ for each of the three $(N-2)$-part subassemblies, and so on. Therefore,

$$\begin{aligned} D &= 1\cdot(N-1)+2\cdot(N-2)+3\cdot(N-3)+\cdots+(N-1)\cdot 1 \\ &= \sum_{i=1}^{N-1} i\cdot(N-i) \\ &= \frac{(N+1)\cdot N\cdot(N-1)}{6} \end{aligned}$$

2. Strongly connected assemblies:

(a) Balanced tree AND/OR graph: the number of prospective decompositions that must be analyzed is $(2^{(N-1)}-1)$ for the initial assembly, $(2^{\mathrm{int}(\frac{N-1}{2})}-1)+(2^{\mathrm{int}(\frac{N-2}{2})}-1)$ for all subassemblies, $(2^{\mathrm{int}(\frac{N-1}{4})}-1)+(2^{\mathrm{int}(\frac{N-2}{4})}-1)+(2^{\mathrm{int}(\frac{N-3}{4})}-1)+(2^{\mathrm{int}(\frac{N-4}{4})}-1)$ for all subsubassemblies, and so on. Therefore,

$$D = \sum_{i=0}^{\mathrm{int}(\log_2 N)} \sum_{j=0}^{2^i-1} (2^{\mathrm{int}(\frac{N-j-1}{2^i})}-1)$$

(b) One-part-at-a-time tree AND/OR graph: the number of prospective decompositions that must be analyzed is $(2^{N-1}-1)$ for the $N$-part

[4] We use the notation $\mathrm{int}(x)$ to represent the largest integer that is less than or equal to $x$. For example, $\mathrm{int}(3)=3$ and $\mathrm{int}(3.5)=3$.

subassembly, $(2^{N-2}-1)$ for the $(N-1)$-part subassembly, $(2^{N-3}-1)$ for the $(N-2)$-part subassembly, and so on. Therefore,

$$\begin{aligned} D &= (2^{N-1}-1)+(2^{N-2}-1)+(2^{N-3}-1)+\cdots+(2-1) \\ &= 2^N - N - 1 \end{aligned}$$

(c) Network AND/OR graph: the number of prospective decompositions that must be analyzed is $(2^{N-1}-1)$ for the $N$-part subassembly, $(2^{N-2}-1)$ for each of the $\binom{N}{N-1}$ $(N-1)$-part subassemblies, $(2^{N-3}-1)$ for each of the $\binom{N}{N-2}$ $(N-2)$-part subassemblies, and so on. Therefore,

$$\begin{aligned} D &= \binom{N}{N}\cdot(2^{N-1}-1)+\binom{N}{N-1}\cdot(2^{N-2}-1)+\cdots \\ &\qquad\cdots+\binom{N}{2}\cdot(2-1) \\ &= \sum_{i=2}^{N}\binom{N}{i}\cdot(2^{i-1}-1) \\ &= \frac{3^N+1}{2}-2^N \end{aligned}$$

For each of the three possibilities of the resulting AND/OR graph, table 7.2 shows the number of decompositions that must be analyzed for weakly connected assemblies and table 7.3 shows the number of decompositions that must be analyzed for strongly connected assemblies, as a function of the number of parts that make up the product. The figures in table 7.3 are given as a reference since it is very unlikely that there would be a twenty-part assembly in which every part is connected to every other part.

The results above take into account the fact that the type of the resulting AND/OR graph is not known a priori. For example, for the weakly connected assembly whose AND/OR graph is a balanced tree, all the $N-1$ cut-sets of the whole assembly were included in the number of decompositions that are tested, although there is only one cut-set that yields two subassemblies that have the same number of parts.

As discussed in the end of section 7.3, a more efficient implementation of the method for the generation of assembly sequences presented in this chapter will include additional tests aimed at avoiding unnecessary computation. One such test is to check whether the feasibility of a decomposition follows from the feasibility of other decompositions. In the case of strongly connected assemblies in which all decompositions of all subassemblies are feasible, the computation can be significantly reduced if this test is performed before analyzing each decomposition. Since all decompositions of the whole assembly are feasible, all

Table 7.2: The number of decompositions that must be analysed for each type of resulting AND/OR graph, as a function of the number of parts, for weakly connected assemblies.

| Number of parts | Balanced tree | One part at a time | Network |
|---|---|---|---|
| 2 | 1 | 1 | 1 |
| 3 | 3 | 3 | 4 |
| 4 | 5 | 6 | 10 |
| 5 | 8 | 10 | 20 |
| 6 | 11 | 15 | 35 |
| 7 | 14 | 21 | 56 |
| 8 | 17 | 28 | 84 |
| 9 | 21 | 36 | 120 |
| 10 | 25 | 45 | 165 |
| 15 | 45 | 105 | 560 |
| 20 | 69 | 190 | 1,330 |
| 25 | 94 | 300 | 2,600 |
| 30 | 119 | 435 | 4,495 |

Table 7.3: The number of decompositions that must be analysed for each type of resulting AND/OR graph, as a function of the number of parts, for strongly connected assemblies.

| Number of parts | Balanced tree | One part at a time | Network |
|---|---|---|---|
| 2 | 1 | 1 | 1 |
| 3 | 4 | 4 | 6 |
| 4 | 9 | 11 | 25 |
| 5 | 20 | 26 | 90 |
| 6 | 39 | 57 | 301 |
| 7 | 76 | 120 | 966 |
| 8 | 145 | 247 | 3,025 |
| 9 | 284 | 502 | 9,330 |
| 10 | 551 | 1,013 | 28,501 |
| 15 | 16,604 | 32,752 | 7,141,686 |
| 20 | 525,389 | 1,048,555 | 1,742,343,625 |
| 25 | 16,783,550 | 33,554,406 | 423,610,750,290 |
| 30 | 536,904,119 | 1,073,741,793 | 102,944,492,305,501 |

decompositions of all subassemblies should also be feasible. Therefore, with a simple additional test, the total number of decompositions that must be analyzed is reduced from $(\frac{3^N+1}{2} - 2^N)$ to $(2^{N-1} - 1)$.

## 7.5 Conclusion

A correct and complete algorithm for the generation of all mechanical assembly sequences was presented. The problem of generating assembly sequences was transformed into the equivalent problem of generating *disassembly* sequences. The algorithm operation consists in looking at all the decompositions of the assembly, that is, all the ways the assembly can be *split* into two subassemblies. This is done by generating all cut-sets of the assembly's graph of connections, and checking which cut-sets correspond to feasible decompositions. A decomposition is feasible if it possible to obtain the assembly by joining the two subassemblies. The same process is repeated for the subassemblies, for the subsubassemblies, and so on, until only single parts are left. At the end, the AND/OR graph representation of assembly sequences is returned.

The algorithm also lends itself to an interactive implementation in which a computer program generates questions that are answered by a human expert. Each question addresses the feasibility of a decomposition. It is also possible to have a computer program, instead of a human, to answer the questions directly from a description of the assembly. Our current implementation, which has the restrictions on the types of assemblies discussed in section 7.1, includes programs that answer the questions.

An approach to compute the answer to the question of whether or not it is feasible to obtain a given assembly by joining two subassemblies was presented. This approach is based on the use of a relational model description of the assembly. The model includes three types of entities: parts, contacts, and attachments; it also includes a set of relationships between entities. Both entities and relationships can have attributes. To decide whether or not a given decomposition is feasible, two predicates must be computed, using the data in the relational model:

- The *TASK-FEASIBILITY* predicate which is true if it is feasible to join two subassemblies.
- The *STABILITY* predicate which is true if the parts in each subassembly maintain their relative position and do not break contact spontaneously.

The key assumption in proving the correctness of the algorithm is that it is always possible to decide correctly, based on geometrical and physical criteria (i.e. using the three predicates above), whether or not it is feasible to obtain a given assembly by joining two subassemblies, and whether or not the subassemblies are stable.

The amount of computation involved in generating all mechanical assembly sequences was assessed by determining the number of decompositions that must be analyzed. That amount depends not only on the number of parts and on how they are interconnected, but on the solution AND/OR graph as well. The least amount of computation occurs for weakly connected assemblies in which each subassembly has only one feasible decomposition and that decomposition yields two subassemblies whose number of parts are either equal or differ by one. The maximum amount of computation occurs for strongly connected assemblies in which all decompositions of all subassemblies are feasible. This worst case, however, is very unlikely to occur in practice. Furthermore, additional simple tests discussed in section 7.3 can reduce the amount of computation.

In practice, an evaluation of the alternative assembly sequences generated by the algorithm presented in this paper is needed in order to choose the sequence that will be actually used in the assembly process. Different evaluation functions have been explored including a function based on parts entropy[13], a function based on the complexity of assembly tasks and the stability of subassemblies[10], a function based on number of different sequences in which the assembly tasks can be executed[11], and a function based on the parallel execution of assembly tasks[11].

It is also possible to implement an interactive system in which a computer program generates the alternative sequences, as described in this paper, and a human expert then selects the best one. Still another possibility would be to use an evaluation function for a preselection of good alternative sequences and then have a human expert to make the final choice.

Whenever the amount of computation exceeds the available computational resources, at least two strategies may be followed:

1. The number of parts can be artificially reduced by treating subassemblies as single parts. An analysis of the graph of connections may indicate the clusterings of parts that yield bigger reductions in the amount of computation.

2. The algorithm generates fewer, hopefully the best, sequences using some heuristics to guide the generation of assembly sequence. Such heuristics should be compatible with the evaluation function used to choose among the alternative assembly sequences.

In both strategies, the computation will be reduced at the expense of the completeness, since not all possible sequences will be generated. The devolpment of a procedure to cluster parts into subassemblies to obtain a hierarchical model of the assembly, and the development of good heuristics to guide the generation of assembly sequences are issues for future research.

# References


[1] M. M. Andreasen et al. *Design for Assembly.* Springer Verlag, 1983.

[2] D. F. Baldwin et al. An Integrated Computer Aid for Generating and Evaluating Assembly Sequences for Mechanical Products. *IEEE Trans. Robotics Automat.*, RA-7(1):78–94, Feb. 1991.

[3] G. Boothroyd et al. *Automatic Assembly.* Marcel Dekker, Inc., New York, 1982.

[4] A. Bourjault. *Contribution a une Approche Méthodologique de L'Assemblage Automatisé: Elaboration Automatique des Séquences Opératoires.* Thèse d'état, Université de Franche-Comté, Besançon, France, November 1984.

[5] T. L. De Fazio and D. E. Whitney. Simplified Generation of All Mechanical Assembly Sequences. *IEEE J. Robotics Automat.*, RA-3(6):640–658, December 1987. Corrections *ibid* RA-4(6):705–708, December 1988.

[6] N. Deo. *Graph Theory with Applications to Engineering and Computer Science.* Prentice-Hall, 1974.

[7] A. J. Goldman and A. W. Tucker. Polyhedral convex cones. In *Linear Inequalities and Related Systems*, pages 19–40. Princeton University Press, 1956.

[8] J. M. Henrioud. *Contribution a la Conceptualisation de l'Assemblage Automatisé: Nouvelle Approche en vue de Détermination des Processus d'Assemblage.* Thèse d'état, Université de Franche-Comté, Besançon, France, December 1989.

[9] L. S. Homem de Mello and A. C. Sanderson. Task Sequence Planning for Assembly. In *12th World Congress on Scientific Computation*, volume 3, pages 390–392. IMACS – International Association for Mathematics and Computers in Simulation, July 1988.

[10] L. S. Homem de Mello and A. C. Sanderson. AND/OR Graph Representation of Assembly Plans. *IEEE Trans. Robotics Automat.*, 6(2):188–199, April 1990.

[11] L. S. Homem de Mello and A. C. Sanderson. Evaluation and Selection of Assembly Plans. In *1990 IEEE International Conference on Robotics and Automation*, pages 1588–1593. IEEE Computer Society Press, May 1990.

[12] C. L. Liu. *Introduction to Combinatorial Mathematics.* McGraw-Hill, 1968.

[13] A. C. Sanderson and L. S. Homem de Mello. Planning Assembly/Disassembly Operations for Space Telerobotics. In W. C. Chiou, editor, *Space Station Automation III*, volume 851 of *SPIE Proceedings Series*, pages 109–115. SPIE – The International Society for Optical Engineering, November 1987.

[14] M. A. Wesley et al. A geometric modeling system for automated mechanical assembly. *IBM J. Res. Develop.*, 24(1):64–74, January 1980.

[15] R. Wilson and J. F. Rit. Maintaining Geometric Dependencies in an Assembly Planner. In *IEEE Int. Conf. Robotics Automat.*, pages 890–895. IEEE Computer Society Press, May 1990.

# Chapter 8

# LEGA : a computer-aided generator of assembly plans

**Jean-Michel Henrioud and Alain Bourjault**

In the field of Assembly Automation, Assembly Planning is an important problem which is still handled empirically in most factories. For any product there are many different assembly plans, and it is obvious that choosing one of them has an influence on the cost of the assembly system. Moreover increasing need in flexibility sometimes leads, for a given assembly system, to real time scheduling requiring local changes in the assembly plan.

A first systematic method for the determination of all the assembly plans available for any given product was proposed by A. Bourjault in 1984 [1]. This method was focused in the liaisons between elementary parts, describing an assembly plan as a sequence of liaisons. An interactive software, SAGA (proprietary), was derived from this method ; it involved a set of questions a human expert was asked about the precedence constraints between the liaisons. In practice the number of questions was prohibitive and it was difficult for the operator to visualize the liaison concept. That is why we have developed a new approach, focused on the parts, presented here and which shares several features with the method concurrently developed by L.S. Homem de Mello [9].

Another approach, derived from Bourjault's work has been proposed by T. De Fazio and D.E. Whitney [2], which used a different formalization of assembly constraints leading to a reduction of the dialogue size with the user . Among the related work, we must mention those of R. Jeannes [10] and of A.Delchambre [3] ( also [4],[11],[13]).

The basic points in the present method are :

- a product model, based on a liaison graph similar to the one used in our previous method ;

- a classification and a formalization of assembly constraints ;

- an algorithm allowing a systematic determination of all assembly plans, described by assembly trees (or part trees, according to a previous definition by D.E. Whitney), satisfying the assembly constraints.

A resulting software, LEGA, has been developed in PROLOG. It has been applied to different industrial products, and has proved to be quite more efficient than its predecessor SAGA.

## 8.1 Assembly system

An Assembly System is a system having p input flows $\phi_{i,e}$ (i = 1, ..., p) and q output flows $\phi_{j,s}$ ( j=1,...,q). Each flow $\phi_{i,e}$ is composed of identical objects $C_i$ and each flow $\phi_{j,s}$ is composed of identical objects $P_j$. Relatively to the considered Assembly System each $C_i$ object is a component and each $P_j$ object is an end-product. The assembly of complex products is carried out through several assembly systems, so that a same object may be an end-product for one system and a component for another.

In this chapter we shall consider assembly systems involving up to thirty components. Thus we assume that more complex products have already been split into smaller parts and we consider either the assembly of a part or the assembly of the end product from the preassembled parts.

Moreover we shall only consider implicitly assembly systems having a single output. In fact most industrial assembly systems have several outputs. Generally the objects $P_j$ are very similar, they share the same structure and differ mostly :

- by some components, called options, which may be present or absent (e.g. an arithmetic processor in a computer).

- by different components, called variants, fulfilling the same function (e.g. different engines for a same car model).

In practice the method presented in the following sections may be applied to multiproduct systems. There is no real difficulty with the options (provided they are not mutually exclusive), for it is sufficient to consider the product bearing all the options. As to the variants, problems may arise when their morphologies are really different, but we have not developed yet any general method to deal with all the possibilities.

## 8.2 Product model

The method proposed in this paper for the generation of assembly plans for any product P is based on the modelling of P by a 5- tuple $<\mathbf{C}, \Gamma, \Sigma, \Delta, f>$ where :

* $\mathbf{C}$ is the set of the **components** of P. The concept of component refers to an assembly system and describes the objects entering this system, they may be some subassemblies produced by another assembly system. In the same way P may be a subassembly for another assembly system.The securing components like screws, bolts and nuts, rivets .... are not included in $\mathbf{C}$.

* $\Gamma$ is the set of **liaisons**, between the components of P. We say that there is a liaison between to components x and y iff there is at least one mechanical liaison between x and y inside the product P for at least one orientation of P.

It is convenient to associate to $\Gamma$ the set $\{1, ..., n\}$ ($n = card(\Gamma)$). $[\mathbf{C}, \Gamma]$ defines a graph called liaison graph which is both simple and connected.

* $\Sigma$ is the set of the **attachments** in P. An attachment is all what contributes to secure at least one liaison. An attachment may be :

- one or several securing components : screws, bolts and nuts, rivets...
- some matter : glue, solder
- some energy : fitting, setting, crimping.

* $\Delta$ is the set of the **complementary features** in P, they are some features which have to be included in P and are defined in the requirement list for P. They are mainly : labels, painting, machining, functional tests, tuning, cleaning,...

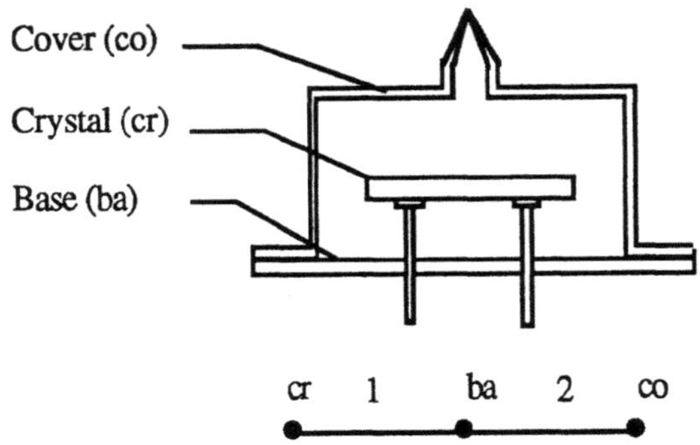


Figure 8.1 : A quartz with its liaison graph

* f is a function which defines, for each attachment or each complementary feature, the set of elements of P which are concerned :

$$f: \quad \Sigma \cup \Delta \quad \rightarrow \quad \mathcal{P}(\mathbf{C}) \times \mathcal{P}(\Sigma \cup \Delta)$$

($\mathcal{P}$(**E**) is the set all of the subsets of **E** )

**Example** : A quartz is presented in figure 8.1. Its model is given by :

* $\mathbf{C}$ = {cr, ba, co}
  . cr is the crystal
  . ba is the base which includes the electrodes
  . co is the cover
* $\Gamma = \{1, 2\}$
  The liaison graph [$\mathbf{C}$, $\Gamma$] is also depicted in figure 8.1
* $\Sigma = \{w_1, w_2\}$
  . $w_1$ is a welding between cr and ba
  . $w_2$ is a welding between co and ba
* $\Delta$ = {el, la, va, sq}
  . el is an electric test
  . la a label printed on the cover which specifies the quality of the quartz as defined by the electric test
  . va is the vacuum created between the cover and the base
  . sq is the squeezing which shuts the tube extending the cover.

* f is defined as follows :

$f(w_1) = (\{cr, ba\}, \Phi)$

$f(w_2) = (\{ba, co\}, \Phi)$

$f(el) = (\{ba, cr\}, \{w_1\})$

$f(la) = (\{co\}, \{el\})$

$f(va) = (\{ba, co\}, \{w_2\})$

$f(sq) = (\{co\}, \{va\})$

Let's notice that in f(la) we have included co, the component which will support la, and also el since its result is an element of la.

**Definition 1** : *A* ***virtual subassembly*** *(X, A) of a product P is an object composed of :*

- *a set X of components of P*
- *the set of all the liaisons which bind the components in X inside P*
- *a set A of attachments and complementary features of P*

*and is such that :*

- *the subgraph of [C, Γ] generated by X is connected*
- *each element a of A is such that :*

$$f(a) \subset (X, A)$$

We shall now refer to virtual subassemblies as VSA.

**Definition 2** : *A virtual subassembly (X, A) for a product P is a* ***subassembly*** *iff there is at least one assembly process for P where the object (X, A) is produced.*

**Examples** : For the quartz previously described :$(\{cr\}, \Phi)$, $(\{cr, ba\}, \Phi)$, $(\{cr, ba\}, \{w_1, el\})$, $(\{cr, ba, co\}, \{w_1, w_2, el, la, va, sq\})$ are subassemblies. The first one is also a component (the crystal) and the last one is the end product ; the second and third ones are subassemblies in the usual meaning. $(\{co, ba\}, \{w_2\})$ is VSA but obviously cannot be a subassembly in any assembly process for the quartz.

## 8.3 Assembly trees

**Definition 3** : *Two VSA $(X_1, A_1)$ and $(X_2, A_2)$ are said to be* ***complementary*** *VSA iff :*

$$X_1 \cap X_2 = \Phi \quad \text{and} \quad A_1 \cap A_2 = \Phi$$

*and there is at least one liaison between one component of $X_1$ and one component of $X_2$.*

Thus two complementary VSA are two objects which are liable to be assembled to produce a new object $(X_1 \cup X_2, A_1 \cup A_2)$.

**Definition 4** : *A* ***geometric operation*** *is the production of a VSA (X, A) realized by mating two complementary VSA $(X_i, A_i)$ and $(X_j, A_j)$ with :*

$$X = X_i \cup X_j \quad \text{and} \quad A = A_i \cup A_j$$

Such an operation will be noted $((X_i, A_i), (X_j, A_j))$.

**Definition 5** : *A* ***non gemetric operation*** *is the production of a VSA (X, A) realized by the adding of either an attachment or a complementary feature p to the VSA (X, A - {p}).*

Such an operation will be noted p(X, A - {p}). A non gemetric operation will be called securing operation or complementary operation according as p is an attachment or a complementary feature.

**Definition 6** : *An* ***assembly tree*** *for a product P is a rooted tree whose :*

- *root is the end product P*
- *nodes are VSA of P*
- *leaves are the components of P*

*and such that every non terminal node may be obtained from its k successors :*

- *by a geometric operation if $k = 2$*
- *by a non geometric operation if $k = 1$*

Two assembly trees for the quartz presented in figure 8.1 are shown in figure 8.2. They are depicted horizontally and each node is represented by a label which is either an elementary part or what has been added to the part(s) associated to its predecessor(s). The horizontal presentation symbolizes the underlying assembly system. The liaisons labelling the nodes resulting from a geometric operation may be omitted for they are redundant.

From the definition of the assembly trees it follows recursively that all VSA associated to the nodes of any tree are subassemblies.

**Fundamental assumption** :

Each assembly plan we shall consider can be represented by an assembly tree.

This fundamental assumption imposes two restrictions to the assembly plans :

1 - The assembly plans cannot include geometric operations mating more than two parts. This is quite sufficient for most assembly processes. Nevertheless some special situations may require the mating of more than two parts. Such situations are studied in [6], where the way of incorporating them a posteriori into the assembly process is described.

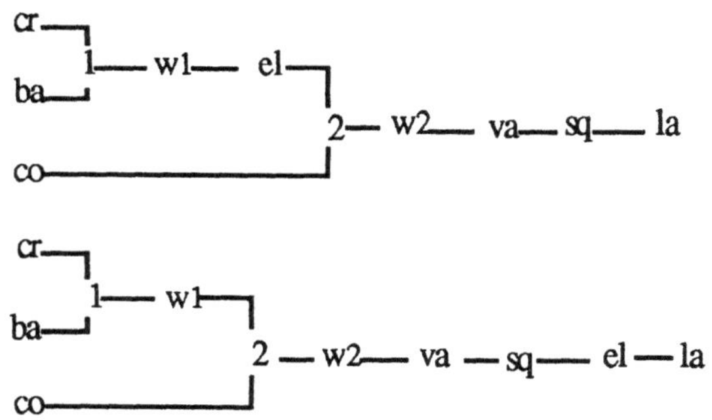


Figure 8.2 : Two assembly trees for the quartz given as exemple.

2 - Definitions 1, 4 and 6 require that when two parts are put together, which is described as a geometric operation, all the liaisons between their components have to be set. It does not matter if they are set simultaneously or sequentially, the point is that no other operation can take place before they are all set. This requirement does not obviously make trouble for rigid products but may prove restricting for non rigid products. In fact, for non rigid products the liaison approach proposed by A.Bourjault [1] is more exhaustive. A detailed discussion of this problem is proposed in [6].

## 8.4 Assembly constraints

A good assembly process has to involve easy operations (at least as easy as allowed by the product design) and has to follow some logic in order to avoid useless reorientations or useless tool changing. We say that assembly planning is subjected to assembly constraints. Some are local and concern the assembly operations, we call them operative constraints while the others are global and concern the assembly process, we call them strategic constraints.

### 8.4.1 Operative constraints

We have divided the operative constraints in three categories. To describe these constraints we shall refer to the two parts of any operation. This refers directly to the geometric operations which involve two parts but also, by extension, to the non geometric ones for which the second part may be some tool or some tool bearing some attachment component.

#### Geometric constraint

When mating two parts there must be at least one collision-free trajectory allowing to bring the two parts in contact. The geometric constraint is absolute, intrinsic to the product and thus is liable to be automatically deduced from a 3D representation of the product.

**Stability constraint**

Each part produced in the course of the assembly process must be stable, that is all its components have to keep their spatial relationships. In fact stability is mostly a relative concept since it depends largely upon data unknown in the assembly planning stage : mainly the motions the part will have to do. Moreover, some loose components may be held by some convenient fixture.

We have defined three stability levels. A part is :

- stable when all its components keep their spatial relationship for any orientation and any move,

- partly stable if there is at least one orientation for which all its components keep their spatial relationship, possibly with help of some holding equipment,

- unstable if neither stable nor partly stable.

The stability constraint, which makes it necessary that no VSA of any assembly tree be unstable is neither absolute nor intrinsic to the product, since it refers to possible holding equipment.

**Material constraint**

The effective realization of any geometric operation requires some equipment handling each involved part, with still the necessity of collision-free trajectory between the two objects which are the base part with its fixture and the secondary part with its handling device (gripper and arm). A material constraint is so neither absolute nor intrinsic to the product.

We have presented in figure 8.3 some illustrations of the operative constraints.

Example A illustrates a geometric constraint, it is impossible to place the black component inside the closed box.

In example B it is difficult to mate components **a** and c because of the tube **b**. The difficulty may turn into an impossibility if **b** is too long or if **c** is fitted tightly in **b**. The decision whether there is or there is not a material constraint is difficult and belongs to the expert.

In example C we suppose that components **a** and **b** are secured to **c** through other components (for instance leads on c). The resulting subassembly will be obviously very difficult to handle and may be considered unstable. There again the decision belongs to the expert.

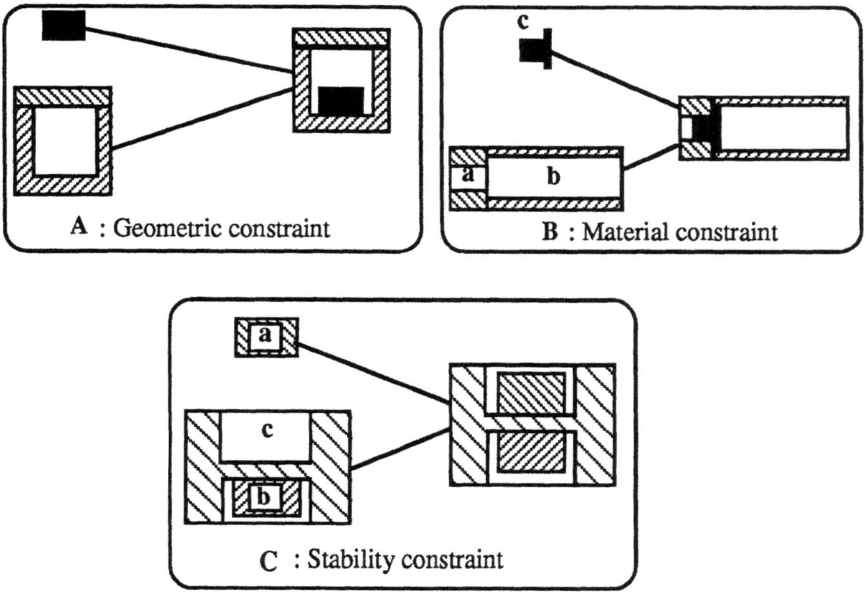


Figure 8.3 : Illustrations of operative constraints.

### 8.4.2 Strategic constraints

According to definitions 4, 5 and 6, each assembly tree respects the operative constraints since all its operations are feasible. The problem is that the number of assembly trees available for any given product is very large (some $10^4$ for products having from 15 to 20 components) and that most of these assembly trees describe obviously awkward assembly processes, involving too many assembly direction changes or tool changes.

In order to avoid the determination of inefficient solutions a set of global constraints may be introduced before the process of assembly trees determination begins. These constraints have to express some reasonable strategies for the assembly process and are to be deduced from the product structure. The principal strategic constraints are described hereafter.

**Imposed subassemblies**

Very often it is possible to define from the end product some subassemblies which are to be produced in the assembly process (for storage purpose, because they are to be used as maintenance spare parts, for functional or stability reason). This leads to a strategic constraint such that only the assembly trees including these subassemblies are to be produced.

**Group of components**

In a manufactured product, some components have similar shapes and similar sizes and are liable to be handled by the same gripper, or they have an identical securing or the same assembly direction. In such cases it may be advisable to group the operations in which they are involved. This leads to a strategic constraint which requires that in each assembly-tree the leaves standing for the components belonging to a same group be consecutive. When some subassemblies are imposed a subassembly may be an element of a group. This concept of group has proved very efficient when applied to several industrial products ; it is equivalent to the concept of cluster developed concurrently by C.J.M. Heermskerk [5]. It should be noticed that it is possible to define subgroups of components inside a same group . Two basic structures for some groups of components, studied in [12] are presented hereafter, which lead to more selective constraints.

**Stacks** : A **stack** is a group of k ($k \geq 3$) parts $C_i$ (components or subassemblies) such that there is a liaison between $C_i$ and $C_{i+1}$ (for i = 1, ..., k - 1) and such that a part $C_i$ may be added to a part S including $C_1$ (respectively $C_k$) iff S includes already $C_{i-1}$ (respectively $C_{i+1}$).

Each time a stack $C_1, ..., C_k$ is declared inside a product P a special strategic constraint is issued so that any resulting assembly tree must include the sub-tree $(...((C_1, C_2), C_3)..., C_k)$, securing and complementary operations being not included.

**Example** : given the product P depicted in figure 8.4 if we define the stack (A,B, C, D, E) there is only one resulting assembly tree $T_1$ presented also in figure 8.4. If we define the stack (B, C, D, E) there are only two resulting assembly trees : $T_1$ and $T_2$. Without any stack, we would find 14 assembly trees.

**Ordered Layers** : A **layer** is a group of k parts $C_1, ..., C_k$ such that each part $C_i$ has a liaison with a same component B. The subgraph associated to the set $(C_1, ..., C_k) \cup \{B\}$ is generally star shaped but the possibility of some liaisons between the $C_i$ is not excluded. When a layer is such that its assembly order is obviously indifferent then it is useless to provide all the possible orders (k!). In such a case the user may choose an arbitrary order by declaring an ordered layer. It is still possible to declare an **ordered layer** when there is some precedence constraints between some of the $C_i$ components if a convenient order is obvious.

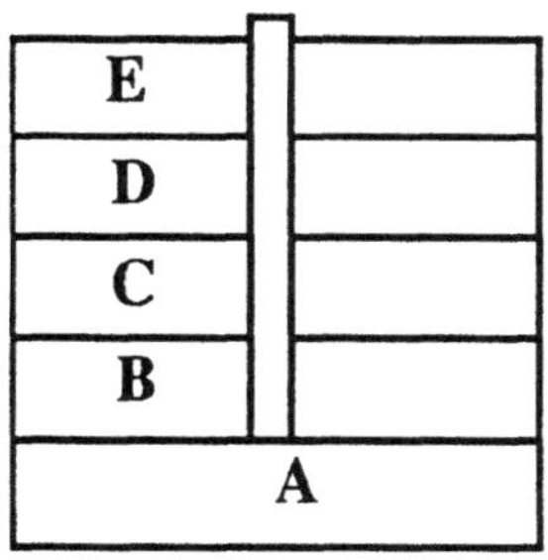


T1 = ((((A,B),C),D),E)
T2 =(A, (((B,C),D),E))

Figure 8.4 : Example of stack strategic constraint.

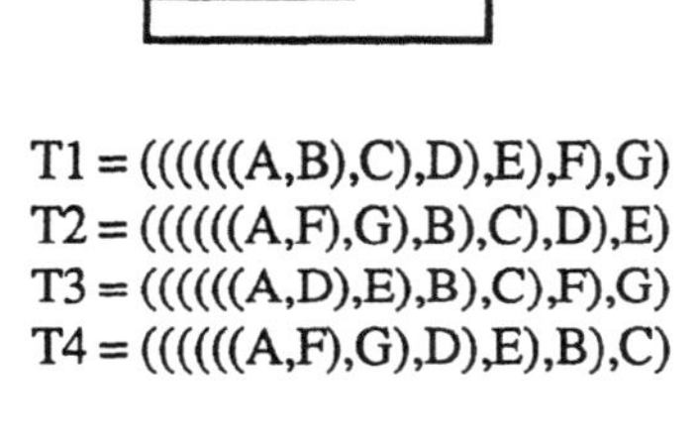


T1 = ((((((A,B),C),D),E),F),G)
T2 = ((((((A,F),G),B),C),D),E)
T3 = ((((((A,D),E),B),C),F),G)
T4 = ((((((A,F),G),D),E),B),C)

Figure 8.5 : Example of layer strategic constraint

Each time an ordered layer $(C_1, ..., C_k)$ is declared inside a product P, a strategic constraint is issued so that each resulting assembly tree must include the subtree $(...((X,C_1), C_2), ..., C_k)$, securing and complementary operations being not included.

**Example** : given the product P depicted in figure 8.5, several possibilities arise for the introduction of strategic constraints :

1. There are two groups of components, according to assembly directions, with layers structures : (B,C,D,E) and (F,G). There are 48 resulting assembly trees .

2. If each layer is declared to be ordered then there are only 2 assembly trees (T1 and T2 in figure 8.5).

3. If (B,C,D,E) is considered as including two sublayers (B,C) and (D,E) then there are 24 assembly trees .

4. If each layer (B,C), (D,E) and (F,G) is declared to be ordered, then there are 4 assembly trees ($T_1$,$T_2$, $T_3$, $T_4$ on figure 8.5).

Layers (B,C,D,E), (B,C) and (D,E) may be declared ordered because their assembly order is indifferent. Layer (F,G) has only one assembly order. Without any strategic constraint the number of assembly trees would be 360 .

**Linear Assembly Trees**

An assembly tree is said to be linear when each node having two successors (the tree being oriented from the root to the leaves) has at least one leaf for successor. So a linear assembly tree describes an assembly process where one single component is added to the current subassembly in each geometric operation.

From our experience of industrial products we have noticed that most of the assembly processes are linear. As we said before we consider assembly systems involving a limited number of components. An assembly process for a whole car is highly parallel but fortunately it is dispatched on several assembly systems which involve mostly linear processes.

So, we have integrated in our method the possibility to restrict the determination of assembly trees to the linear ones. However this constraint is considered weaker than the one dealing with the imposed sub-assemblies. Thus, when some imposed sub-assemblies have been defined and when the restriction to linear assembly trees is chosen, the resulting trees are linear relatively to the imposed sub-assemblies and to the other components.

This concept of linear assembly tree is also proposed by J.D.Wolter [14] inside a classification of assembly plans.

**Advantages and drawbacks of strategic constraints**

Any assembly tree which satisfies the operative constraints describes a feasible assembly process. Let us say that such an assembly tree is valid. Practically, it turns out that the number of valid assembly trees increases dramatically fast with the number of components of the end product. The number of valid assembly trees for products having 20 components ranges from $10^4$ to $10^6$. So it is obviously useless to know all the valid assembly trees unless we can select the best ones.

In order to reduce the set of resulting assembly trees to a few best ones we have proposed a method presented in [6] and [8]. Succinctly, we consider all the different operations involved in all the assembly trees (for products having 20 components there are hundreds of different operations). The complexity of each operation is evaluated, which requires the choice of the base component as well as its orientation. When there are several possible choices, then several different operations (which then are called oriented) are issued and have to be evaluated. A new set of assembly trees, also called oriented, is produced. For each of them an operative cost is computed from its operations and a logistic cost is deduced from the number of assembly direction changes between its operations. These two costs are then combined in a global cost which allows their ranking. Unfortunately this method is quite time consuming for the expert, who has to evaluate some hundreds of operations, and thus cannot be effectively applied until this evaluation process is partly automated.

In the present state of our work the most efficient tool allowing the determination of a reasonable set of assembly trees (e.g. 10 to 50) is the introduction of strategic constraints.This tool has proved very efficient and it had been applied to different industrial products (14 to the present day). It requires some time to the user to study the product to assemble before the process of assembly tree generation begins, but it is quite reasonable. Moreover, this is a creative task, more rewarding than the one involved in the interactive determination of the assembly trees presented in the following section.

Thus we may conclude that the introduction of strategic constraints is a practical and efficient tool for assembly planning.The only drawback is that the choice and definition of assembly constraints is left to the user, so that subjectivity is introduced inside a systematic process. The method quoted above, with a numerical evaluation of each resulting assembly tree would be better, but there is still a lot of work to be done towards a system that is adequate for practical industrial use.

To conclude this section, we must emphazise that for a given product, the user may hesitate between mutually exclusive strategic constraints. Then he will have to define several sets of constraints, each set leading to a corresponding set of assembly trees. When such a situation arises, the different sets of resulting assembly processes will have to be kept to be compared when the resulting assembly systems are sufficiently defined (equipment costs and performances) to allow a precise comparison.

## 8.5 Generation of all Assembly Trees

### 8.5.1 Basic formulation

Informally, the approach we have developed for the systematic generation of assembly trees is a decomposition one. Starting from the end product P we search all the feasible operations having P for result. The set of these operations defines a set of subassemblies. For each resulting subassemblies (X, A) we search again all the feasible operations having (X,A) for result. And we proceed recursively until the resulting subassemblies are reduced to components of P.

This algorithm is quite similar to the one defined by L.S. Homem de Mello [9]. To find the geometric operations which have a subassembly (X, A) for result we search all the cut sets of the graph associated to (X, A) that is all the pairs $X_1$, $X_2$ such that :

$$X_1, X_2 \neq \Phi \text{ and } X_1 \cap X_2 = \Phi \text{ and } X_1 \cup X_2 = X$$

and such that the graphs associated to $X_1$ and $X_2$ are connected. But, according to the model proposed here, we have also to search all the non geometric operations having (X, A) for result. So we have to look, for each p belonging to A, if the operation p(X, A-{p}) is feasible.

More formally, we define now an assembly tree in the following way, where $V_P$ is the set of all the VSA for a given product P.

**Definition 7** : *T is an* ***assembly tree*** *for (X, A)* $\in V_P$ *iff :*

1. $X = \{c_i\}$, $c_i \in C$ *and* $A = \Phi$

   *then* $T = c_i$

*or* 2. $\exists (X_i, A_i), (X_j, A_j) \in V_P$ *such that*

- $T_i$ *is an assembly tree for* $(X_i, A_i)$
- $T_j$ *is an assembly tree for* $(X_j, A_j)$
- *The geometric operation* $((X_i, A_i), (X_j, A_j))$ *is feasible*

   *then* $T = (T_i, T_j)$

*or* 3. $\exists p \in A$

- *T' is an assembly tree for (X, A-{p})*
- *The non geometric operation p (X, A-{p}) is feasible*

   *then* $T = p(T')$.

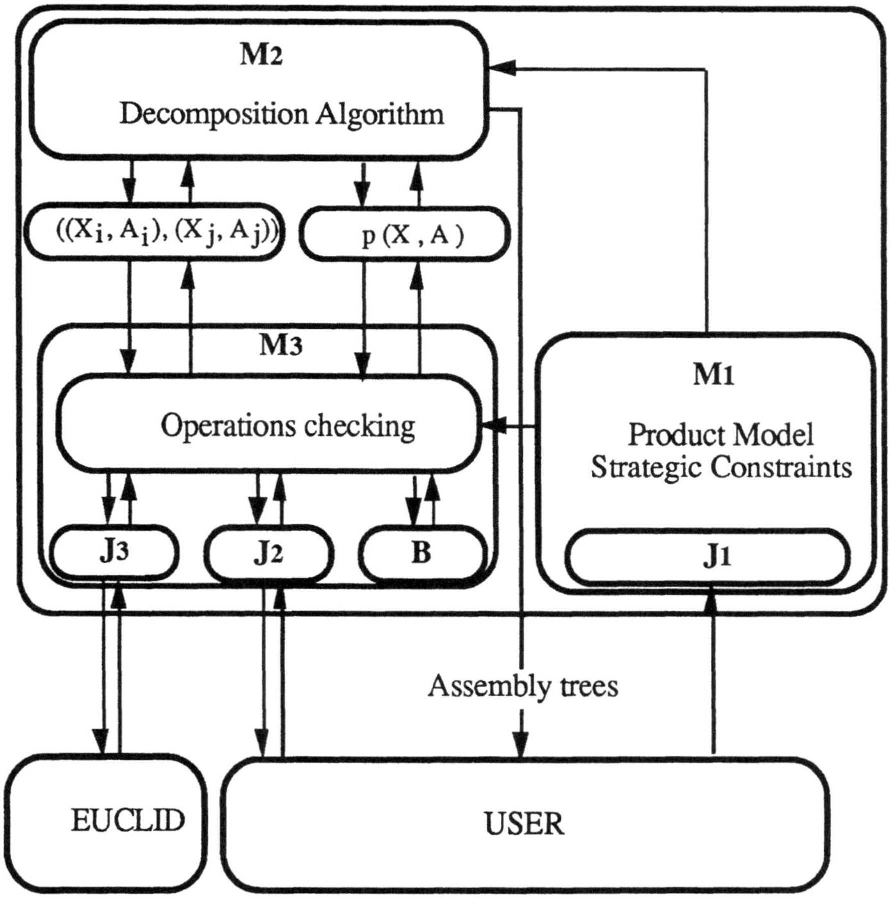


Figure 8.6 : Organisation of LEGA .

Definition 8.7 is composed of three disjunctive clauses which lead directly to three Prolog clauses. These clauses are the basic part of our present program for generating assembly tree. This prototype program, named LEGA ("Logiciel d'Elaboration des Gammes d'Assemblage") is written in Prolog II. Its architecture is depicted in figure 8.6, which includes three main modules :

* **M1** describes the product model and the set of strategic constraints. All data in **M1** are provided by the user. **M1** is able to check the coherence of this data. **J1** is the user's interface.

* **M2** provides the decomposition algorithm. It is mainly composed of the clauses resulting from Definition 7. **M2** provides only decompositions consistent with the strategic constraints defined in **M1**. **M2** provides candidate operations (geometric and non geometric) resulting from the product decomposition.

* **M3** checks all the operations issued by **M2** first by searching a database **B** which contains the already known operative constraints, and then, when no answer is found in **B**, by questioning the user. The user's answer is used to enrich **B**. **J2** is the user's interface. **M3** includes also an optional interface **J3** with an external program, connected to a CAD software containing a 3D solid geometric model of the product,which allows an automatic checking of the geometric constraint. This connexion has been successfully tested in association with CAD software EUCLID.

* Module "operations checking" includes a set of rules allowing the deduction of new operative constraints from the ones already registered in **B** and from the product model (see section 8.6).

### 8.5.2 Modified algorithm

The third condition given in Definition 7, which allows to check all the feasible non geometric operations having for result subassembly (X, A), is quite inefficient in practice. This condition leads to check for every p in A the feasibility of operation p (X, A -{p}). In fact, it appears that most of the non geometric operations have to take place as soon as the conditions for their realization (defined by f function in the product model) are fullfilled. Thus a securing operation is usually carried out as soon as the components to be secured are in a subassembly and this for stability purpose. In the same way a checking operation is carried out as soon as the components necessary to the function to be checked are in a subassembly. To add more components before the checking would lead to reject or to rework a more complex subassembly. That is why we have added to LEGA the following rule:

**Rule** :Any subassembly (X, A) is authorized to enter a geometric operation ((X,A), ( )) iff :

$$\forall p[p\in \Sigma\cup\Delta\wedge P_{\mathbb{C}}(f(p)) \subset X\wedge P_{\Sigma\cup\Delta}(f(p)) \subset A \Rightarrow p\in A]$$

where $P_{\mathbb{C}}$ and $P_{\Sigma\cup\Delta}$ denote the projections respectively on $\mathbb{C}$ and $\Sigma\cup\Delta$.

Informally, any subassembly is allowed to enter a geometric operation only once it has been provided with all its attachments and all its complementary features.

Nevertheless, there are obviously a few exceptions to this rule . So when the user thinks that it would be interesting to delay the introduction of some attachment or some complementary feature p, he has to declare it in the product modelling stage. This leads to the issue of a formula free (p).

## 8.6 Formalization of operative constraints

### 8.6.1 The CG relation

**Definition 8** : *the* ***CG relation*** *for a product P is a binary relation on Vp such that :*

$$((X_i, A_i), (X_j, A_j)) \in CG \qquad \textit{iff :}$$

*1. $(X_i, A_i)$ and $(X_j, A_j)$ are two complementary VSA*

*and 2. Any geometric operation $((X_P, A_P), (X_q, A_q))$ such that :*

$$X_i \subset X_p \quad \text{and} \quad X_j \subset X_q$$

*is unfeasible.*

**Properties** :

The three properties below follow directly from definition 8

1. $\forall (X_i, A_i), (X_j, A_j) \in V_P\ [CG((X_i, A_i), (X_j, A_j)) \Leftrightarrow CG((X_j, A_j), (X_i, A_i))]$

2. $\forall (X_i, A_i), (X_j, A_j), (X_P, A_P), (X_q, A_q) \in V_P$
   $[CG((X_i, A_i), (X_j, A_j)) \wedge X_i \subset X_P \wedge X_j \subset X_q \Rightarrow CG((X_P, A_P), (X_q, A_q))]$

3. $\forall (X_i, A_i), (X_j, A_j), (X_P, A_P), (X_q, A_q) \in V_P$
   $[\neg CG((X_i, A_i), (X_j, A_j)) \wedge X_P \subset X_i \wedge X_q \subset X_j \Rightarrow \neg CG((X_P, A_P), (X_q, A_q))]$

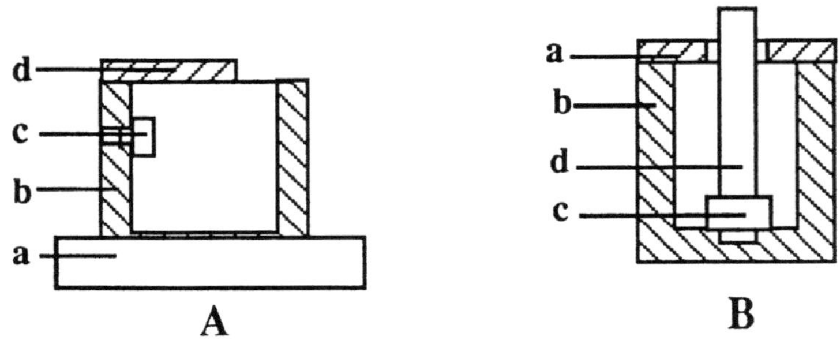


Figure 8.7 : Two examples illustrating material constraint .

In LEGA, each time a candidate geometric operation $((X_i,A_i), (X_j,A_j))$ issued by the decomposition algorithm is rejected because of the geometric constraint, then a clause CG $((X_i, A_i), (X_j, A_j))$ is created in database **B**. The reason is that since there is no free collision trajectory allowing to bring the two parts $(X_i, A_i)$ and $(X_j, A_j)$ in contact, the problem will be the same or even worse if we add more components to these parts.

When a candidate geometric operation $((X_i, A_i), (X_j, A_j))$ satisfies the geometric constraint but does not satisfy the material constraint, the user is allowed, under his responsibility, to force in LEGA a clause CG $((X_i, A_i), (X_j, A_j))$. This may be done when there is obviously no room between the two parts for any possible gripper. Thus there is a geometric constraint between the objects constituted by the involved subassemblies and any possible handling device.

This situation is illustrated by example A in figure 8.7. There is no geometric constraint for the operation $((\{a, b, d\}, \Phi), (\{c\},\Phi))$ but this operation may be judged impossible or too difficult by the user. Obviously the impossibility will stand or even increase if more components are added to any of the two involved parts.

However the possibility to express a material constraint by means of relation CG must be used carefully, for it may lead to some mistakes as, for instance, in the example B in figure 8.7. Operation $((\{a, b\}, \Phi), (\{c\}, \Phi))$ may be judged impossible while operation $((\{a,b\}), \Phi), (\{c, d\}, \Phi))$ may be a rather easy one.

### 8.6.2 The CS relation

**Definition 9** : *the* ***CS relation*** *for a product P is a property (or unary relation) on* $V_P$ *such that :*

$$(X_i, A_i) \in CS \quad \textit{iff the subassembly } (X_i, A_i) \textit{ is unstable.}$$

We have shown in [6] that some automatic deductions about the stability of the subassemblies issued by the decomposition algorithm could be derived from the product model and from the CS clauses already created in database **B**.

### 8.6.3 The CM relation

**Definition 10** : *the* ***CM relation*** *for a product P is a binary relation on $V_P$ such that :*

$((X_i, A_i), (X_j, A_j)) \in CM$ *iff*

*1.* $((X_i, A_i), (X_j, A_j)) \notin CG$

*and* *2.* $(X_i, A_i) \notin CS \wedge (X_j, A_j) \notin CS$

*and* *3.operation* $((X_i, A_i), (X_j, A_j))$ *is unfeasible*

An example of use of CM relation is given by product B in figure 8.7. If the user wants to reject operation $((\{c\}), (\{a, b\}))$ we have seen that it was not correct to use relation CG, so we would instead create a clause : CM $((\{c\},\Phi), (\{a, b\},\Phi))$

### 8.6.4 The AG relation

**Definition 11** : *the* ***AG relation*** *for a product P is a property on $(\Sigma \cup \Delta) \times V_P$ such that :* $(p, (X, A)) \in AG$

*iff any non geometric operation p (Y, B) such that* $X \subset Y$ *is unfeasible.*

**Properties** :

1. $\forall\, p \in \Sigma \cup \Delta, (X_i, A_i) \in V_P\ [AG(p, (X_i, A_i)) \wedge X_i \subset X_j \Rightarrow AG(p, (X_j, A_j)]$

2. $\forall\, p \in \Sigma \cup \Delta, (X_i, A_i) \in V_P\ [\neg AG(p, (X_i, A_i)) \wedge (X_j \subset X_i)$

$$\Rightarrow \neg AG(p, (X_j, A_j)]$$

In LEGA, each time a candidate non geometric operation $p(X_i, A_i)$ issued by the decomposition algorithm is rejected because of a geometric constraint, then a clause AG $(p, (X_i, A_i))$ is created in database **B**.

An example is given in figure 8.8 where it is impossible to screw the components a and b when the component c is mounted on b, so we have : AG $(s, (\{a, b, c\}, \Phi))$

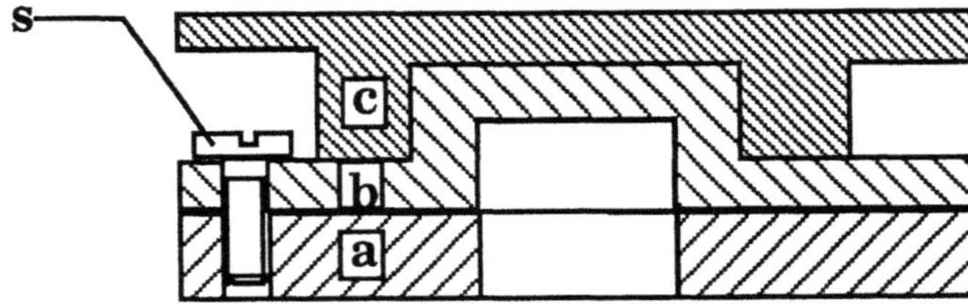


Figure 8.8 : Geometric constraint for a securing operation.

### 8.6.5 The AM relation

**Definition 12** : *the* ***AM relation*** *for a product P is a property on* $(\Sigma \cup \Delta) \times V_P$ *such that :* $(p, (X_i, A_i)) \in AM$ *iff*

1. $(p, (X_i, A_i)) \notin AG$

*and* 2. *The non geometric operation* $p(X_i, A_i)$ *is unfeasible.*

In practice the AM relation is not used frequently. Its main use is to correct an insufficient definition of the f function in the product model. For instance, consider the quartz presented in section 8.1. Labelling la has to be set in the cover co. If the user does not include the electric state in the defintion of f (la) and declares f(la) = ({co}, Φ), then the decomposition process will suggest the operation la ({co}, Φ) which may be realized before the electric testing.This operation is impossible because the label cannot include the quality of the resulting quartz. In this case this operation may be rejected by using a clause AM (a clause CG would obviously eliminate every assembly process).

## 8.7 User's part in assembly tree generation

As described in section 8.5, LEGA produces a set of operations whose feasibility has to be checked. Whenever the feasibility of an operation cannot be decided from the database **B** and the set of rules included in LEGA, the user is questioned. According to the operation category (geometric or not) two different dialogues may be issued which are presented in figure 8.9.

The first one, issued for a geometric operation decides its feasibility and, when the operation is unfeasible, the nature of the operative constraint involved.The operation is noted (S,T) where S and T are the two involved subassemblies. Moreover, when there is a stability problem, the unstable part is defined.

The second one, issued for a non geometric operation is quite similar but, since there is only one subassembly involved, when there is a stability problem there is not need to ask which part is unstable.The operation is noted p(S).

In both dialogues we have presented the clauses deducible from the user's answers. In fact, the work in LEGA is more complex than the simple creation of these clauses. Properties of CG and CS relations are used to eliminate redundant clauses in **B**.

Also the knowledge included in **B** allows LEGA, sometimes, to skip some questions in the dialogue. Thus when at least one part $(X_i, A_i)$ or $(X_j, A_j)$ is known to be stable (because one is reduced to a component or because all its components are secured), no question is asked about the unstable part. Moreover, if the two parts are known to be stable and if there is a clause :

$$\neg CG\ ((X_p, A_p), (X_q, A_q))$$

with $$((X_i \subset X_p) \wedge (X_j \subset X_q)) \vee ((X_i \subset X_q) \vee (X_j \subset X_p))$$

then the answer that the operation $((X_i, A_i), (X_j, A_j))$ is unfeasible implies that a material constraint is involved, a geometric constraint being excluded.

In practice, the number of dialogues issued in the assembly tree generation of a product or subassembly having 20 components, without any strategic constraint, is of some hundreds. Moreover the waiting time between two dialogues increases as the data that is added to database **B** ( which is empty at the beginning) enables LEGA to do more and more automatic deductions without the user's help. Thus, the generation of all valid trees with LEGA is not interesting in practice because it requires too much time from the user and produces too many assembly trees.

The connection to a CAD software reduces substantially the number of questions but increases the computing time. Thus the good use of a CAD software is to have LEGA working in two stages. A first stage completely automatic which defines the whole set of geometric constraints in **B**, and a second stage which starts again the whole process interactively with the user.

However LEGA is most efficient when supplied with strategic constraints.Typically only a few of them are needed to reduce considerably both the number of questions and the number of resulting assembly trees. In the next section we shall present a simplified version of LEGA which has proved to be very effective for most of the products we have studied, and has been considered more practical by engineers working on assembly.

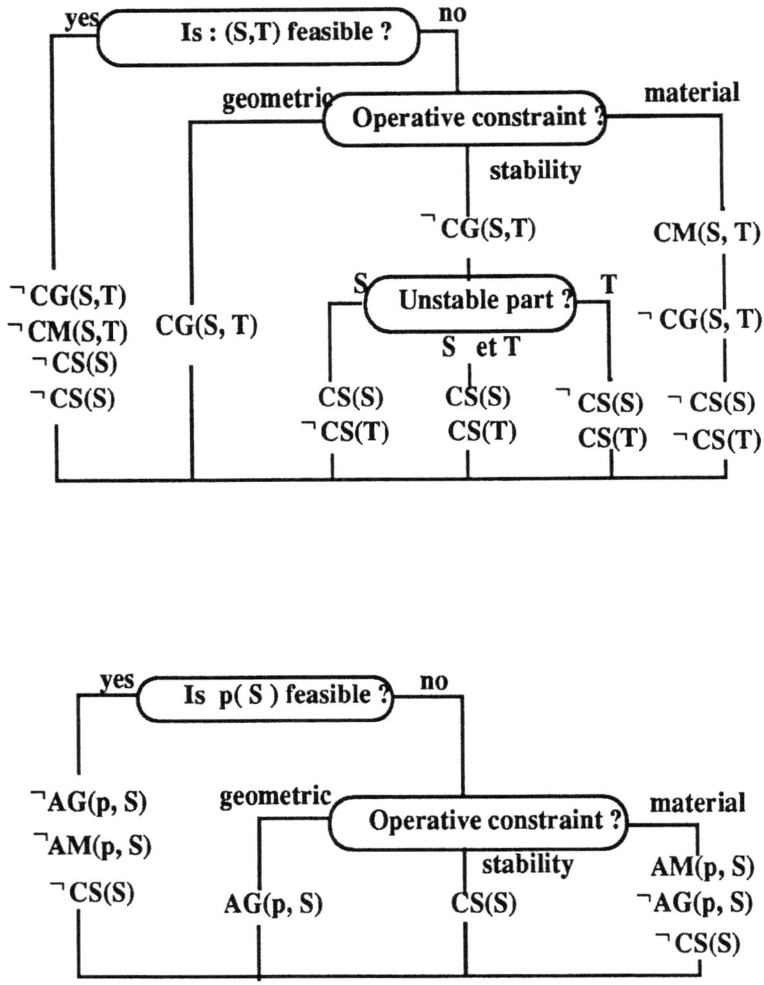


Figure 8.9 : Dialogues issued between LEGA and the user .

## 8.8 Simplified version of LEGA

As we have stated in section 8.5, when two parts are assembled, if they are to be secured, usually they are secured immediately. That is why we have added a corresponding general rule in LEGA. Thus, let p be an attachment securing two subassemblies $(X_i, A_i)$ and $(X_j, A_j)$, that means, with $f(p) = (X, \Phi)$ :

$$X \cap Xi \neq \Phi \quad \text{and} \quad X \cap X_j \neq \Phi \quad \text{and} \quad X \subset X_i \cup X_j$$

LEGA will produce successively the two following operations :

1. is $p(X_i \cup X_j, A_i \cup A_j)$ feasible ?

2. is $((X_i, A_i), (X_j, A_j))$ feasible ?

In fact it is far better for the user to consider globally these two questions and to find out at the same time whether it is possible to mate and to secure these two subassemblies. So we have introduced a new kind of operation.

**Definition 13** : *An* ***assembly operation*** *is the production of a VSA (X, A) realized by mating and securing two complementary VSA $(X_i, A_i)$ and $(X_j, A_j)$*
*with :* $X = X_i \cup X_j$ *and* $A = A_i \cup A_j \cup \alpha$
*where* $\alpha = [p/p \in \Sigma \ : P_C(f(p)) \subset X \wedge P_C(f(p)) \not\subset X_i \wedge P_C(f(p)) \not\subset X_j]$

Accordingly we have developed a simplified version of LEGA which deals mainly with assembly and complementary operations. Nevertheless, it is still possible to declare that some attachment may be delayed by means of predicate free. For such attachments the concept of assembly operation will only appear in the trees where their setting follows immediately the mating of two corresponding parts. In other trees the user will still have to deal separately with the geometric operation and the securing operation.

To close this section it is important to state that for many products having few complementary features and when their placing in the assembly process appears to be easy, it is more convenient for the user not to include them in the product model. They will be introduced later in the definition of the assembly system. Such a way to deal with complementary features may not be satisfactory from a theoretical point of view but it greatly simplifies the user's work. Thus, ultimately LEGA may be simplified to the point where the only operations involved are the assembly ones as defined above. This implies the merging of relation CG and AG on the one hand and relations CM and AM on the other hand, and a slight change in the dialogue between LEGA and the user. This simplified version of LEGA has been presented in [7] while a complete description of the full version was given in [6].

## 8.9 Conclusion

The method proposed in this paper allows to find all the valid assembly sequences for any product to assemble. The resulting assembly sequences, represented by assembly trees, include not only the assembly operations (mating and securing of parts), but also every complementary operation imposed by product definition.The resulting program, LEGA, which is still a prototype, has been written in Prolog which proved quite suitable to the formalization we have chosen both for the decomposition algorithm and for the assembly constraints.

The current version of LEGA is still quite prohibitive for products having more than ten components, both because it requires too much time from the user and produces too many valid assembly trees. We are studying a partially automatized evaluation of the operations involved in the different resulting trees allowing to rank them, but obviously there is still a lot of work to be done.

The key to turn LEGA into an efficient generator of assembly plans was to define some strategic constraints which may be introduced by the user in the product modelling stage. These constraints allow, for most products, a considerable reduction of the number of resulting assembly trees and of the user's work. The drawback is that their choice depends upon the user's expertise and that there is no proof that they select the best trees. Anyway this impoverishment in the theoretical aspect is compensated by a real increase in efficiency. It is now possible to generate assembly trees for products having about twenty components in reasonable time, and even to deal with more complex products by splitting them into subassemblies. It is important to emphasize that only a few strategic constraints are needed and, however subjective they are, they are registered so that the user's decision may be easily discussed.

# References


[1] A. Bourjault. *Contribution à une approche méthodologique de l'Assemblage Automatisé : Elaboration Automatique des Séquences Opératoires*. Thèse d'Etat, Université de Franche-Comté, December 1984.

[2] T.L. De Fazio and D.E. Whitney. *Simplified Generation of all mechanical assembly sequences.* IEEE J. Robotics and Automation, Vol. RA.3, 640-658. December, 1987.

[3] A. Delchambre. *Conception assistée par ordinateur des gammes opératoires d'assemblage*. Thèse de Doctorat, Université libre de Bruxelles,May 1990.

[4] B. Frommherz and J. Hornberger. *Automatic Generation of Precedence Graphs.* Proceedings of 18th ISIR, Lausanne, 453-466, IFS Publications, Avril 1988.

[5] C.J.M. Heermskerk. *A Concept for Computer Aided Process Planning for Flexible Assembly.* Ph.D.Thesis, Delft University of Technology, May 1990.

[6] J.M.Henrioud. *Contribution à la conceptualisation de l'assemblage automatisé: nouvelle approche en vue de la détermination des processus d'assemblage.* Thèse d'Etat, Université de Franche-Comté, December 1989.

[7] J.M. Henrioud and A.Bourjault. *Détermination des arbres d'assemblage*, APII vol. 24, 547-564, November 1990.

[8] J.M. Henrioud, F. Bonneville and A. Bourjault. *Evaluation and Selection of Assembly Plans*, APMS'90,Espoo, Finland, 212-219, August 1990.

[9] L.S. Homem de Mello. *Task Sequence Planning for Robotic Assembly*. Ph.D.Thesis, Carnegie Mellon University, May 1989.

[10] R.Jeannes. *Méthodologie d'analyse des produits pour leur reconception en vue du montage automatique.*Thèse de Doctorat, ENSAM, 1986.

[11] W. Jentsch F. Kaden. *Automatic Generation of Assembly Sequences.* Art.Int. and Inf. Control systems of Robots,Tchecoslovaquie, 197-200, 1984.

[12] M. Meunier. *Eléments méthodologiques pour la conception des systèmes flexibles d'assemblage.* Thèse de Doctorat, Université de Franche-comté, June 1989.

[13] H. Sekigushi et al. *Study on Automatic Determination of Assembly Sequences*, Annals of the CIRP, Vol.32/1,,371-374, 1983.

[14] J.D. Wolter. *On the Automatic Generation of Assembly Plans.* Proceedings of the 1989 IEEE Int. Conf. on Robotics and Automation Scottsdale, 62-68.

# Chapter 9

# Maintaining geometric dependencies in assembly planning

**Randall H. Wilson and Jean-François Rit**

Used as a tool to get manufacturability feedback in early design stages, an assembly sequence planner would be a boon to designers. By pressing a button, an engineer could receive an assembly plan or set of plans for the product being designed, along with estimates of production time and cost. The design could then be adjusted to make it easier to build. But to be most useful, such a tool must be both autonomous and fast. If the designer has to do a great deal of geometric reasoning for the planner or wait days for the system to do it automatically, the assembly planner will be used infrequently, and the impact on the design will be minimal.

In this chapter we describe GRASP, an assembly planner designed for use in a concurrent design environment. GRASP generates assembly sequences directly from a geometric model of the target assembly, and includes a method that makes assembly planning more efficient by establishing a tight connection between physical reasoning and sequence planning. In our approach, the geometric reasoner constructs an expression describing the conditions under which the result of a geometric calculation will still be valid. These expressions are then evaluated at later steps, often avoiding costly geometric computation. In our experiments the method has caused large speed increases when planning for real assemblies.

In our treatment of assembly sequence planning, we make some simplifying definitions. First, parts are placed directly into their final positions relative to each other, and no connection between two parts is ever changed after it is established. No assembly operation mates more than two subassemblies. In addition, each intermediate subassembly must be connected.

We also make assumptions about the geometry of the parts in the assembly: the parts are rigid, they have exact geometry, and their positions have no tolerances. This has the disadvantage that the planner cannot accurately reason about springs, snap-fit connections, or other flexible parts, but it allows the planner, with a geometric model of the final assembly, to describe any partial assembly with just a list of the parts contained therein. Although the methods we discuss could be generalized further, the solid modeler we use reasons about 3D polyhedral objects, with a limited added ability to reason with cylindrical, conical, and spherical surfaces. Moreover, our planner can only mate two subassemblies with a single translation.

Finally, in this chapter we only allow assembly operations that mate a single part with a subassembly; Woo [21] and Wolter [20] call this the *linear* case, while De Fazio and Whitney [7] call it "precluding a plurality of unconnected subassemblies." Chapter 10 examines the general case, when two subassemblies can be placed together in an assembly operation.

## 9.1 GRASP

We have implemented an assembly sequence planner called GRASP[1]. GRASP takes as input just the 3D geometric models of a set of parts comprising a target assembly, automatically builds a model of the assembly including the parts and contact relations between them, and finally generates an AND/OR graph representing a set of possible assembly sequences.

### 9.1.1 Building the Assembly Model

Most of the geometric information in GRASP is contained in the description of the goal assembly, which is built and maintained using Vantage [1], a 3D polyhedral modeling system. A user defines the geometry and position of each part using constructive solid geometry, then the modeler creates a boundary

[1]GRASP stands for "Geometric Reasoning Assembly Sequence Planner." It has nothing to do with grasp planning.

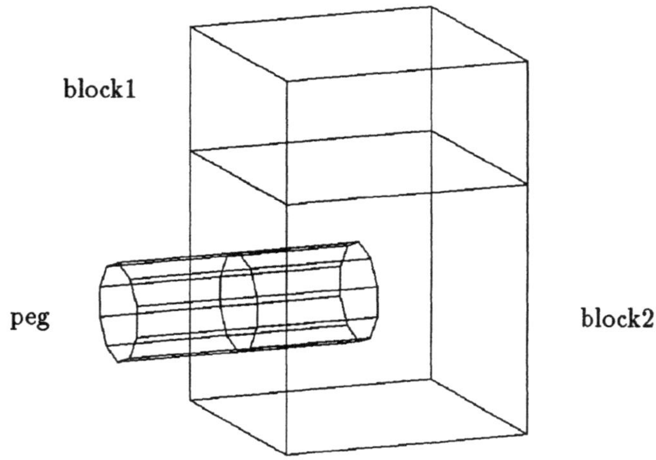


```
(block2 (name "Block 2")
        (assembly blocks-and-peg)
        (b-rep block2z)
        (part-number 2)
        (contacts ((peg (cylinder-contact (1.0 0.0 0.0)
                                          (0.0 20.0 20.0))
                        (planar-contact (-1.0 0.0 0.0)))
                   (block1 (planar-contact (0.0 0.0 -1.0))))))
```

Figure 9.1: GRASP's contact representation

representation from the CSG tree. The boundary representations for the parts of the target assembly are the only input to GRASP.

From the part models, the procedure *COMPLETE-ASSEMBLY* deduces all the contacts between faces of parts and records the contact information in a connection graph, where each node corresponds to a part and each edge is a contact between two parts. The connection graph, much like Homem de Mello's relational model [8], forms the basis of the planning process. In a real model of an assembly, contact information could be ambiguous due to tolerances or small distances between parts, requiring explicit human clarification. However, since we assume no tolerances on our parts, GRASP constructs its relational model of the assembly autonomously.

To find contacts between parts, GRASP checks each pair of surfaces from different parts for a possible contact. For every pair of surface types, a special-purpose routine determines whether there is contact between an instance of each; for example, two planar faces are in contact if they are coplanar, have opposing normals, and intersect in their common plane. Routines have been

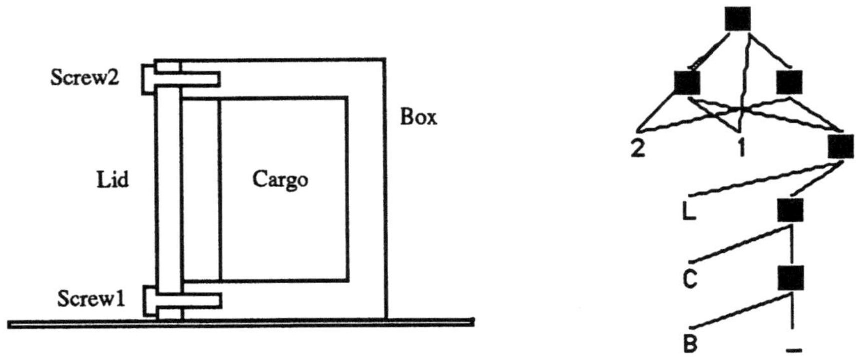


Figure 9.2: A crate assembly and its AND/OR graph

written to check for face-face, face-cylinder, and cylinder-cylinder contacts. Although Vantage approximates curved surfaces with faces, it remembers the parameters of the real surfaces, allowing accurate detection of curved contacts.

The contacts for each part and parameters describing each contact—the inward-pointing normal for a planar contact, the direction and position of the axis for a cylindrical contact—are stored in the connection graph of the assembly. Figure 9.1 shows the Vantage drawing of a three-part assembly and the representation of one part and its connections. Once the connection graph has been constructed, sequence planning can begin.

### 9.1.2 Building the AND/OR Graph

GRASP adopts Homem de Mello and Sanderson's AND/OR graph representation of the space of possible assembly sequences (Chapter 6). Associated with each node in the graph is a partial assembly that might be reached in the process of building the final product. An AND-arc represents the operation of putting two child assemblies together to make the parent, while OR-arcs give different ways of creating the same parent assembly. Thus each AND-subtree of a full AND/OR graph represents a partial assembly order for the product. Figure 9.2 shows a simple assembly and the AND/OR graph GRASP generated for it.

Figure 9.3 gives the main algorithm of GRASP, which follows loosely the one given by Homem de Mello and Sanderson in Chapter 7. It builds the AND/OR graph from the top down, beginning at the goal assembly. This corresponds to the *disassembly heuristic* of taking the assembly apart, then reversing the order to find an assembly plan. Since GRASP only models free flying rigid parts with no uncertainty, removing a part is exactly the reverse of placing it.

Before expanding a node in the AND/OR graph, GRASP checks for instability

Procedure *GENERATE-AND/OR-GRAPH*(goal-assembly)
  open ← {goal-assembly}
  until empty(open)
    $S$ ← pop(open)
    If *STABLE*($S$)
      For each part $p \in S$
        If *CONNECTED*($S - p$) and *MOVABLE*($p, S$)
          expand($S, p, S - p$)
          unless expanded($S - p$)
            push($S - p$, open)

Figure 9.3: Main algorithm of GRASP

of the corresponding assembly using techniques described below. If it is stable, the planner considers removing each part in the assembly. GRASP verifies that removing the part will leave the rest of the assembly connected, then calls the procedure *MOVABLE* to determine whether a geometrically feasible path exists to remove the part from the assembly.

### 9.1.3 Local Motion

*MOVABLE* performs two types of geometric calculations to find a path to remove a part. The first is a local-motion check based on computing the part's local translational freedom. If a part can be removed, it can be moved a very small distance; conversely, if no small motion is allowed by the part's contacts, it cannot be removed.

A part's local translational freedom is the set of directions in which the part can translate an infinitesimal distance from its current position, given the geometry of the rest of the assembly considered as a solid. We can compute the local translational freedom of a part $p$ by analyzing $p$'s contacts with other parts (figure 9.4). Each planar face in contact restricts the translational freedom of $p$ to a half-space on one side of that plane; each cylindrical contact restricts $p$ to translate along the axis of the cylinders. The intersection of all the spaces of freedom is the part's local freedom. In general, the local translational freedom of a part in three dimensions takes the form of a convex cone as in figure 9.5. If the cone is not null, then we say that the part is *locally free* in that assembly.

### 9.1.4 Global Motion

If a part is not fully constrained by its contacts with the rest of the assembly, GRASP tries to sweep the part in some of the legal directions from its current

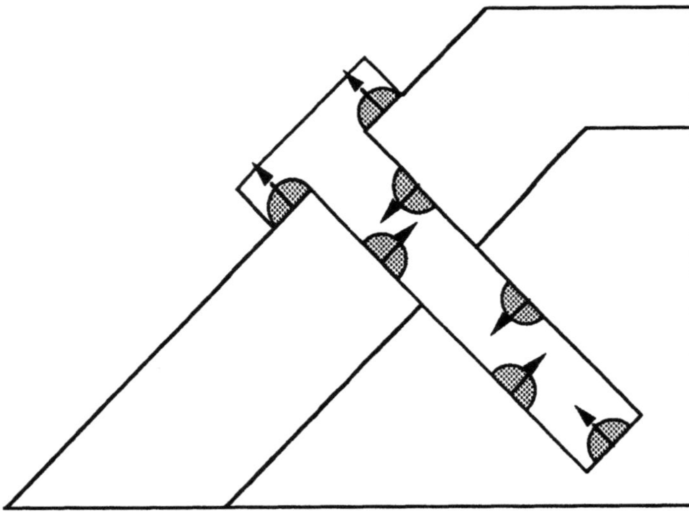

Figure 9.4: Local freedom computation

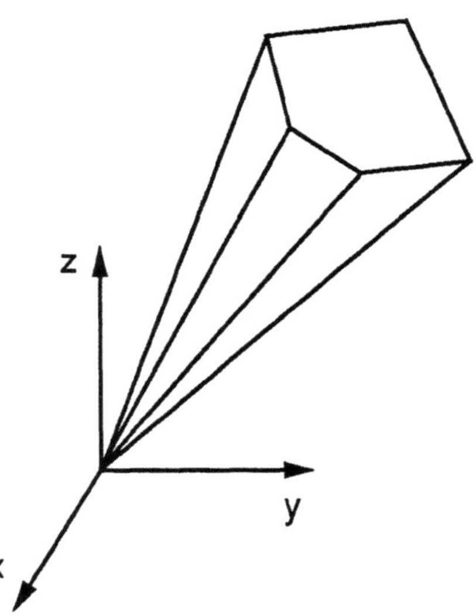


Figure 9.5: A 3D convex cone of removal directions

position to infinity. The directions to sweep are chosen heuristically based on the shape of the local freedom cone. For instance, for a half space, the normal of the plane facing into the half-space and four perpendicular directions in the bounding plane of the half space are chosen as sweeping directions. For a cone such as the one in figure 9.5, GRASP sweeps along vectors parallel to the edges of the cone.

To sweep a part, the faces of each possible interfering part are compared pair-wise with the faces of the translating part to check for collision. If the two faces intersect when projected into the plane perpendicular to the vector of translation, and the face being swept is "behind" the interfering face at one of the points of intersection, then a collision exists. This sweeping check is quite expensive to calculate, but it is necessary to ensure the global validity of assembly operations.

If one of the chosen directions is free from collision with all other parts present in the assembly, it constitutes a valid insertion path for the part. Note that even when all sweeping directions result in collision, a straight path along another vector or a bent path might exist. To find such insertion paths, a more sophisticated motion planner [12, 18] could be called as discussed in Section 9.5.2, but this could be expensive. In Chapter 7, Homem de Mello and Sanderson mention the creation of *virtual contacts* to express global constraints between two parts in the assembly model. Since such constraints must come from either human input or a calculation similar to sweeping, we chose sweeping as a simple solution.

To minimize sweeping computations, GRASP saves the result of each sweep for later retrieval. Previous-sweeps[$p_1$, $p_2$] is a two-dimensional array holding a list of pairs ($d$, *collides*). When a part $p_1$ needs to be swept against part $p_2$ in direction $d_1$, previous-sweeps[$p_1$, $p_2$] is searched for a pair whose first element is $d_1$. If one exists, *collides* is *true* if $p_1$ hits $p_2$ in direction $d_1$, and *false* if not. If no pair ($d_1$, *collides*) is found, $p_1$ is swept against $p_2$ as above, and the result is stored in the table. We call this method *sweep caching*, and in our experiments it has accelerated planning considerably.

### 9.1.5 Stability

GRASP has a limited ability to reason about the environment in which assembly will happen. The user can describe fixed objects that are not part of the target assembly, for instance a table and fixtures. A node in the AND/OR graph is then considered an undecomposable subassembly if it consists of only one part or only fixed objects. In this mode GRASP also does a fast stability check on each assembly. If the local translational freedom of any one part includes a downward component, the part is free to slide (figure 9.6). This algorithm is not exact; figure 9.6c shows a simple unstable assembly that the method does not identify because two parts will slide together. The general

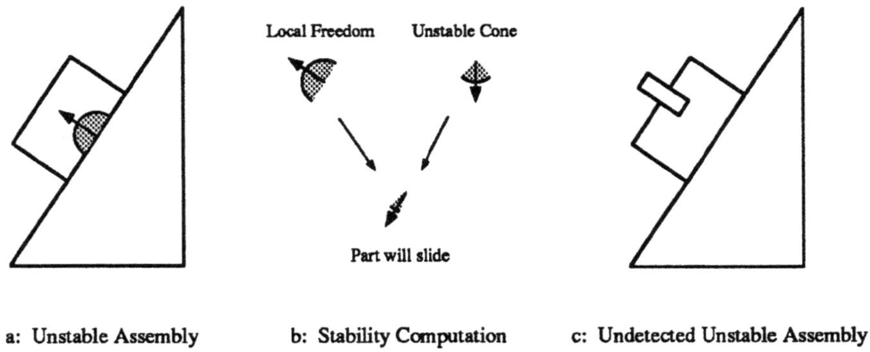


Figure 9.6: Single-part stability computation

stability problem is very difficult [5], but in practice the local stability check has proved to be quite valuable, pruning much of the search space and leaving few unstable assemblies.

Figure 9.7 shows GRASP planning for the assembly of an electric bell, discussed further in Section 9.4.2. It has generated the partial AND/OR graph seen in the upper left, and the graphics window shows the current operation it is adding to the graph: placement of the battery into its case.

## 9.2 Maintaining Movability Dependencies

The obvious algorithm given above for generating the AND/OR graph just checks the movability of each part at each node in the AND/OR graph. However, this repeats a great deal of computation about the geometry of the assemblies. Because there is little change in geometry between assemblies and their children in the AND/OR graph, most of the geometric reasoning about the movability of parts in the parent assembly should still be valid for the same parts in each child subassembly. For instance in the assembly of figure 9.2, screw2 is movable whether screw1 is present or not, while the cargo is not movable as long as the box and the lid are there, independent of screw1 and screw2. Essentially, we would like to exploit regularities in the geometry of assemblies and their subassemblies to reduce the geometric computation necessary to plan assembly sequences.

### 9.2.1 Principles of Dependency Maintenance

In the algorithm above there is a very weak link between the geometric reasoning modules and the symbolic reasoner constructing the AND/OR graph. For each query about the movability of a part in an assembly, the geometric

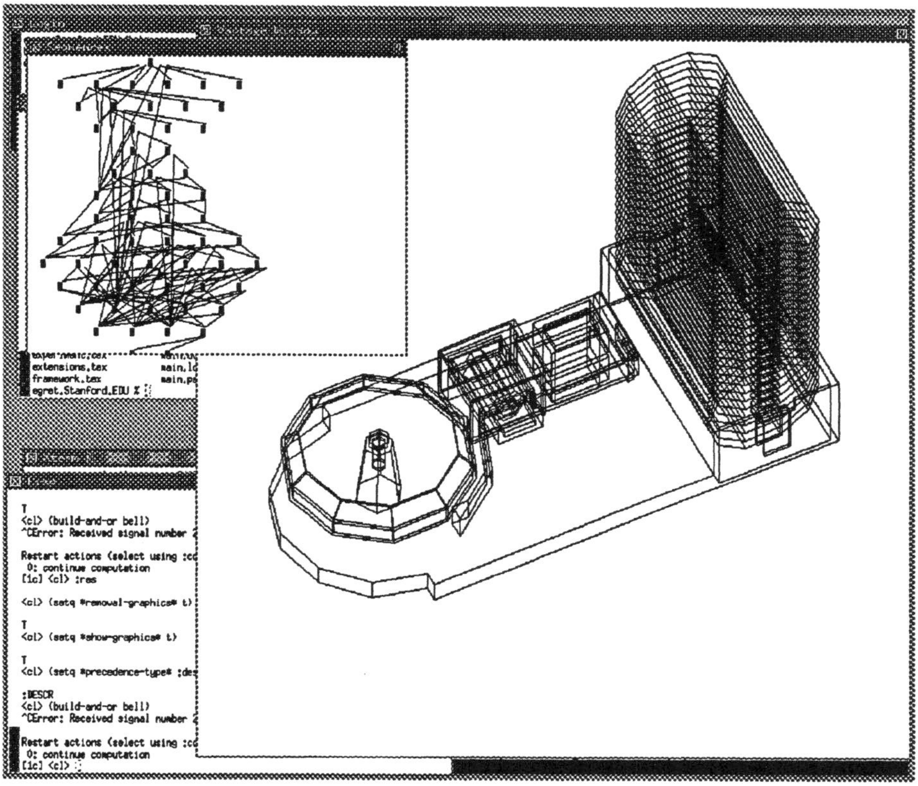


Figure 9.7: GRASP in operation

reasoner only answers that yes, the part is movable, or no, it is not. GRASP significantly reduces the number of geometric reasoning steps by having the geometric module return an expression stating the conditions under which the given part would be movable. When this *precedence expression* is still valid in descendants of the current assembly, evaluating it is usually much faster than performing a full geometric check.

In general, a precedence expression cannot fully describe the conditions of movability for a part. It only gives some sufficient and some necessary conditions. In order to manage this easily, we use a direct extension of the classical propositional calculus where each expression can take the value *true*, *false*, or *maybe*. The truth tables of this calculus are given in figure 9.8. The main property that we use is the following: given a proposition $M$ (for movability) that has $(N_i)_{i\in I}$ as necessary conditions and $(S_j)_{j\in J}$ as sufficient conditions, then the expression

$$(maybe \wedge \bigwedge_I N_i) \vee \bigvee_J S_j \tag{9.1}$$

| OR | T | ? | F |
|---|---|---|---|
| T | T | T | T |
| ? | T | ? | ? |
| F | T | ? | F |

| AND | T | ? | F |
|---|---|---|---|
| T | T | ? | F |
| ? | ? | ? | F |
| F | F | F | F |

| NOT | |
|---|---|
| T | F |
| ? | ? |
| F | T |

Figure 9.8: Truth tables for GRASP's three-valued propositional calculus

| $\mathcal{N}$ | $\mathcal{S}$[a] | $(? \wedge \mathcal{N}) \vee \mathcal{S}$ |
|---|---|---|
| T | T | T |
| T | F | ? |
| F | T | T[b] |
| F | F | F |

[a]In GRASP, conditions are always $T$ or $F$. When no necessary or sufficient conditions are present, $\mathcal{N}$ and $\mathcal{S}$ are set respectively to $T$ and $F$.

[b]In GRASP, only proven necessary or sufficient conditions are used. Thus this case — which should return a contradiction — does not appear.

Figure 9.9: Necessary and sufficient conditions represented in a single formula.

will be true whenever one of the $S_j$ is true, false whenever one of the $N_i$ is false, and *maybe* in all other cases (figure 9.9). Therefore, expression 9.1 embodies conditions on $M$, the movability of a part. This is the general form of a precedence expression. In the following, we will use the notation $M(p, A)$ to represent the movability of a part $p$ from an assembly $A$.

Passing down movability properties from an assembly to its subassemblies can be implemented by passing down precedence expressions. To prove that the precedence expression $M(p, S)$, inherited by a subassembly $S$ of $A$, still denotes movability in $S$, we only need to have implications of the form:

$$\begin{aligned} \forall S \subseteq A, \forall p \in S, \qquad & \mathcal{S} \Rightarrow M(p, S) \\ & \neg\mathcal{N} \Rightarrow \neg M(p, S) \end{aligned} \tag{9.2}$$

where $\mathcal{N}$ and $\mathcal{S}$ are necessary and sufficient conditions on the movability of $p$ in $A$. Several types of expressions of this form are given in Section 9.3.

We thus replace the *MOVABLE* procedure, called in the main algorithm, by a more sophisticated version given in figure 9.10. *MOVABLE* must return *true* if a part $p$ is removable from an assembly $S$ and *false* if it is not. It also has the side effect of setting the precedence expression of a part, which is *maybe* by default. Real geometric computation only occurs in the *PRECEDENCE* procedure, which is called when the movability of the part cannot be deduced symbolically from the inherited precedence expression.

Notice that in general, each node in the AND/OR graph has many parents, and so the choice of parent assembly from which to inherit is arbitrary. It would be possible to combine the precedence expressions from the different parents for the child node's expression, sometimes saving more geometric com-

Procedure $MOVABLE$(p,S)
    $A \leftarrow A\text{-}PARENT\text{-}OF(S)$
    $M(p,S) \leftarrow M(p,A)$
    If $M(p,S)$ evaluates to
        **False**: return False
        **True**: return True
        **Maybe**: $M(p,S) \leftarrow PRECEDENCE(p,S)$
            If $M(p,S)$ evaluates to True
                return True
            Else return False

Figure 9.10: Procedure *MOVABLE*, taking advantage of precedence expressions

putation than with the single-inheritance method. However, it is not clear that the savings would outweigh the extra overhead and complexity of combining precedence expressions.

## 9.3 Precedence Expressions

We tested three kinds of precedence expressions, called *simple*, *contact*, and *local* precedence expressions. They are increasingly complex and accurate in describing symbolically when a part is movable. They are all built from atomic propositions $P_i$, each of which represents the assertion that part $i$ is present in the assembly under consideration.

### 9.3.1 A Simple Sufficient Condition

When parts are removed from an assembly $A$, the free space for the remaining parts is broadened. Thus if a part is movable in $A$ it is guaranteed to be movable in any subassembly $S$ of $A$:

$$\forall S \subseteq A, \forall p \in S, M(p,A) \Rightarrow M(p,S) \qquad (9.3)$$

The *simple* type of precedence expression takes advantage of this. Homem de Mello [8, page 168] mentions the possibility of performing a check similar to the one that *simple* precedence expressions achieve, but does not elaborate.

From equation 9.3, whenever a part $p$ is movable in an assembly $A$, *true* is a sufficient condition for $M(p,S)$. Therefore, from expression 9.1, the precedence expression is set trivially to *true*. When subassemblies inherit their precedence expressions from $A$, geometric computation will not have to be done for parts that were movable in $A$. On the other hand, the movability computation will need to be redone in subassemblies of $A$ for each unmovable part. For instance, after expanding the root node in the crate assembly in figure 9.2, the simple precedence expression for screw2 will be *true*, and will not need to be recomputed in the subassembly with screw1 removed.

### 9.3.2 A Necessary Condition on the Parts in Contact

In the next version, *contact* precedence expressions, the geometric module supplies the planner with a list of parts that constrain a part $p$ in the assembly $A$. This is a list of all parts $p'$ in $A$ such that either $p'$ is in contact with $p$ or in at least one chosen direction of sweeping, $p$ collides with $p'$. In subassemblies of $A$, movability does not need to be recomputed when all of these parts are still in the subassembly:

$$\forall S \subseteq A, \forall p \in S, \neg M(p,A) \wedge \bigwedge_{q \in \mathcal{C}(p,A)} q \in S \Rightarrow \neg M(p,S) \tag{9.4}$$

where $\mathcal{C}(p,A)$ is the set of parts in $A$ in contact with or swept into by $p$.

From equations 9.2 and 9.4, we can infer that when $p$ is not movable in $A$, $PRECEDENCE(p,A)$ must return an expression

$$M(p,A) = maybe \wedge \neg \bigwedge_{q \in \mathcal{C}(p,A)} P_q$$

In subassemblies $S$ this expression will evaluate to *false* as long as all of the original contacting parts $\mathcal{C}(p,A)$ are present. When one of them is removed, the truth value of a $P_q$ will become false, causing $M(p,S)$ to evaluate to *maybe*. Geometric computation will then have to be done for $p$.

In addition, $PRECEDENCE(p,A)$ returns *true* when $p$ is movable, so the simple sufficient condition of the previous section is maintained.

For instance, after expanding the root node $A$ in the crate assembly, the contact precedence expression for the cargo in $A$ will be

$$M(cargo,A) = maybe \wedge \neg(P_{box} \wedge P_{lid})$$

because the cargo is constrained to move left by the box, and sweeps into the lid in that direction. In subassemblies resulting from removing one screw, the cargo will still be unmovable but no geometric calculation will be done.

### 9.3.3 Necessary and Sufficient Conditions

In the final and most complicated version, called *local* precedence expressions, the geometric reasoner returns a precedence expression more closely stating the conditions under which a part might be movable. The parts in contact with a part $p$ in $A$ are grouped such that all the parts in a group constrain the freedom of $p$ in the same way, either along the same plane or in parallel cylindrical contacts. Moreover, the parts swept into along one direction are also grouped together. In subassemblies of $A$, $p$ will not be movable unless all of the parts in one such group are missing. Furthermore, if all swept-into parts along a vector are missing, a sweep in that direction must be valid, and so $p$ is guaranteed to be movable:

$$\forall S \subseteq A, \forall p \in S,$$

$$\neg M(p,A) \quad \wedge \left[\bigwedge_{f \in \mathcal{F}(p,A)} \bigvee_{q \in f} q \in S\right] \wedge \left[\bigwedge_{d \in \mathcal{D}(p,A)} \bigvee_{q' \in d} q' \in S\right] \Rightarrow \neg M(p,S) \tag{9.5}$$

$$\neg M(p,A) \quad \wedge \neg \left[\bigwedge_{d \in \mathcal{D}(p,A)} \bigvee_{r \in d} r \in S\right] \Rightarrow M(p,S) \tag{9.6}$$

$\mathcal{F}(p,A)$ is a set indexed by facets of the freedom cone of $p$ in $A$, where $f$ is the set of parts constraining a given facet, and $\mathcal{D}(p,A)$ is a set indexed by directions of sweep, $d$ being the set of parts swept into when $p$ moves along one given direction.

Thus when a part is not movable, we have one necessary (9.5) and one sufficient (9.6) condition, so the full expression 9.1 applies and $PRECEDENCE(p,A)$ returns an expression

$$M(p,A) = \left[maybe \wedge \neg \bigwedge_{f \in \mathcal{F}(p,A)} \bigvee_{q \in f} P_q\right] \vee \neg \bigwedge_{d \in \mathcal{D}(p,A)} \bigvee_{r \in d} P_r$$

For example, the local precedence expression for the cargo after the expansion of the root node $A$ will be[2]

$$M(cargo,A) = [maybe \wedge \neg(P_{box})] \vee \neg(P_{lid})$$

and the local precedence expression for the lid after expanding $S$ will be

$$M(lid,A) = maybe \wedge \neg[(P_{screw1} \vee P_{screw2}) \wedge (P_{box})]$$

Since the lid is completely constrained by contacts, no sweep term is included in the local precedence expression. Notice that using contact precedence, GRASP would recompute the movability for the lid after removing one screw.

Local precedence expressions subsume contact ones. To see this, consider that the contact expression $C$ for a part $p$ lists the $P_i$ that are ground propositions

[2] Actually, GRASP does not simplify its precedence expressions, and so $P_{box}$ is listed three times because the box contributes three facets to the local freedom cone of the lid.

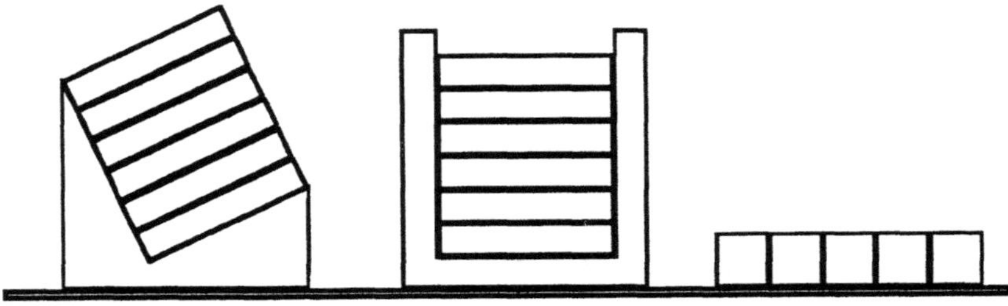


Figure 9.11: Three simple types of assemblies

in the local precedence expression $L$ for $p$. The truth value of $L$ can only change when the truth value of a $P_i$ changes to *false*, which would change $C$ to *maybe*. As a result, $L$ will be recalculated the same or fewer times than $C$ would have been.

## 9.4 Evaluation

We introduced precedence expressions in order to accelerate the process of assembly sequence planning. We have evaluated the three types of precedence expressions theoretically and experimentally and found that they shortened planning time considerably.

### 9.4.1 Theoretical Complexity

Because of the complex ways in which the geometry of an assembly can affect the size of its AND/OR graph, it is difficult to find meaningful bounds on the computation required to build it. For instance, given an assembly with $N$ parts, the number of nodes in the AND/OR graph can range from $2N-1$ when there is only one legal sequence, to $2^N-1$ when all sequences are legal. Below we analyze the complexity for three types of assemblies and for each type of precedence expressions. We assume that the number of calls $GC$ to the geometric reasoner is the overriding factor for the total running time of the algorithm to generate the AND/OR graph.

Consider the situation in which all parts are free to move in the initial assembly, but only one sequence satisfies stability considerations, as in Figure 9.11a. With $N$ parts, the AND/OR graph has $N-1$ non-terminal nodes. The time required to generate the graph using each type of precedence expression is:

**None** At each step in the generation of the AND/OR graph, all of the parts in the subassembly being considered must be checked for movability. Therefore, without precedence expressions $GC = \sum_{i=0}^{N-1} N - i = \frac{N(N+1)}{2} = O(N^2)$

**Simple** Using simple precedence expressions, the accessibility will be computed for the original N parts, finding each expression to be *true*. These expressions will be inherited downward, and no more geometric reasoning will be necessary. Thus $GC = O(N)$.

**Contact and Local** The complexity is the same as in the simple case.

In assemblies like the one in Figure 9.11b with $N - 1$ plates inside a box, only one sequence is valid because just one part is removable in each subassembly. Again the complexity depends on the type of precedence used:

**None** The obvious algorithm will again require $GC = O(N^2)$ calls to the geometric reasoner.

**Simple** In this case simple precedence gains us nothing. In each node, only one part is movable, so no *true* precedence expressions will be inherited. As a result, $GC = O(N^2)$.

**Contact** Since each plate $p_i$ is constrained by the box and parts $p_{i-1}$ and $p_{i+1}$, when we remove part $p_{i+1}$ only the contact precedence expressions of the box and part $p_i$ will evaluate to *maybe*, forcing a geometric call. Thus the number of calls will be 2 at each step except the last where only the box remains, so $GC = N + 2N - 1 = O(N)$.

**Local** The local precedence expressions for the plates will result in the same behavior as in the contact case. However, since the last plate $p_{N-1}$ contributes to each of the constraints on the box, the box's precedence expression will not evaluate to *maybe* until $p_{N-1}$ is removed, so $GC = N + N - 1 = O(N)$.

Finally, consider an assembly in which all sequences of assembly are valid, such as in Figure 9.11c. Without precedence expressions, the accessibility of every part in each of the $2^N - 1$ nodes would be computed. Using simple (or any other) precedence expressions, the accessibility would be found to be *true* for each part in the final assembly. This information would be inherited down the tree, making the total number of geometric calls N, even though there are $2^N - 1$ nodes in the AND/OR graph.

### 9.4.2 Experimentation

GRASP is implemented in Common Lisp, and runs on a DECstation 5000 under Allegro Common Lisp. We have planned the construction of about 10

| Precedence Type | Geometric Calls | Sweeps | Time in Seconds |
|---|---|---|---|
| none | 14 | 11 | 9.0 |
| simple | 12 | 9 | 8.8 |
| contact | 9 | 6 | 8.5 |
| local | 6 | 5 | 8.9 |

Table 9.1: Planning times for the crate assembly

assemblies with GRASP, including two-dimensional assemblies using the prototype previously written [19]. Table 9.1 shows the number of geometric calls required for the prototype to generate the full AND/OR graph for the crate assembly in figure 9.2 using each kind of precedence expression. The courageous reader can check it by hand to help understand the method. Note that the time to generate the graph was greater using local precedence expressions than with contact expressions; the geometry is so simple that the time required to create complex precedence expressions outweighs the savings.

A more interesting example is the *assembly from industry* (figure 9.12) with which De Fazio and Whitney [7] illustrate their method for generating assembly sequences. Figure 9.13 shows the assembly's *liaison diagram* as given there. It is a model of a transmission with 11 parts, 21 when the geometry of the bolts is explicitly represented. It is symmetric around an axis of revolution, and as such its geometry can be fully modeled in the two dimensions of the GRASP prototype. Figure 9.14 shows the assembly sequences for the transmission without bolts in De Fazio and Whitney's graphical representation of all valid liaison sequences. Single-part assemblies are not included in the diagram, and assemblies are shown by filling the box corresponding to each liaison that has been established in that assembly (liaisons 1-6 in the first row, etc.). For example, the leftmost assembly in the third row down has all connections established except for 4, 5, 16, 17, and 18; this corresponds to the assembly with all parts except K and L.

The set of sequences shown in figure 9.14 is not quite the same as the ones given in [7]. These differences are a result of GRASP generating its AND/OR graph from geometry alone, while De Fazio and Whitney compute their sequences from *precedence constraints* incorporating human geometric and mechanical insight. For example, De Fazio and Whitney find six possible ways to start the assembly process; GRASP finds eight (the bottom row of figure 9.14). One of these, the assembly consisting of parts C and D, cannot result in a finished assembly because the bolts connecting C to A are not accessible when C and D are connected. Because the bolts are not represented explicitly, GRASP cannot take this into account. However, when GRASP is run on the full model including bolts, it does not find any sequences using the assembly of only C and D.

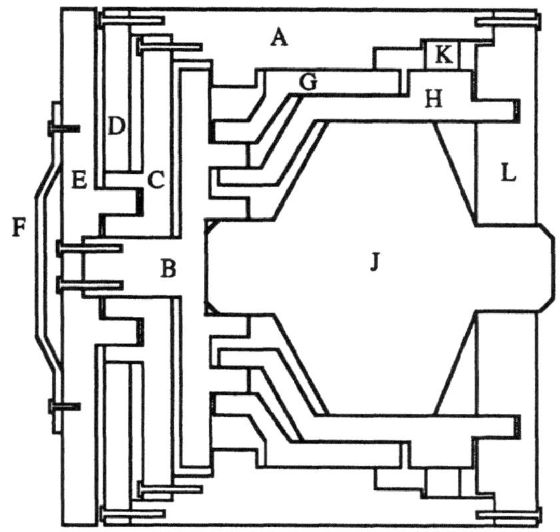


Figure 9.12: De Fazio and Whitney's transmission

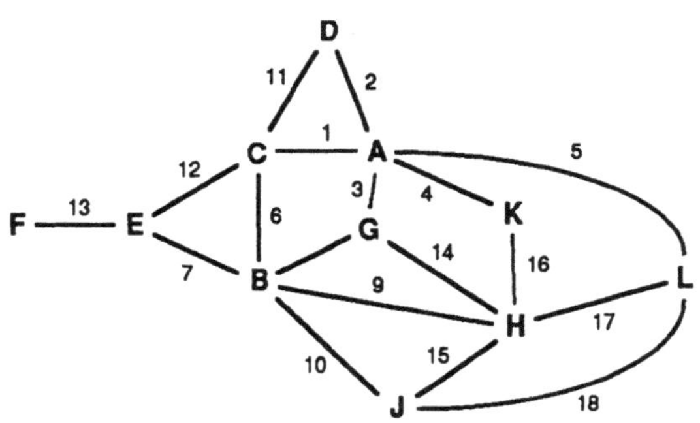


Figure 9.13: Liaison diagram for the transmission

| Precedence Type | Geometric Calls | Sweeps | Time in Seconds |
|---|---|---|---|
| none | 2508 | 26343 | 1151 |
| simple | 2035 | 20526 | 943 |
| contact | 669 | 6645 | 445 |
| local | 121 | 1193 | 99 |

Table 9.2: Planning times for the transmission, with bolts

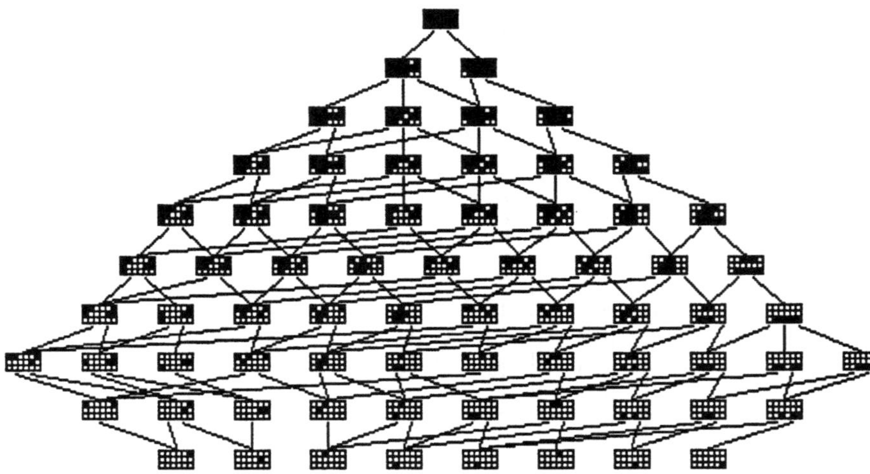

Figure 9.14: Assembly sequences for the transmission

| Precedence Type | Geometric Calls | Sweeps | Time in Seconds |
|---|---|---|---|
| none | 11125 | 581 | 578 |
| simple | 6377 | 559 | 325 |
| contact | 1369 | 559 | 113 |
| local | 638 | 559 | 60 |

Table 9.3: Planning times for the electric bell

The number of geometric calls and the time required for GRASP to generate the AND/OR graph for the transmission, with bolts as separate parts, is shown in table 9.2. Sweep caching is not used in the 2D prototype, and consequently the number of sweeps and total planning time is quite large. The resulting AND/OR graph has 295 subassembly nodes and 668 AND-arcs.

Figure 9.15 shows an assembly with which we have tested the full three-dimensional version of GRASP. It is an electric bell kit with 22 parts, not including the flexible wires that GRASP cannot reason about. Two contacts in the real bell are threaded, but for these experiments GRASP models them as pegs. The AND/OR graph representing all the linear sequences of assembly for the bell has 1389 nodes and 5486 AND-arcs. Table 9.3 shows the number of geometric calls and the time required to generate the AND/OR graph for the bell assembly.

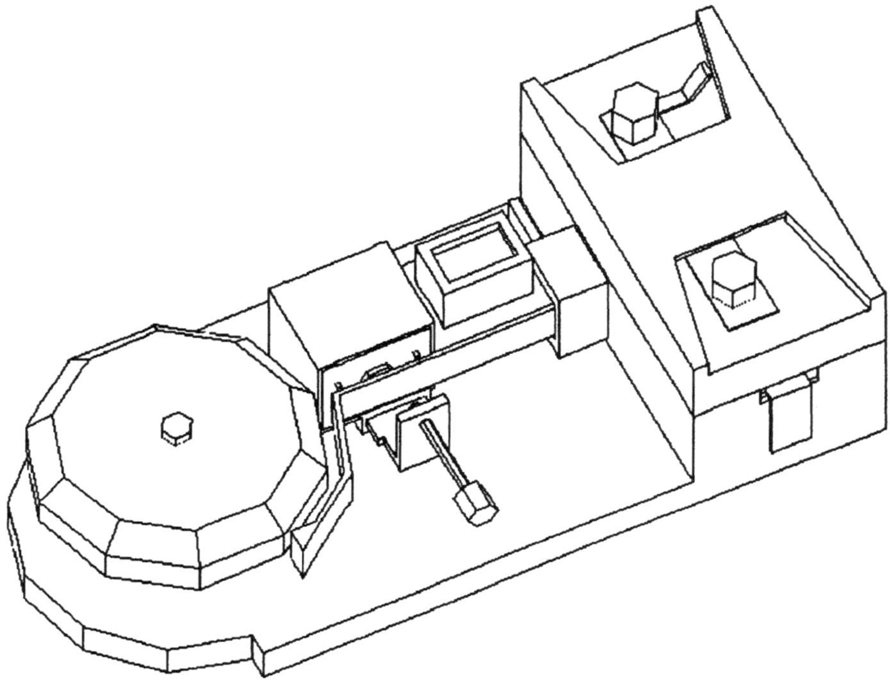

Figure 9.15: The electric bell

## 9.5 Possible Extensions

We have used precedence expressions to encode the conditions under which a single part will have the same local freedom or set of swept-into parts. However, precedence expressions are general enough to encode other geometric results, and with some modifications could be used in planners that have significant differences from GRASP.

### 9.5.1 Non-linear Sequence Planning

All the algorithms we have given are easily extensible to the non-linear case, where two subassemblies can be mated in the same operation. A quick justification is that a subassembly can be considered as a part for the purpose of movability, as long as it remains stable throughout the removal motion.

As far as dependency maintenance is concerned, the general mechanism and the *MOVABLE* procedure are identical. Furthermore, all the necessary and sufficient conditions (equations 9.3, 9.4, 9.5 and 9.6) are valid for a subassembly $S$ just as for a part $p$. The important thing to note is that they still depend on

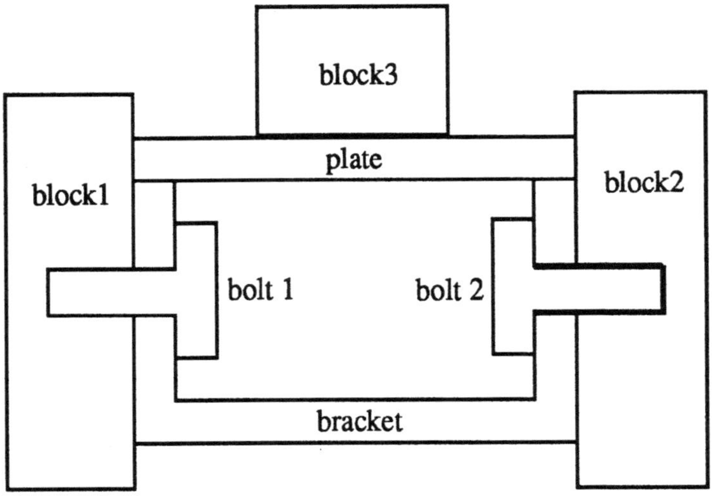


Figure 9.16: A bracket assembly

the presence of *parts* in contact or being swept. In other words, the precedence expressions are identical, with each atom denoting the presence of a part. Thus, after expanding the root node in the crate example of figure 9.2, the local precedence expression for the subassembly {*box,cargo,screw1*} would be:

$$M(\{box, cargo, screw1\}, A) = maybe \wedge \neg[(P_{lid} \vee P_{screw2}) \wedge (P_{lid})]$$

However, the theoretical validity of maintaining dependencies for subassemblies does not mean that the method is practical. Managing and evaluating precedence expressions is notably more costly in the non-linear case because the number of precedence expressions is larger. A node with $n$ parts can have as many as $2^n - 2$ expressions in the non-linear case (one for each subassembly) compared to at most $n$ in the linear case. Chapter 10 describes a more practical approach to speeding up non-linear sequence planning.

## 9.5.2 General Path Planning

A part's movability could be computed in other ways than contact analysis and sweeping. For example, a part may not be fully constrained by its contacts with other parts, yet the planner can find no straight path to disassemble it, as is the case with the bolts in figure 9.16. We might then call a more powerful path planner [3, 18] to try to find a curved path for the part.

In a sequence planner using a global path planner, precedence expressions could be incorporated by adding some functionality to the path planner. Checking for the movability of a part $p$ in assembly $A$ under this scheme, there are three cases to express:

- Contacting parts fully constrain the part. In this case local precedence expressions can be used.
- The global path planner finds a (possibly curved) path. We then set $M(p, A) = true$.
- The path planner cannot find a path. To construct the precedence expression for this case, the motion planner used must be modified to find a set of parts that together constrain $p$. The changes to the path planner will depend heavily on the planning technique employed.

  For instance, if the planner builds an adjacency graph based on a cell decomposition of the configuration space of $p$ [13, chapters 5-6], it may return a list of the parts contributing boundaries to the connected component that contains the starting position of $p$. If the motion planner is based on a local exploration of the configuration space [3] it may return the list of parts with which $p$ collided during the search process.

  A motion planner with such abilities would allow us to construct a contact precedence expression

  $$M(p, A) = maybe \wedge \neg \bigwedge_{c \in \mathcal{C}(p,A)} P_c$$

  where $\mathcal{C}(p, A)$ is the set of parts the motion planner returns as constraining $p$ in $A$. Furthermore, if we use sweeping as a first check for global motion, or our motion planner is able to find sets of parts whose removal would allow a path for $p$, we can build an expression

  $$M(p, A) = \left[ maybe \wedge \neg \bigwedge_{c \in \mathcal{C}(p,A)} P_c \right] \vee \neg \bigwedge_{d \in \mathcal{D}(p,A)} \bigvee_{r \in d} P_r$$

  where $\mathcal{D}(p, A)$ is a set indexed by possible paths or directions of sweep and $d$ is the set of parts $p$ collides with along one blocked path. In subassemblies, if all of the blocking parts $d$ are missing in one direction then that path is valid, while if any of the parts directly blocking a curved path are missing, then such a path might exist.

  For example, bolt1 in figure 9.16 would be given a precedence expression like

  $$M(bolt1, A) = [maybe \wedge \neg(P_{block1} \wedge P_{bolt2} \wedge P_{plate} \wedge P_{bracket})] \\ \vee \neg(P_{bolt2} \vee P_{bracket} \vee P_{block2})$$

## 9.6 Relation to Other Work

The assembly planning problem has been addressed by several authors. This paper builds upon the work of some of them, while the techniques found here may coordinate well with the ideas of others.

Bourjault [6] proposes a procedure to enumerate all the sequences of assembly for a product through a series of structured questions to a human. He uses a *subset* rule, somewhat like simple precedence expressions, and its contrapositive the *superset* rule to reduce the number of questions to the human. De Fazio and Whitney [7] drastically reduce the number of questions asked of the human by requiring each answer to state the situations in which one connection can be established. Baldwin [2] implements and compares these and other methods. Because a human is an integral part of these approaches, the emphasis has been on reducing the number of questions to the user and increasing the utility of each answer. It would be possible to have the human enter a reason for each decision, and use this somewhat like a precedence expression; however, it is not clear that doing so would be a more efficient method of entering constraints on assembly sequences.

Homem de Mello and Sanderson [9] introduce the AND/OR graph representation of the space of assembly sequences. Homem de Mello [8] also gives a rigorous method for local motion analysis in three dimensions. GRASP is built upon their framework and basic techniques of assembly planning. However, Homem de Mello and Sanderson do not concentrate on questions of speed and reducing computation. In this chapter we demonstrate that maintaining geometric results can improve the efficiency of their methods considerably.

The problem of finding assembly sequences has been addressed by other authors, but few attempt to completely automate geometric reasoning. Wolter [20] assumes that geometric reasoning has already resulted in a list of directions in which each part might move, along with lists of other parts that interfere with those motions. Ko and Lee [11] present a method for assembly planning but give no experimental results. Miller and Hoffman [15] use ray casting techniques much like GRASP's sweeping to derive assembly sequences, reasoning especially about fasteners. Their system finds one sequence of assembly for a mechanical mouse in 6-9 minutes, which it appears could be shortened using techniques such as precedence expressions.

A great deal of research has been conducted into the physical reasoning needed to accomplish single steps in an assembly plan. Motion planning to assemble single parts or subassemblies has been studied by a number of authors [12, 14, 16, 18]. Palmer [17] showed that in general deciding whether an assembly is stable is an NP-complete problem. Grasp [10] and fixture planning [4] will also be crucial to the development of competent assembly planners. Since these authors generally consider one step in the planning process, they do not address the issue of saving computation between steps. However, we believe precedence expressions are general enough that they can be applied to many different geometric reasoning methods.

# Conclusion

The task of assembly planning is very dependent on geometry. As a result, capable automatic assembly planners will need to incorporate powerful yet fast geometric reasoning methods. In addition, the assembly planning problem imposes special requirements on the algorithms used. In this work we have explored the close links between assembly planning and geometric reasoning and proposed a compromise solution that allows a computer to generate assembly sequences strictly from the geometry of the target assembly, with no human input.

This compromise includes checking the movability of each part in a target assembly by analyzing the part's contacts to find its local translational freedom, then sweeping the part along chosen directions in the resulting cone to ensure global validity of the paths found. Though not as expensive as calling a motion planner, these computations are numerous and costly. However, it is possible to exploit the similarity between the geometry of an assembly and that of its subassemblies to replace, in many cases, a geometric calculation by a symbolic one, using previously computed results. Encoding results in precedence expressions for later use yields large improvements in running times and the number of calls to the geometric reasoner in our experiments on real assemblies.

While our experimental results can only justify the use of precedence expressions with the current geometric reasoning techniques, the method is much more general. It could be used to save computation when assembly operations are not limited to single-part insertions, or when it is necessary to call a more time-consuming motion planning algorithm. Furthermore, precedence expressions pinpoint the kinds of information that are both obtainable with existing geometric reasoning methods and profitable once put in symbolic form. As a link between the two levels, precedence expressions are a prototype of the richer forms of communication between geometric and symbolic reasoning methods that will be necessary to solve real-world planning problems efficiently.

# References

[1] P. Balakumar, J.-C. Robert, R. Hoffman, K. Ikeuchi, and T. Kanade. *VANTAGE: A Frame-Based Geometric Modeling System - Programmer/User's Manual V1.0.* The Robotics Institute, Carnegie Mellon University, 1989.

[2] D. F. Baldwin. Algorithmic methods and software tools for the generation of mechanical assembly sequences. Master's thesis, Massachusetts Institute of Technology, 1990.

[3] J. Barraquand and J.-C. Latombe. Robot motion planning: A distributed representation approach. *To appear in International Journal of Robotics Research*, 1991.

[4] J. J. Bausch and K. Youcef-Toumi. Kinematic methods for automated fixture reconfiguration planning. In *Proceedings of the IEEE International Conference on Robotics and Automation*, pages 1396–1401, 1990.

[5] N. Boneschanscher, H. van der Drift, S. J. Buckley, and R. H. Taylor. Subassembly stability. In *Proceedings of the National Conference on Artificial Intelligence*, pages 780–785, 1988.

[6] A. Bourjault. *Contribution à une approche méthodologique de l'assemblage automatisé: élaboration automatique des séquences opératoires.* PhD thesis, Faculté des Sciences et des Techniques de l'Université de Franche-Comté, 1984.

[7] T. L. De Fazio and D. E. Whitney. Simplified generation of all mechanical assembly sequences. *IEEE Journal of Robotics and Automation*, RA-3(6):640–658, December 1987. Errata in RA-4(6):705-708.

[8] L. S. Homem de Mello. *Task Sequence Planning for Robotic Assembly.* PhD thesis, Carnegie Mellon University, 1989.

[9] L. S. Homem de Mello and A. C. Sanderson. AND/OR graph representation of assembly plans. Technical Report CMU-RI-TR-86-8, Robotics Institute - Carnegie-Mellon University, 1986.

[10] J. Jones and T. Lozano-Perez. Planning two-fingered grasps for pick-and-place operations on polyhedra. In *Proceedings of the IEEE International Conference on Robotics and Automation*, pages 683–688, 1990.

[11] H. Ko and K. Lee. Automatic assembling procedure generation from mating conditions. *Computer Aided Design*, 19(1):3–10, February 1987.

[12] A. Koutsou. *Planning Motion in Contact to Achieve Parts Mating.* PhD thesis, University of Edinburgh, 1986.

[13] J.-C. Latombe. *Robot Motion Planning.* Kluwer Academic Publishers, Boston, November 1990.

[14] T. Lozano-Perez. Spatial planning: A configuration space approach. *IEEE Transactions on Computers*, C-32(2):108–120, 1983.

[15] J. M. Miller and R. L. Hoffman. Automatic assembly planning with fasteners. In *Proceedings of the IEEE International Conference on Robotics and Automation*, pages 69–74, 1989.

[16] D. K. Pai and B. R. Donald. On the motion of compliantly-connected rigid bodies in contact, part I: The motion prediction problem. Technical Report 89-1047, Computer Science Department - Cornell University, October 1989.

[17] R. S. Palmer. *Computational Complexity of Motion and Stability of Polygons.* PhD thesis, Department of Computer Science - Cornell University, 1989.

[18] J.-M. Valade. Geometric reasoning and automatic synthesis of assembly trajectory. In *Proceedings of the International Conference on Advanced Robotics*, pages 43–50, 1985.

[19] R. H. Wilson and J.-F. Rit. Maintaining geometric dependencies in an assembly planner. In *Proceedings of the IEEE International Conference on Robotics and Automation*, pages 890–895, 1990.

[20] J. D. Wolter. *On the Automatic Generation of Plans for Mechanical Assembly.* PhD thesis, The University of Michigan, 1988.

[21] T. C. Woo. Automatic disassembly and total ordering in three dimensions. In *Conference on Intelligent and Integrated Manufacturing Analysis and Synthesis*, pages 291–303, 1987.

# Chapter 10

# Efficiently partitioning an assembly

**Randall H. Wilson**

The previous chapter introduces the assembly planner GRASP and describes a method to efficiently build an AND/OR graph representing the linear assembly plans for a product. However, the best assembly sequence might not be linear, and indeed some assemblies cannot be built one part at a time. This chapter introduces an algorithm to facilitate generating non-linear assembly sequences quickly.

Specifically, I present an algorithm that efficiently solves the *physical partitioning* problem. Physically partitioning an assembly consists of finding all ways to divide the parts of an assembly into two subassemblies, such that the operation of bringing them together is physically feasible. In Chapter 9 the partitioning problem is simplified to the question of whether any one part can be placed in the assembly; here the general case of placing a subassembly is considered.

Other researchers [2, 3] have described complete algorithms based on a generate-and-test approach that can be very slow for many assemblies. Here I introduce a new algorithm to partition assemblies, prove its correctness and completeness, and show a worst-case time bound that is polynomial in the number of features in the assembly and the number of feasible decompositions.

GRASP, the assembly planner described in Chapter 9, has been extended to include the new partitioning algorithm, and experimental results are shown for real assemblies.

As in the previous chapter, the following presentation assumes that all parts have a rigid, exact, certain geometry; that all parts are assembled along a single translation directly to their final relative positions and not moved later; and that no assembly operation mates more than two subassemblies.

## 10.1 Get-Feasible-Decompositions

An obvious method to physically partition an assembly is to generate all divisions of the parts of the assembly into two sets, then test the assembly operation bringing each pair of subassemblies together according to a set of constraints. If the assembly has $n$ parts, there are $2^n$ partitionings to test, making this approach prohibitive. However, the efficiency of the algorithm can be improved by integrating feasibility considerations into the generating stage, so that many infeasible decompositions are never generated.

In Chapter 7, Homem de Mello and Sanderson present an algorithm to physically partition assemblies, *GET-FEASIBLE-DECOMPOSITIONS*, that generates all cut-sets of the assembly's connection graph, then tests the assembly operation corresponding to each cut-set for physical realizability using the predicate *FEASIBILITY-TEST*. Since by definition a cut-set divides a graph into two connected components, the procedure only explicitly generates those partitionings that have two connected subassemblies. In other words, *GET-FEASIBLE-DECOMPOSITIONS* requires both subassemblies to be connected and includes this constraint in the generation process. Other criteria, including geometric interference, stability of the two partitions, and mechanical feasibility, are checked in *FEASIBILITY-TEST*.

Specifically, Homem de Mello and Sanderson show how to compute the local translational freedom of one subassembly with respect to the other. Recall from Chapter 9 that for a part with planar contacts, the set of possible infinitesimal translations forms a convex cone in three dimensions (figure 10.1), and if the cone is empty the part cannot be removed. To calculate the range of motion for a subassembly, we can consider it as a single part and calculate its cone of removal directions in the same way. Each small motion inside the cone is a *removal trajectory*, and if at least one exists, we say the two subassemblies are *locally free* from each other.

Although *GET-FEASIBLE-DECOMPOSITIONS* is more efficient than the obvious approach, many assemblies have a large number of cut-sets, and as a result it can still be very slow. When automating the method, Baldwin could not plan for assemblies with more than 11 parts because of this fact [2, page 96]. Wolter [8] limited assembly plans to single-part insertions because of the complexity

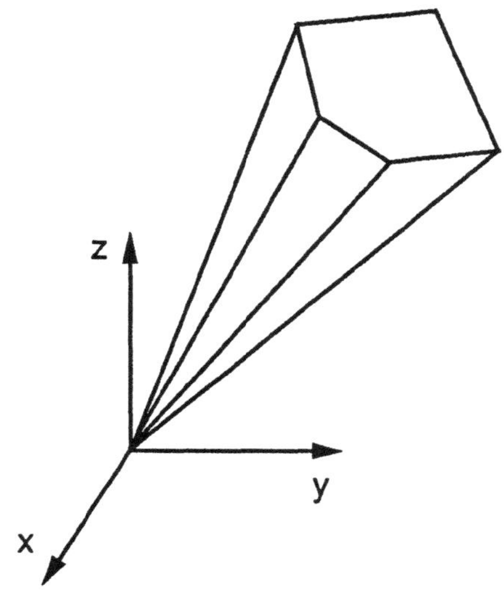


Figure 10.1: A 3D convex cone of removal directions

of generating decompositions, and GRASP originally focused on linear plans (as in Chapter 9) for the same reason. A faster method of physically partitioning an assembly will make non-linear assembly sequence planning feasible for much more complex assemblies. The algorithm presented in the next section integrates geometric interference constraints as well as connectedness criteria into the generation process, thereby achieving greater efficiency.

## 10.2 Partitioning Assemblies

For simplicity in the following description, assume that all parts meet only in planar contacts and that only translations are allowed to separate subassemblies. Section 10.6 shows how the method can be generalized to correctly and efficiently reason about cylindrical and threaded contacts, and motions including rotation.

The new algorithm to physically partition an assembly, *PARTITION*, gains efficiency by incorporating two necessary constraints on assembly decompositions into the generation process:

**Connectedness** The two subassemblies to be separated must each constitute a connected subassembly.

**Local freedom** A removable subassembly must be free to move a small distance with respect to the rest of the assembly.

These two constraints are encoded in the procedure *GROW-SUBASSEMBLY*, described below, which takes a small subassembly as input and adds to it all the parts that *must* be removed with it from the full assembly. Since *PARTITION* includes these necessary constraints in the generation process, it generates only the partitionings in which one subassembly is locally free from the rest of the assembly. As a result, far fewer decompositions must be tested against the predicate *FEASIBILITY-TEST*.

To accomplish this, *PARTITION* divides the physical partitioning problem into three parts:

- find a set of *critical directions* that is sufficient to allow all the locally free decompositions of the assembly to be disassembled
- for each critical direction, use *GROW-SUBASSEMBLY* to find all the subassemblies that are locally free from the rest of the assembly in that direction
- test each locally free decomposition with *FEASIBILITY-TEST*.

### 10.2.1 Finding Critical Directions

The procedure *CRITICAL-DIRECTIONS* takes an assembly $A$ as input and returns a list of possible removal trajectories $T$ such that for every decomposition of $A$ into $S$ and $A-S$ where $S$ is locally free from $A-S$, at least one of the trajectories in $T$ is included in the local freedom cone of $S$ with respect to $A-S$. In other words, if an assembly can be partitioned into two subassemblies that are locally free from each other, then one of the directions in $T$ will allow the subassemblies to be separated.

Computing this set of critical directions $T$ is quite simple. Any two subassemblies that meet in planar contacts will be separable, if at all, in a set of translations given by a 3D convex cone, as in figure 10.1. Since each face on a convex cone results from one or more planar constraints, every edge of a cone is at the intersection of two planes. Thus the set of critical directions $T$ includes two trajectories (a vector and its inverse) parallel to the intersection of each pair of planar contacts that exists in the assembly; duplicate vectors are removed. In the special case where all the contacts in an assembly are in parallel planes, *CRITICAL-DIRECTIONS* returns a single vector parallel to the plane, allowing the parts to slide in that plane. As a result, for every cone that could result from the planar contacts in the assembly, there exists a trajectory in $T$ that is a translation along one of its edges. Therefore the set $T$ is sufficient to allow removal of all locally free subassemblies of the assembly.

In general, with $n$ planar contacts in an assembly there are $O(n^2)$ removal directions in $T$. However, in practice many of these turn out to be parallel, and in most of our experiments *CRITICAL-DIRECTIONS* returns only a small number of trajectories. In fact, many assemblies can be built using only motions parallel to the major axes, and the step of generating critical directions might be skipped if it is known *a priori* that the target is such an assembly.

### 10.2.2 Generating Decompositions

The procedure *PARTITION* proceeds as follows. First, it calls *CRITICAL-DIRECTIONS* to enumerate a sufficient set of trajectories to remove all subassemblies from the assembly $A$. For each trajectory $t$, it starts one *seed partition* $S$ for each part in $A$, and calls *GROW-SUBASSEMBLY*$(S,t)$. *GROW-SUBASSEMBLY* adds to the subassembly $S$ all the parts that *must* be removed with $S$ if it is to be moved a small distance along trajectory $t$ away from a connected complement $A-S$. Thus when *GROW-SUBASSEMBLY* returns, one of two conditions must be true: (1) $S$ contains over half the parts in $A$, in which case *PARTITION* goes on to another seed partition (the smaller half will be found as a free subassembly in the reverse direction if it exists), or (2) $S$ is a subassembly that is locally free from $A$. If the latter is true and $S$ hasn't already been found in direction $t$, $S$ is put on the list of free subassemblies and new seed partitions are generated, each consisting of $S$ and one other part in contact with $S$. Then the procedure chooses another seed partition, and the process repeats. After generating all the subassemblies that are locally free from the rest of the assembly, the decomposition corresponding to each subassembly is checked against *FEASIBILITY-TEST* for global validity. Figure 10.2 gives a more precise description of *PARTITION*.

*GROW-SUBASSEMBLY* uses two functions to compute the generating constraints that add parts to a partial subassembly. The first, *LOCALLY-COMPATIBLE*, allows generation with regard to the local freedom constraint. If the current subassembly includes a part $p_1$ and a contact between $p_1$ and another part $p_2$ keeps $p_1$ from moving along the selected trajectory, then $p_2$ must be added to the subassembly to allow it to move. Therefore *LOCALLY-COMPATIBLE*$(p_1, p_2, t)$ returns *true* if and only if the local freedom of part $p_2$ with respect to $p_1$ includes trajectory $t$. It simply finds the contacts between $p_1$ and $p_2$ and returns *true* if all of them are compatible with the motion of $p_1$ in direction $t$. A planar contact interferes with a translation $t$ when the dot-product of $t$ with the outward normal of the contacted face of $p_2$ is negative.

The second function, *OTHER-PARTITIONS*, allows the use of connectedness as a generating constraint. *OTHER-PARTITIONS*$(S,A)$ returns a list of the connected components $C_i$ of the connection graph of assembly $A$ once the current subassembly $S$ has been removed. If $A-S$ is not connected, then there will be more than one $C_i$. It is clear that no superset $S'$ of $S$ will have a connected complement unless $S'$ contains all but one of the $C_i$. Furthermore, for $S'$ to encompass half of the parts of $A$ or less, the one $C_i$ not included in $S'$ must have half or more of the parts in $A$. If $A$ has $n$ parts, there can only be one connected component $C_i$ with at least $n/2$ parts. If each of the $C_i$ contains

```
Procedure PARTITION(full-assembly)
     feasible-decompositions ← {}
     all-directions ← CRITICAL-DIRECTIONS(full-assembly)
(1)  for each dir ∈ all-directions
          d-locally-free ← {}
          seed-stack ← {}
(2)       for each part ∈ full-assembly
               push({part}, seed-stack)
(3)       until empty(seed-stack)
               subassembly ← pop(seed-stack)
(4)            subassembly ←
                    GROW-SUBASSEMBLY(subassembly, dir, full-assembly)
(5)            if length(subassembly) ≤ length(full-assembly) / 2 and
                         subassembly ∉ d-locally-free
                    push(subassembly, d-locally-free)
                    for each part2 ∈ full-assembly - subassembly
                         if part2 contacts subassembly
                              push(subassembly ∪ {part2}, seed-stack)
(6)       for each S ∈ d-locally-free
               if {S, full-assembly - S} ∉ feasible-decompositions and
                         FEASIBILITY-TEST(S, full-assembly)
                    push({S, full-assembly - S}, feasible-decompositions)
     return feasible-decompositions
```

Figure 10.2: Procedure *PARTITION*

fewer than $n/2$ parts, then for no superset $S'$ of $S$ will $A - S'$ contain over half the parts of $A$. Thus, whenever $A - S$ has several connected components $C_i$, *GROW-SUBASSEMBLY* adds all the $C_i$ except the one with the most parts to $S$. In the case where the largest $C_i$ has fewer than $n/2$ parts, the resulting subassembly will be rejected after *GROW-SUBASSEMBLY* returns.

Note that because only parts in contact with $S$ are ever added to it, $S$ is always a connected partition and does not need to be checked. In addition, because the connected components $C_i$ only contact parts that are already in $S$, they cannot have any contacts with other parts that need to be checked with *LOCALLY-COMPATIBLE*. Figure 10.3 gives a more formal description of the procedure *GROW-SUBASSEMBLY*.

### 10.2.3 Checking Global Interference

Once locally free connected subassemblies are identified, each decomposition must be checked using the predicate *FEASIBILITY-TEST*. The global feasibility

```
Procedure GROW-SUBASSEMBLY(subassembly, dir, full-assembly)
    part-stack ← subassembly
    until empty(part-stack)
        part1 ← pop(part-stack)
        for each part2 contacting part1
            if part2 ∉ subassembly and
                    ¬ LOCALLY-COMPATIBLE(part1, part2, dir)
                push(part2, subassembly)
                push(part2, part-stack)
    others ← OTHER-PARTITIONS(subassembly, full-assembly)
    others ← others - largest(others)
    for each partition ∈ others
        subassembly ← subassembly ∪ partition
    return subassembly
```

Figure 10.3: Procedure *GROW-SUBASSEMBLY*

of motions in GRASP is computed by heuristically choosing a set of directions in the local freedom cone of the subassembly, and sweeping the subassembly against the rest of the assembly. One way to do this would be to construct a solid model of the subassembly and sweep it against the rest of the parts. In experiments, building this model turns out to be more expensive than sweeping all the individual parts. Chapter 9 explains GRASP's sweeping algorithms in more detail.

## 10.3 An Example

Figure 10.4 shows the crate assembly from Chapter 9 during various stages of the procedure *PARTITION*. The three dimensional assembly is shown in figure 10.4a, and a side view in 10.4b. For this example we will consider the "screws" as square pegs.

Each planar contact in the crate assembly is parallel to either the XY-plane, the YZ-plane, or the XZ-plane. The pairwise intersections between these planes yield vectors parallel to each of the major axes. Figure 10.4a shows the six vectors returned by *CRITICAL-DIRECTIONS*.

Starting with the trajectory +z, each part of the crate forms a seed subassembly. Consider the seed subassembly $S$ consisting of just the lid (figure 10.4c). *GROW-SUBASSEMBLY* now finds the parts in contact with the lid, which are screw1, screw2, and the box. *LOCALLY-COMPATIBLE*(lid, box, +z) returns *true*, since the dot-product of +z with +x, the normal of the planar contact, is 0; thus the box is not added to $S$. However, the bottom planar contact of

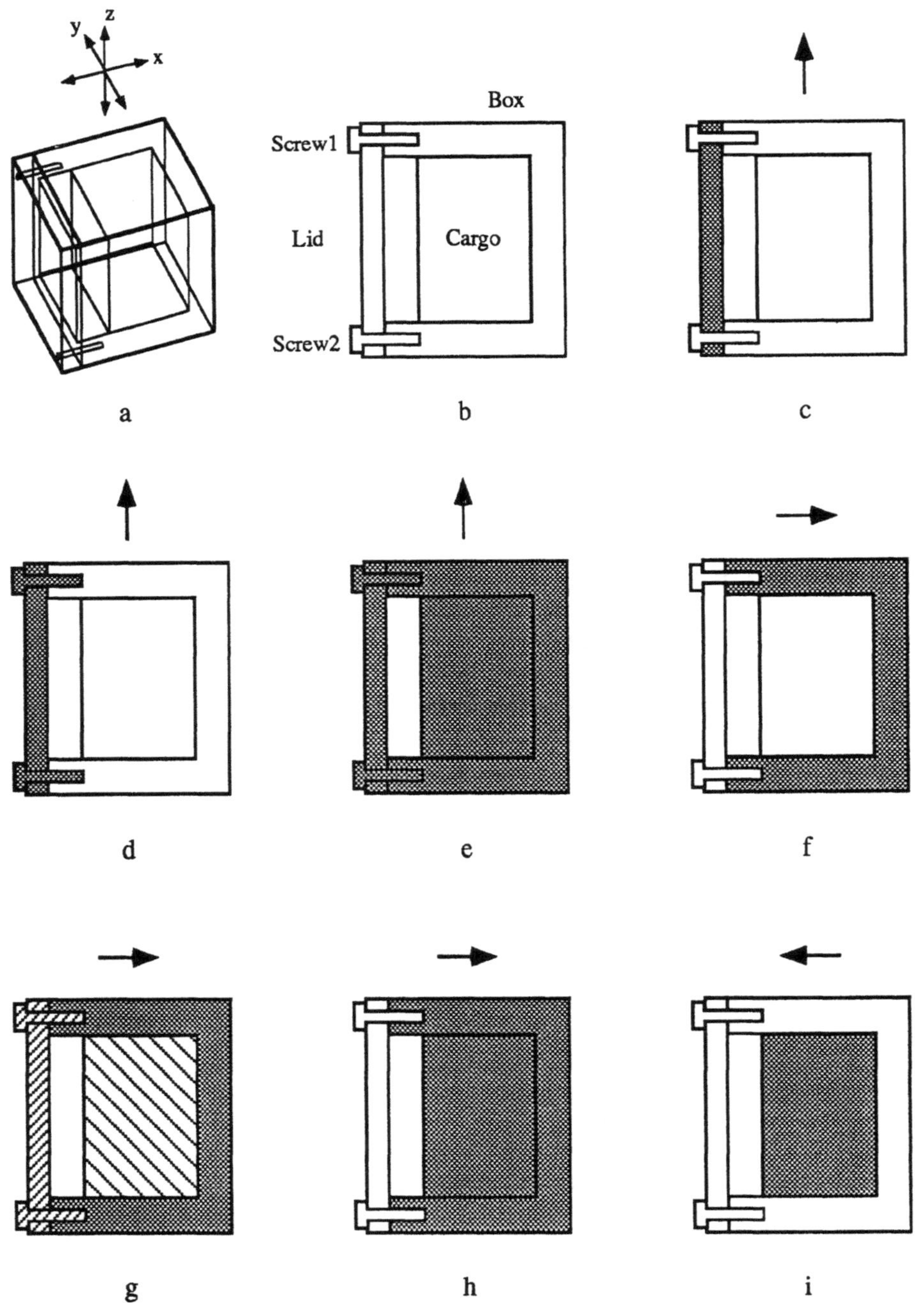


Figure 10.4: The crate during procedure *PARTITION*

the lid with screw1 prevents the lid from moving in the +z direction, and so *LOCALLY-COMPATIBLE*(lid, screw1, +z) returns *false* and screw1 is added to $S$, as is screw2 (10.4d).

Since screw1 has been added to $S$, its contacts with other parts are now examined. The box prevents screw1 from moving along the +z trajectory, so the box is added to $S$. Screw2 is considered, but all of its contacted parts are now in $S$ already. Finally, the bottom planar contact between the box and the cargo prevents the box from moving upward, and so the cargo is added to $S$ (10.4e). Since $S$ now contains the whole crate assembly, there are no connected components returned by *OTHER-PARTITIONS*($S$, Crate), and *GROW-SUBASSEMBLY* returns.

At this point $S$ contains more than half of the parts in the crate, so it is not entered in d-locally-free as a removable subassembly, and no seed subassemblies are generated from it. A similar process will happen with the seed subassemblies starting from each of the other parts of the crate: each will be grown to include the whole assembly. Thus no locally-free subassembly can be removed in the +z direction from the crate, and the same is found for the −z, +y, and −y trajectories.

Now consider the +x trajectory and the seed subassembly consisting of just the box (figure 10.4f). No contacts constrain the box in the +x direction, so *OTHER-PARTITIONS*({lid, box}, Crate) is called, and returns the connected components {lid, screw1, screw2} and {cargo} (10.4g). The component with fewer parts is added to $S$ (10.4h), and then $S$={box, cargo} is entered as a locally-free subassembly. The resulting seed partitions {box, cargo, lid}, {box, cargo, screw1}, and {box, cargo, screw2} each constitute over half the assembly, so they are eventually eliminated.

The last three decompositions of the crate result from the −x direction, since the seed partitions {screw1}, {screw2}, and {cargo} are all locally-free along the −x trajectory. However, when *FEASIBILITY-TEST* tries to find a global path to remove {cargo}, it finds that the cargo collides with the lid (10.4i). Therefore the three globally removable connected subassemblies found are {screw1}, {screw2}, and {box, cargo}.

## 10.4 Soundness and Completeness

The soundness of *PARTITION* follows immediately. Feasible-decompositions starts out empty, and only partitionings that satisfy *FEASIBILITY-TEST* are added to it. Therefore, as long as *FEASIBILITY-TEST* is accurate, *PARTITION* will only return decompositions that are physically feasible.

The partial completeness of *PARTITION* is also easy to show. The only loop that might not have finite iterations is that on line (3), which loops until the seed-stack is empty. The list of seeds starts with a finite number of assemblies, and each time through the loop, one is removed from it. Each seed added to the stack has more parts than the one removed. But since a subassembly is limited to one-half the number of parts in the full assembly, this process must end. Therefore *PARTITION* will terminate.

For *PARTITION* to be complete, the function *LOCALLY-COMPATIBLE* must have the ***locality*** property:

$$\forall t \forall S_1 \forall S_2 \; LC(S_1, S_2, t) \iff \forall S_3 \forall S_4 \, (S_3 \subseteq S_1 \wedge S_4 \subseteq S_2 \Rightarrow LC(S_3, S_4, t)) \tag{10.1}$$

Equation 10.1 states that a subassembly can be removed along a trajectory from a set of parts if and only if that trajectory allows removal from any subset of those parts. This property is true, specifically, for interference checking between rigid parts. If subassembly $S_1$ can be removed from $S_2$ using a certain path, then taking parts from $S_2$ will never cause an interference to appear, since the free space of $S_1$ will monotonically increase; furthermore, if all the individual parts do not interfere with $S_1$'s path, their sum will not either.

Note that the locality property does not hold true for many constraints on assembly sequences. For instance it very often happens that an assembly is stable but removing a part will leave it unstable. Fine motion planning and flexible parts also violate the locality property.

A special case of equation 10.1 is the fact that if a part $p$ interferes with the motion of a subassembly $S$ in a direction, then any set of parts including $p$ will also interfere with the motion of $S$:

$$\forall t \forall S_1 \forall p \forall S_2 \; \neg LC(S_1, \{p\}, t) \wedge p \in S_2 \Rightarrow \neg LC(S_1, S_2, t) \tag{10.2}$$

Therefore, when any part $p$ interferes with the removal of a subassembly $S$ along a trajectory, then any removable superset of $S$ must include $p$.

To show completeness of *PARTITION*, first remove the call to *GROW-SUBASSEMBLY* in line (4), and add a connectedness check to *FEASIBILITY-TEST*. The procedure that remains generates all possible decompositions of an assembly and checks each against *FEASIBILITY-TEST*, once for each critical direction of assembly. Each part in the assembly is a seed assembly, and every connecting part is added to each of these, and so on, thus generating every connected set of parts in the assembly. Each subassembly is checked for removal from the rest of the assembly using *FEASIBILITY-TEST*. Subassemblies for which the rest of the assembly is not connected will be caught by *FEASIBILITY-TEST*. This is equivalent to generating all cut-sets of the connection graph and testing them, so this modified algorithm is complete.

When line (4) is placed back in *PARTITION*, some of the subassemblies above are no longer generated. Take any subassembly $S$ that was generated by the modified algorithm and passed *FEASIBILITY-TEST*. At least one trajectory

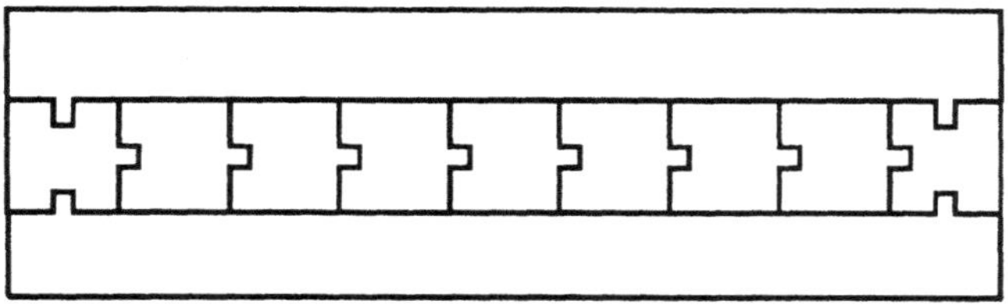

Figure 10.5: An assembly with 2 feasible decompositions

$t$ in the set of critical directions allows separation of $S$ from $A - S$ (Section 10.2.1). Along trajectory $t$, $S$ is locally-compatible with $A - S$, so by equation 10.2, all of the subassemblies of $S$ are locally-compatible with the parts of $A - S$. As a result, *GROW-SUBASSEMBLY* will never insist on adding parts from $A - S$ to the seed subassembly that built into $S$, and so this cut-set is not removed from the ones generated. Furthermore, because $A - S$ is connected and includes at least half the parts in $A$, $A - S$ will always be in the largest connected partition found by *OTHER-PARTITIONS*, and none of its parts will be added to $S$ for that reason. Therefore no parts of $A - S$ are added to $S$ by *GROW-SUBASSEMBLY*, and the decomposition into $S$ and $A-S$ is not removed from the ones generated. Since $S$ was taken to be any feasible decomposition of $A$, the procedure generates the same list of decompositions as the modified version above. *PARTITION* is therefore complete.

## 10.5 Complexity

Because in general the connection graph for an assembly with $n$ parts can have $O(2^n)$ cut-sets, the worst-case time complexity of *GET-FEASIBLE-DECOMPOSITIONS* is $O(2^n)$. For instance, the sandwich assembly in figure 10.5 has an exponential number of cut-sets, but only two feasible decompositions.

The geometry of assemblies with the same number of parts can vary a great deal, so another measure of assembly complexity is the number of mating features $m$ in the assembly. Since each center part in the sandwich assembly has a constant number of contacts with other parts, the number of parts is $n = cm$ for some constant $c \leq 1$. Therefore the worst-case time complexity of *GET-FEASIBLE-DECOMPOSITIONS* is at least $O(2^{cm})$.

Now consider *PARTITION*. Assume that there are $m$ mating features in the assembly, $n \leq m$ parts, and the assembly has $s$ decompositions that both satisfy *LOCALLY-COMPATIBLE* in at least one direction and have internally connected subassemblies. *CRITICAL-DIRECTIONS* will find one direction for every pair of planar contacts in the assembly, or $O(m^2)$ total directions in the worst case. Loop (1) will be executed once for each direction. Line (3) will loop once for each subassembly on the seed stack; since each seed is put on the stack either by line (2) or as the sum of a locally-free decomposition and one connecting part, loop (3) will execute at most $n + sn$ times for each direction. Checking each contact takes constant time, and each contact includes exactly two parts, so *LOCALLY-COMPATIBLE* will be called at most $2m$ times in one call to *GROW-SUBASSEMBLY*. Using a depth-first marking algorithm, *OTHER-PARTITIONS* also takes $O(m)$ time, so *GROW-SUBASSEMBLY* is $O(m)$. Line (5) requires looking through a list of up to $s$ subassemblies. In total, each time through loop (3) will take $O(m + s)$ time. Finally, loop (6) might take $s^2$ time to make sure duplicate decompositions are not in the final list. Combining these nested loops, *PARTITION* could take time

$$m^2(sn(m+s)+s^2) = snm^3 + s^2nm^2 + s^2m^2 = O(s^2m^3 + sm^4)$$

in the worst case, plus the time to evaluate *FEASIBILITY-TEST* on $s$ decompositions. In fact, in practice the time to compute the global constraints in *FEASIBILITY-TEST* become the dominant factor in running time (see Section 10.7).

In our implementation, d-locally-free and feasible-decompositions are kept as hash tables, making lookup times essentially constant. The effective complexity then becomes

$$m^2(sn(m)+s) = O(sm^4)$$

and if we assume that the number of mating features is proportional to the number of parts (which is not always the case), then *PARTITION* has time complexity $O(sn^4)$.

Note that if the target assembly has an exponential number $s$ of locally free subassemblies, *PARTITION* takes exponential time. For instance, if the center parts in the sandwich assembly (figure 10.5) were not interlocking, there would be $O(2^n)$ locally-free decompositions to test. Both *GET-FEASIBLE-DECOMPOSITIONS* and *PARTITION* must test all the partitionings they generate; the former generates all cut-sets and the latter all locally free decompositions. Because every locally free decomposition corresponds to a cut-set, *PARTITION* will always test the same number or fewer decompositions than *GET-FEASIBLE-DECOMPOSITIONS*. In our experiments, local freedom has proven a very strong constraint on decompositions, and as a result *PARTITION* generates a short list of partitionings to test.

## 10.6 Extensions to the Basic Algorithm

The version of *PARTITION* described above only considers parts that meet each other in planar contacts, and only uses those contacts and connectedness as generating constraints. The description has been limited to these simple assembly geometries partly to make the discussion more understandable to the reader. However, it is possible to include other types of assemblies and constraints in the algorithm without much modification. I have analyzed and implemented several extensions to *PARTITION*, allowing GRASP to handle cylindrical contacts, no connectedness constraint, and helical motions, while global sweeping constraints and general rotational motions have been investigated.

### 10.6.1 Cylindrical Contacts

To extend *PARTITION* to handle full cylindrical contacts such as a peg in a hole, the two geometric procedures *CRITICAL-DIRECTIONS* and *LOCALLY-COMPATIBLE* must be modified. *CRITICAL-DIRECTIONS* adds a trajectory parallel to the axis of the cylinder, and its inverse, to the list of translations, since any decomposition breaking the peg-in-hole contact will need to move along the axis of the cylinder. In *LOCALLY-COMPATIBLE* if a contact $c$ is a cylindrical contact, $c$ interferes with translation $t$ unless the axis of the cylinder is parallel to $t$. Cylinder-face contacts are much like planar contacts. With these changes, *PARTITION* finds all translational partitionings of assemblies with both planar and cylindrical contacts.

### 10.6.2 Unconnected Subassemblies

*PARTITION* need not include the connectedness of subassemblies as a generating constraint. Although assembly planners usually assume subassemblies are connected, the constraint forbids some valid assembly sequences. For instance, one might want to connect two subassemblies by a fixture, mate them with another subassembly, and then remove the fixture: imagine putting together the wheels and axles of a wagon, then lowering the body onto the two axles simultaneously. Without the connectedness constraint, however, there may be a much larger number of feasible decompositions. This change has been implemented in GRASP, but few experiments have been done.

### 10.6.3 Threaded Contacts

*PARTITION* as given above cannot easily be applied to most mechanical products, since it only allows translational motions to separate parts. Many assemblies have threaded connections, where the two parts to be mated must follow a helical trajectory relative to each other, and thus cannot be assembled with translations only. A simple example consists of a nut and a bolt. However, *PARTITION* can be extended to handle screw contacts, and any other type of connection that allows only a finite set of mating trajectories to assemble the two connected parts.

To correctly partition assemblies with threaded contacts, the trajectories considered by the algorithm must be extended to encompass both translations and motions with an element of rotation about an axis parallel to the translation. *LOCALLY-COMPATIBLE* thus must be able to discern whether another contact will prevent such helical motions. This is more complicated when screwing motions are allowed. In addition, a threaded contact is incompatible with any motion between the two parts except for a helical motion with the same axis and pitch as the contact.

Furthermore, *CRITICAL-DIRECTIONS* must add the finite set of removal trajectories associated with each such contact to the list of those considered. Since any feasible partitioning of an assembly that breaks a screw contact $s$ must follow one of the feasible trajectories allowed by $s$, this augmented list will be sufficient to find all partitionings of the assembly using these new trajectories.

Once a locally-free subassembly is generated along a helical motion, the global validity of the operation must be checked. This requires that a part be swept along a curved path, which is more difficult than sweeping along a translation. GRASP conservatively approximates this calculation by rotating the subassembly around the axis of the helical motion to compute an object that is then translated along the axis to check for collisions.

Note that in certain cases rotational motions might allow additional partitionings of an assembly that has only planar contacts, and these decompositions will not be found by the augmented *PARTITION* (see figure 10.7). Instead, in cases of rotational motions separating partitions with planar contacts, *PARTITION* will find the same set of decompositions as *GET-FEASIBLE-DECOMPOSITIONS* does. Section 10.6.5 discusses how both algorithms can be extended to handle infinitesimal rotational motions in a complete way.

### 10.6.4 Global Constraints

Since the basic partitioning algorithm only considers contacts, it may return a large number of decompositions that do not satisfy global constraints. In particular, *FEASIBILITY-TEST* sweeps parts to check for the existence of a straight motion to remove each locally-free subassembly; however, the requirement of a single translation to separate subassemblies can be included as a generating constraint in the procedure. *PARTITION* can be modified to find all the pairs of subassemblies that can be completely separated by a single translation, in polynomial time on the number of such decompositions and the number of features in the assembly. In fact, because GRASP currently only sweeps a subassembly along a few heuristically-chosen directions, the extended procedure would find decompositions that are missed by the current geometric reasoner.

In an assembly $A$, the set of directions an subassembly $S$ can translate with

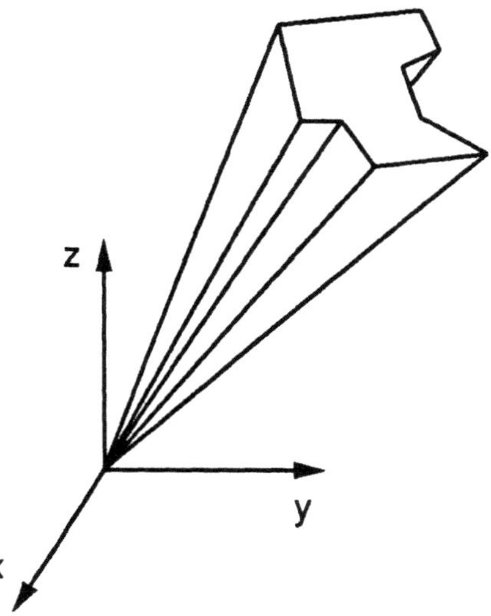


Figure 10.6: A non-convex cone of removal translations

respect to its complement $A - S$ an infinitesimal distance is always included within a convex cone, as in figure 10.1. To find the set of directions that $S$ can translate indefinitely, we can project the obstacles corresponding to the parts of $A - S$ back onto the unit sphere to obtain a cone of straight removal directions, which is in general not convex (figure 10.6). Each face of this cone that is not in common with the local translational freedom cone will arise either from a vertex of $S$ and an edge of $A - S$, or from an edge of $S$ and a vertex of $A - S$. Each vertex-edge pair defines a plane constraint, and all non-convex cones arising from partitionings of $A$ will be made up of these planes. Thus we can find a set of *global critical directions* that includes the edges of all non-convex cones arising from an assembly. If there are $m$ features, there are $m^2$ planar constraints, and $m^4$ global critical directions. When generating assemblies, the function *LOCALLY-COMPATIBLE* is replaced by *GLOBALLY-COMPATIBLE*, which sweeps parts as well as checking contacts. As a result, the modified procedure will generate all globally feasible decompositions of the target assembly, and no more.

Arkin et al [1] present a procedure to find globally valid separation translations for polygons in the plane using the notion of a monotone path between obstacles, and show how to use this to physically partition two-dimensional assemblies of polygons. Their method has since been extended to three dimensions in a way very similar to the above [5]; however, it is unclear how easily their algorithm could be extended in other ways, such as using a connectedness generating constraint, cylindrical contacts, or motions with a rotational component.

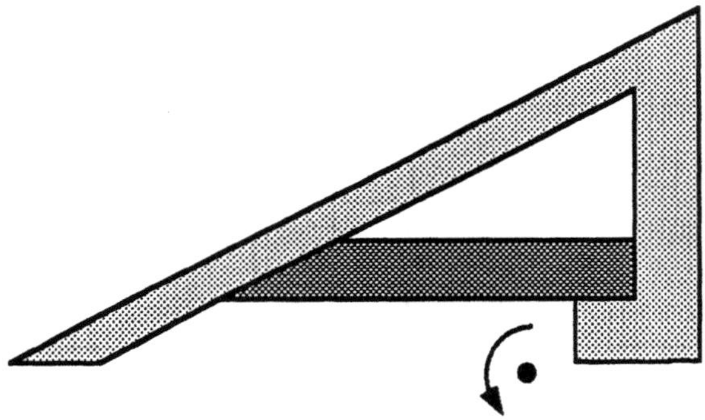

Figure 10.7: A rotation to remove a part that cannot translate

### 10.6.5 General Rotations

That a subassembly have local translational freedom is a necessary constraint on assembly decompositions only when the global motions for assembling parts are limited to translations. Consider for instance the planar assembly in figure 10.7. Because no translations exist for the inside part, its local translational freedom cone is null. However, a rotation around the point shown can free the part. *PARTITION* will not generate this decomposition, even when augmented with the methods of Section 10.6.3, because this rotation is not included in the list of removal directions for any one contact. It is possible to extend the procedure to find all such decompositions while preserving the polynomial time bound. This method is preliminary and not implemented, and it will only be sketched here.

The positions of a rigid part $p$ translating and rotating in three dimensions can be represented as points in a configuration space $\mathcal{C} = R^3 \times SO(3)$. A motion direction at a position $q \in \mathcal{C}$ is a tangent at $q$ to a trajectory going through $q$. A direction can be represented by a 6D vector of the tangent space [4]. Moreover, a set of contacts constrains the set of motion directions into a 6D local freedom cone which is linear and convex for planar contacts [6].

This formalism can be immediately included in *GET-FEASIBLE-DECOMPOSITIONS*. The 6D local freedom cone for the contacts of the decomposition corresponding to each cut-set of the connection graph is calculated, and if it is null the decomposition is not feasible. This version of the cut-set method is truly complete for rigid parts: it will find all possible decompositions of the target assembly.

*PARTITION* can also be extended to use 6D cones. Each cone is defined as the intersection of several half-spaces delimited by hyperplanes corresponding to well chosen contact points. It can be shown that the lines generated by combining all groups of five hyperplanes form a set of critical directions. Thus with $m$ contact points, there are $O(m^5)$ directions in which to partition the assembly. The function *LOCALLY-COMPATIBLE* tests whether any such rotation and translation is incompatible with a single contact, although this will be a much more complicated calculation.

With the extension to 6D freedom cones, *PARTITION* operates in the worst case in polynomial time on the number of point-plane contacts and feasible decompositions. However, note that the degree of the polynomial and the complexity of the geometric reasoning involved would make it advantageous to use this method only when the number of parts is quite large. In addition, the combination of 6D freedom cones with the global sweeping constraints of the previous section would not be straightforward, since non-instantaneous motions cannot be embedded in a vector space and the freedom cones will no longer have planar sides. Indeed, checking for collisions along a helicoidal path is already much harder than the translational case.

### 10.6.6 Maintaining Geometric Dependencies

As discussed in Chapter 9, the direct extension of *precedence expressions* to *GET-FEASIBLE-DECOMPOSITIONS* is straightforward yet not very useful in practice. A large number of cut-sets might be generated, few of which correspond to feasible decompositions, and a precedence expression must be maintained and evaluated for each one. On the other hand, because *PARTITION* does not explicitly build the local freedom cone for each decomposition, dependency maintenance cannot be applied to this aspect of the partitioning process easily.

However, *PARTITION* still calls *FEASIBILITY-TEST* to assess the constraints on assembly operations that are not used as generating constraints in the algorithm. Precedence expressions could be used to hold the results of other expensive calculations in *FEASIBILITY-TEST* and the conditions under which they will still be valid. These precedence expressions would only be stored or evaluated once a subassembly is found to be locally free, thus keeping the number of expressions manageable. Such a capability has not been implemented in the current version of the planner; as a result, GRASP finds either linear assembly sequences using precedence expressions, or non-linear assembly sequences using *PARTITION*. Further investigation should investigate combining the two techniques.

## 10.7 Experimentation

The algorithm *PARTITION* has been implemented in the assembly sequence planner GRASP, which is described in Chapter 9. The experiments described here were performed using a version of the algorithm extended to include the

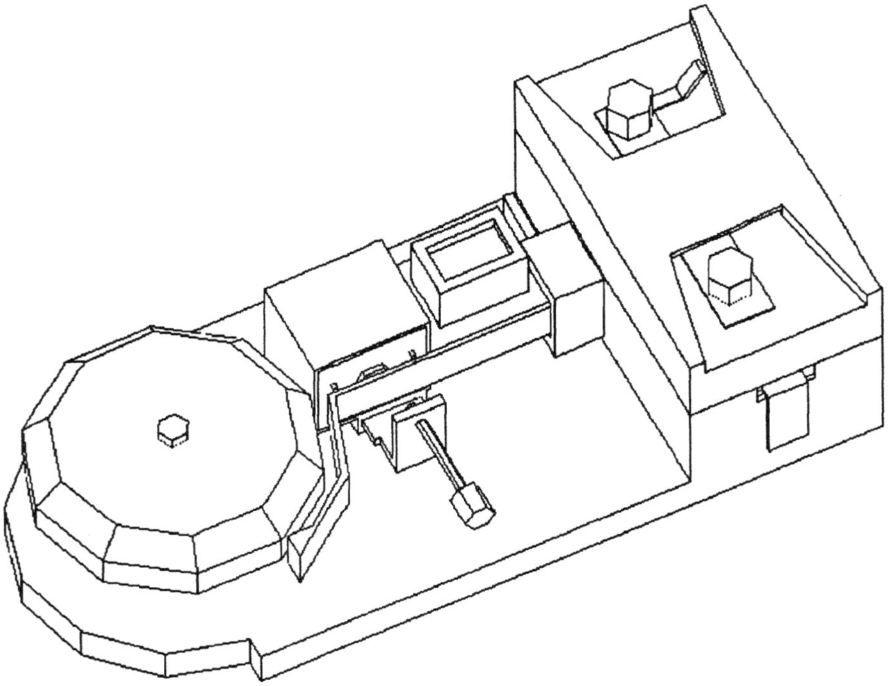

Figure 10.8: The electric bell

methods of Sections 10.6.1 and 10.6.3. There are a number of cylinder-cylinder and cylinder-plane contacts in the 22-part electric bell (figure 10.8), as well as two threaded contacts.

Table 10.1 compares the running times of *GET-FEASIBLE-DECOMPOSITIONS* and *PARTITION* when run on the electric bell. It gives the time required to partition the bell according to local motion and connectedness constraints only, to partition including global motion checking, and to build the full AND/OR graph. With non-linear assembly sequences allowed, the AND/OR graph for the bell has 1,710 nodes and 12,447 AND-arcs. Note that the total AND/OR graph generation time is comparable for the two algorithms. This results from *PARTITION* being slower than the cut-set method when partitioning the many small assemblies close to the leaves of the graph.

With larger assemblies the cut-set algorithm quickly becomes intractable. For instance, GRASP has planned for the assembly of a skin-machine product composed of 36 parts. Using *PARTITION*, GRASP finds one assembly plan for the skin-machine in about a minute. Planning for the same assembly, *GET-FEASIBLE-DECOMPOSITIONS* was stopped after two days without partitioning the root node of the AND/OR graph.

| Decomposition Algorithm | First Partitioning | Swept First Partitioning | AND/OR Graph |
|---|---|---|---|
| *GET-FEASIBLE* | 54 | 62 | 3894 |
| *PARTITION* | 1.1 | 9.1 | 1032 |

Table 10.1: Time to plan for the electric bell, in seconds

# Conclusion

The problem of automatic assembly sequence planning requires two relatively distinct types of reasoning: largely symbolic reasoning about sets of assembly sequences and more physical methods to evaluate the geometric and mechanical feasibility of individual assembly operations. This dichotomy leads naturally to a generate-and-test methodology for solving the assembly planning problem, in which assembly operations and sequences are proposed and then critiqued by the geometric reasoning module.

The physical partitioning problem—finding all ways to divide an assembly into two subassemblies that can be mated obeying physical constraints—lies on the boundary between the symbolic and geometric sides of assembly planning. A generate-and-test approach can end up evaluating a large number of infeasible decompositions for some assemblies; *PARTITION* addresses this problem by including the geometric constraint of local motion freedom in the generation process. By tightly merging the symbolic and geometric reasoning required in physical partitioning, the algorithm achieves much greater efficiency. As a result, it will prove an invaluable technique to help make automatic assembly sequence planning practical for real world assemblies with many parts.

# References

[1] E. M. Arkin, R. Connelly, and J. S. B. Mitchell. On monotone paths among obstacles, with applications to planning assemblies. In *Proceedings of the ACM Symposium on Computational Geometry*, pages 334–343, 1989.

[2] D. F. Baldwin. Algorithmic methods and software tools for the generation of mechanical assembly sequences. Master's thesis, Massachusetts Institute of Technology, 1990.

[3] L. S. Homem de Mello. *Task Sequence Planning for Robotic Assembly.* PhD thesis, Carnegie Mellon University, 1989.

[4] J.-C. Latombe. *Robot Motion Planning.* Kluwer Academic Publishers, Boston, November 1990.

[5] J. S. B. Mitchell. Personal communication, December 1990.

[6] M. S. Ohwovoriole. *An Extension of Screw Theory and its Application to the Automation of Industrial Assemblies.* PhD thesis, Stanford University, April 1980.

[7] R. H. Wilson. Efficiently partitioning an assembly. In *Proceedings of the IASTED International Symposium on Robotics and Manufacturing*, 1990.

[8] J. D. Wolter. *On the Automatic Generation of Plans for Mechanical Assembly.* PhD thesis, The University of Michigan, 1988.

# Chapter 11

# On the automatic generation of assembly plans

**Jan D. Wolter**

One of the primary goals of robotics research has been the development of flexible manufacturing systems that allow a manufacturer to bring new product designs into production rapidly. This has led to the development of programmable machine tools, manipulators, and workholding systems that can be adapted to new manufacturing tasks simply by loading new software. However the full benefits of such tools can only be attained if reliable software can be rapidly generated. The ideal system would be one which could automatically program itself to produce a new product given only a description of the product. It is in the pursuit of this goal that the assembly planning problem arises.

Assembling a mechanical structure is typically seen as being performed by a series of operations, such as the insertion of a peg into a hole. Under this model, the first stage in programming an assembly system must be to identify the operations necessary to manufacture the given assembly, and to specify the sequence in which they are to be performed. The selection of such a sequence of operations is called the *assembly planning problem*.

This paper describes the XAP/1 assembly planning system. It will begin by describing in detail the problem solved by XAP/1. Although several assembly planning systems have been developed, each has been based on a slightly different definition of the assembly planning problem. To clarify the relationship between XAP/1 and other planners, four key questions will be discussed in section 11.1:

- What kinds of operations are allowed?
- How much detail is included in the plans generated?
- How is the input assembly described?
- How much optimization is done on the plans generated?

It will be seen that it is XAP/1's strong orientation toward optimization that distinguishes it from other assembly planners, and to a large extent determines the design of the planner.

Section 11.2 will describe some theoretical bounds on the computational complexity of the assembly planning problem. Although finding a solution to an assembly planning problem can be quite difficult in general, polynomial algorithms exist for many special cases. In particular, given the inputs used by XAP/1, it is possible to find some valid plan in polynomial time. However, the problem solved by XAP/1, that of finding an optimal plan, is NP-hard.

Section 11.3 will describe the operation of the XAP/1 planner. Most assembly planners plan sequentially, starting with the finished product and removing parts until it is completely decomposed. XAP/1, on the other hand, plans by posting constraints and is able to make decisions in any order. This allows it to plan opportunistically, making obvious decisions first and rapidly narrowing the search space. Its search is guided by advice from a collection of plug-in criterion modules.

Section 11.4 presents some example results and studies the performance of the planner experimentally. XAP/1 generates plans for reasonably sized assemblies very quickly, but its time complexity does grow exponentially with the number of parts. This section will conclude by describing work in progress on the XAP/1 planner.

## 11.1 Problem Definition

This section will describe the assembly planning problem in terms of its inputs and outputs. Some of the important differences between the problem definition used here and those used in other assembly planners will be detailed.

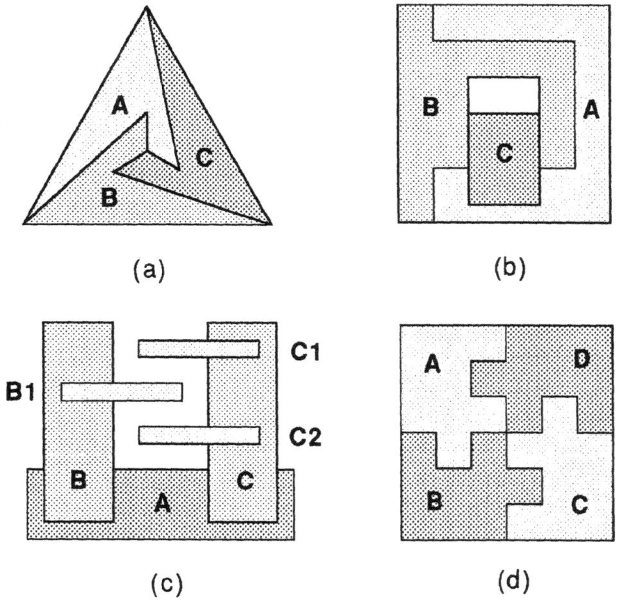


Figure 11.1: Two dimensional assemblies which cannot be built by (a) a sequential plan, (b) a monotone plan, (c) a contact-coherent plan, and (d) a linear plan.

### 11.1.1 Legal Operations

All assembly planning systems place some constraints on the types of operations that may be used. The most common is the sequentiality assumption. A plan is *sequential* if it can be decomposed into a sequence of operations such that each operation involves moving one set of parts along one common trajectory. Thus, a sequential plan is any plan that could, in theory, be executed by a one-handed robot. Each operation would consist of grasping a set of parts, moving them, and then releasing them. Sequential plans are not sufficient for the construction of all assemblies. The two-dimensional assembly in figure 11.1a, for example, requires that part A and part B be simultaneously inserted along distinct trajectories, and so would require a two-handed robot. All assembly planners developed to date produce sequential plans. This is because finding mating trajectories that involve the coordinated motion of many parts is difficult, and because such mating operations are rarely needed in practice.

Another common assumption is monotonicity. A plan is *monotone* if no operation ever separates any pair of parts that were already in their goal positions relative to each other, and every operation leaves all moved parts in their goal positions relative to some unmoved part. The monotonicity restriction excludes operations which place parts in temporary positions. For example, the assembly in figure 11.1b could only be built if part C was first temporarily inserted fully into part B, and then slid to its goal position only after part A had been installed. Such a plan would not be monotone because it leaves C in a temporary position while A is inserted. Generating non-monotone plans is challenging because of the difficulty of finding appropriate temporary positions for parts. Only Hoffman's assembly planning system has significant capabilities in this respect[10,9]. In practice, non-monotone plans are most commonly needed for assemblies with moving parts, which may be moved to different positions for different assembly operations.

We call the set of parts which are placed in their final relative positions by an operation a *partial assembly*. Many planners restrict the kinds of partial assemblies which may be produced through a coherence restriction. Given any connected graph $G$ whose vertices correspond one-to-one with the parts of the assembly, a plan is *coherent* for $G$ if every partial assembly that occurs in the plan forms a connected subgraph in $G$.

One commonly used form of coherence graph is a *contact graph*. A contact graph has an edge connecting the nodes for any two parts that touch each other in the goal assembly. Plans which are coherent for a contact graph only allow operations which place the moved parts into contact with some other part. Not all assemblies can be built by contact-coherent plans. Figure 11.1c is an example of assembly which cannot. There is only one way to split the complete assembly into two sets of parts and that is by removing part A from the rest. But doing so leaves an unconnected partial assembly, $\{B, B1, C, C1, C2\}$, so the plan is not contact-coherent. In practice, non-coherent plans are often used when fixtures are available that can accurately position parts not in contact[15].

Contact graphs were first used by Jentsch and Kaden[13]. Homem de Mello and Sanderson's "relational model graphs" can be viewed as augmented contact graphs[11]. The "liaison graphs" used by Bourjault[1] and by De Fazio and Whitney[4,5] are similar to contact graphs, but are more loosely defined. All these systems generate only coherent plans. Of course, this restriction can be eliminated by using a complete graph as the coherence graph, however this would entail a significant performance penalty for most systems.

This paper considers the generation of plans which are sequential and monotone, but without coherence constraints. In the generation of optimal plans, the XAP/1 planner applies an additional constraint, that of linearity. A plan is *linear* if no more than one part is moved at a time. Such plans will contain no subassemblies and will be constructed entirely in a single fixture. There are many common assemblies in which the use of subassemblies is necessary (as in figure 11.1d) or desirable. To plan such systems with XAP/1, the user would have to supply a breakdown of the parts into subassemblies. The system could then generate a separate linear plan for each subassembly.

Assembly $S_0$:
(1) Insert part $P_1$ with trajectory $\beta$.
(2) Insert part $P_2$ with trajectory $\gamma$.
(3) Insert subassembly $S_1$ with trajectory $\beta$.
(4) Insert part $P_5$ with trajectory $\beta$.

Subassembly $S_1$:
(1) Insert part $P_4$ with trajectory $\alpha$.
(2) Insert part $P_3$ with trajectory $\alpha$.
(3) Insert part $P_6$ with trajectory $\alpha$.

(a)

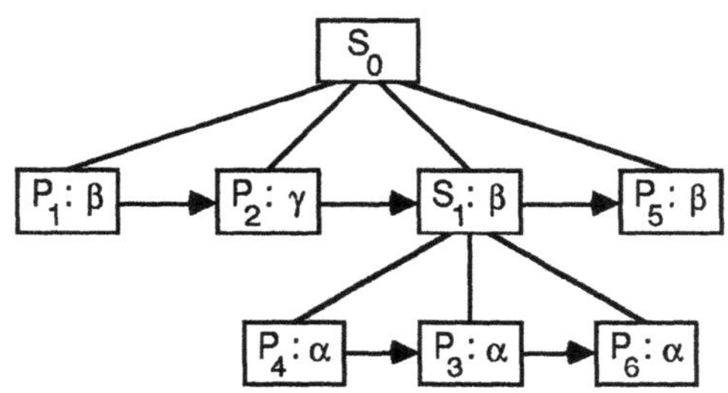


(b)

Figure 11.2: A plan to build a six part assembly using one subassembly. This can be represented as (a) a sequence of insertions for each subassembly or (b) a subassembly tree diagram.

### 11.1.2 Level of Detail

Assembly planning is the first step toward producing a complete production plan for the product. This complete plan might include executable robot programs, workcell layouts, and fixture designs. Normally the assembly planner will not decide on all these details. Instead it will make only high-level decisions, and leave the lower-level details to be determined by lower-level planners.

The amount and type of detail to be included in the plans produced by an assembly planner is a critical design issue. The planning problem becomes more difficult as plans become more detailed, and the number of different kinds of decisions the planner must be able to make increases. However, to have any value, a planner must produce plans which are detailed enough so they can be meaningfully evaluated. A planner cannot find "good" plans, unless it knows enough about those plans to be able to judge them.

The planners described by Homem de Mello and Sanderson and by De Fazio and Whitney produce plans composed of *assembly operations* which combine two sets of parts. Though neither system stores mating trajectories for these operations, both check if a feasible one exists. They do not, however, decide which of the two sets of parts is to be held by a fixture. Homem de Mello and Sanderson produce a partially-ordered tree of assembly operations, while De Fazio and Whitney produce a totally-ordered linear sequence. These models of a plan include all the detail necessary to allow the geometric feasibility of a plan to be determined.

The XAP/1 planner is based on slightly lower-level operations, called *insertion operations*. An insertion operation consists of inserting a part or subassembly into a fixture. A plan consisting of insertion operations may be drawn as a *subassembly tree*, as shown in figure 11.2. Each internal node corresponds to a *subassembly*, where a subassembly is defined as a set of parts that is assembled in one fixture, and then inserted as a unit into a larger assembly. The children of a subassembly are the ordered sequence of parts and subassemblies that are inserted to create it. Thus, for the example in figure 11.2b, subassembly $S_0$ is built by inserting first part $P_1^*$, followed by part $P_2$, subassembly $S_1$ and part $P_5$ so these are the children of node $S_0$.

Subassembly trees contain considerably more detail than the plans produced by Homem de Mello and Sanderson and De Fazio and Whitney. For an $n$-part assembly, each of the plans produced by those systems will correspond to between $\frac{1}{4}(2\sqrt{2})^n$ and $2(3^n)$ different subassembly trees. The subassembly tree representation makes it possible to uniquely determine such facts as the number of subassemblies used, and the order in which the parts are inserted into a fixture. It is possible to extract a fixture specification from a subassembly tree plan—one that tells for each fixture what parts it must hold, what directions those parts will be inserted from, and what assembly forces it must restrain those parts against once they are placed. Since plans described in terms of assembly operations do not distinguish between held and moved parts, this kind of information cannot be extracted from them.

Of course, since the XAP/1 planner is currently limited to the generation of linear plans, it produces trees of only one level. Thus it is restricted to producing plans which are simply totally-ordered sequences of insertion operations.

### 11.1.3 System Inputs

Ultimately, an assembly planner should be able to generate plans directly from a CAD model of the goal assembly. Several systems already support such interfaces, including those described by Hoffman and by Homem de Mello and Sanderson. Both of these systems query the geometric models directly during the planning process.

XAP/1, like De Fazio and Whitney's system, does its planning based entirely on a symbolic description of the problem. All the information that makes up this description is to be computed before the actual planning begins, and the geometric model is not accessed during planning. This is in contrast to the planners described by Hoffman and by Homem de Mello and Sanderson, which make geometric queries during the plan generation.

The XAP/1 planner represents the problem with two kinds of information. First, a set $\Theta_P$ of insertion trajectories are proposed for each part $P$. Second, sequencing constraints are generated by checking which parts would block the insertion of a given part by a given trajectory. If moving part $P$ along trajectory $\theta \in \Theta_P$ causes $P$ to collide with another part $Q$ then we have a constraint of the form:

if $P$ is inserted along $\theta$, then it must precede $Q$.

All constraints in XAP/1 take this form.

These trajectory proposals and sequencing constraints are generated manually and input to the current system. However, insertion trajectories could be proposed automatically by recognizing common relationships between nearby parts in a CAD model. Some might be generated based on the geometry of the assembly, while others might be found using some higher-level knowledge about the assembly. Generally the number of trajectories proposed for each part can be quite small—perhaps between two and a dozen.

Geometrically, trajectories perpendicular to planar contact surfaces or parallel to the axis of cylindrical contact surfaces might be proposed. These will often work well where the parts do not interlock in complex ways. More sophisticated geometrical techniques are known that can be used to find straight-line trajectories to separate pairs of adjacent parts in two or three dimensions[19,16]. In the relatively rare cases where two parts interlock in very complex ways, a general geometric path planner could be applied to separate them.

Perhaps more promising than purely geometrical techniques for proposing trajectories are techniques which use higher-level knowledge about common structures in assemblies. Knowledge about fasteners is especially useful. For example, if a structure where a nut and bolt hold together a set of parts can be recognized, then we can propose spiral trajectories for the threaded parts, and straight trajectories parallel to the axis of the bolt for both the threaded parts and all the parts they hold together. Similarly other types of fasteners suggest certain trajectories might be used both for the fasteners and the parts they fasten. This would allow us to find trajectories for many common deformable parts, such as C-rings, cotter pins and rivets.

Knowledge about other types of part features, such as tabs, slots and holes can also be used to guess mating trajectories. Some research has been done on methods to identify such features from geometric models[2,8,14]. It seems likely, however, that identifying such features in assemblies would be easier than identifying them on individual parts. For example, for a depression in the surface of a part to be classified as a slot, there should be another part which has a tab that fits into it. There is also research in progress aimed at the development of functional CAD systems which would make this kind of feature-based planning much easier[20].

### 11.1.4 Optimization

All plans must be geometrically feasible, in the sense that it must be possible to perform the operations without causing any intersection between parts. However there are usually many feasible plans, so the planner must use some other basis to choose the best of them. Ideally it would always choose a plan with minimal set-up and production costs, but in practice this is difficult to achieve because the actual costs can only be computed if the plan is very detailed. Since assembly planners produce only high-level plans, we must be content to evaluate those plans on the basis of some estimate of how likely they are to have an inexpensive implementation.

Homem de Mello and Sanderson evaluate their plans for flexibility and parallelism. A flexible plan is simply one that has many possible implementations. Such a plan is certainly more likely to have a good implementation. Highly parallel plans can reduce production time when multiple manipulators are present. It is worth noting, however, that plans with high parallelism will also tend to use large numbers of subassemblies. In a single manipulator environment this is a disadvantage because each additional subassembly will require one additional insertion operation. This situation is typical of nearly all evaluation criteria in all assembly planners: different criteria must be applied in generating plans for different manufacturing environments.

The XAP/1 planner addresses this by implementing criteria as independent "plug-in" modules that can be installed as needed and that can be given different weights for different applications. Because the XAP/1 planner generates more detailed plans than most other planners, it is also able to evaluate plans with respect to a wider range of more realistic criteria than other planners. Currently criterion modules have been implemented for the following three criteria:

1. **Directionality**. We would prefer to insert all parts, as much as possible, from a single direction. This simplifies the fixtures, requires a less dexterous robot, and avoids extra operations to reorient the work piece.

2. **Fixture Complexity**. We would like to sequence the operations so that the partially built assemblies hold themselves together as much as possible. For example, if we wish to place ten washers on a peg, it is

better to place the washers on the peg one by one than to hold the ten washers in place while inserting the peg.

3. **Manipulability**. We would like to perform the more difficult operations with the more easily handled parts. For example, if we wish to attach a bolt to an engine block, we would prefer to fixture the engine block while screwing in the bolt, rather than fixture the bolt while screwing on the engine block.

Note that none of these criteria could easily be implemented in a planner that describes the plan only in terms of assembly operations, because they cannot be assessed without knowledge of the fixturing and manipulation requirements of the plan.

In practice, the number of different criteria which are relevant to a particular planning problem may be large. It is the need to be able to plan efficiently with a large and variable set of conflicting criteria that drove the design of the XAP/1 planner.

## 11.2 Computational Complexity

In this section we will discuss the computational complexity of the assembly planning problem. In certain special cases, it has been shown that an assembly plan can be found rapidly, if we do not require optimality. Dawson[3] has shown that for any set of star-shaped parts in $k$-dimensional space there exists a non-sequential assembly plan which can be generated in linear time. Guibas and Yao[7] give an $O(n \log n)$ time algorithm to find a linear, monotone, monodirectional plan for any two-dimensional assembly of convex parts of total complexity $n$. For three dimensions, however, Dawson[3] has shown that there are assemblies of convex parts which cannot be built by any sequential plan (i.e., no set of parts can be moved without disturbing some others).

On the other hand, it can be shown that finding a non-monotone assembly plan is PSPACE-hard even in two-dimensions. Only exponential algorithms are known for PSPACE-hard problems. Hopcroft, Schwartz and Sharir have shown that moving a set of parts from a given starting position to a given goal position is PSPACE-hard[12]. Though no initial position is specified in assembly planning problems, it is possible to design assemblies which must pass through a specific intermediate position, and moving from that intermediate position to the final position requires the solution of a problem of this type. Using this observation, it is possible to modify Hopcroft, Schwartz and Sharir's proof to produce a proof that finding a non-monotone assembly plan is PSPACE-hard[21].

As described in section 11.1.3, the XAP/1 planner is given planning problems described by sets of proposed trajectories for each part and a set of constraints

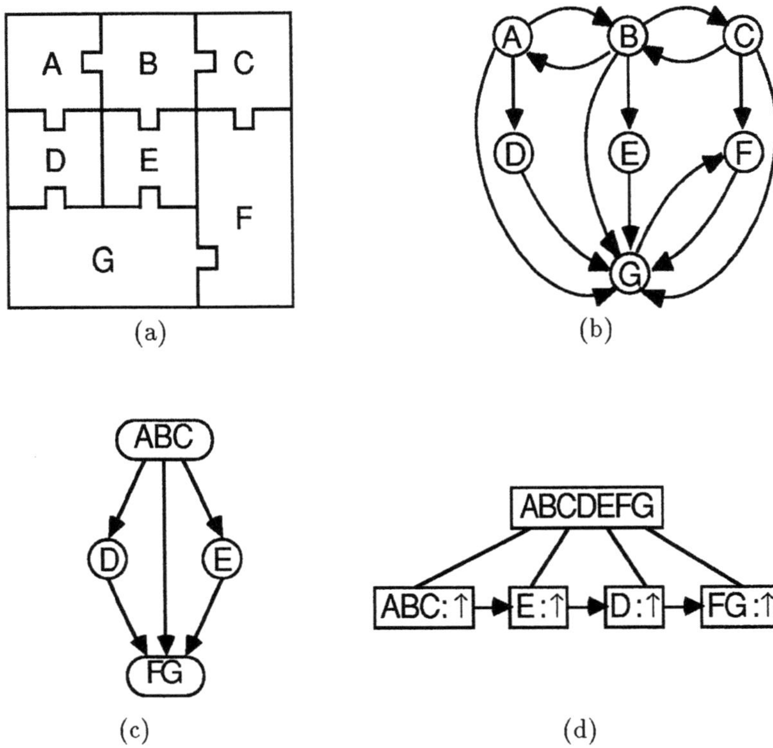


Figure 11.3: Assembly decomposition: (a) is the assembly; (b) is the precedence graph for insertion of parts from below; (c) is the precedence graph after collapsing strongly connected components; (d) is the subassembly tree showing the resulting decomposition of the assembly.

on each trajectory. Given this information, it is possible to find a monotone solution, if one exists, quite quickly. The algorithm can be described quite easily, since it is based on well-known graph algorithms. For simplicity, we will assume that the same set of trajectories is proposed for every part. Then we can draw a different precedence graph for each proposed trajectory. The graph for the trajectory $\theta$ would have one node for each part, and would have an edge directed from part $p$'s node to part $q$'s node if there is a constraint that says that part $p$ must be inserted before $q$ when trajectory $\theta$ is used. For example, the precedence graph for insertion from below in the assembly shown in figure 11.3a is shown in figure 11.3b. If part $C$ were inserted from below, it would collide with parts $B$, $F$ and $G$, so there are arrows pointing to those parts from part $C$. Clearly these precedence graphs can be constructed for a set of $n$ parts, with $t$ trajectories and $c$ constraints in $O(nt + c)$ time.

The algorithm will construct a subassembly tree from the root downward, by dividing each subassembly into an ordered sequence of parts and subassemblies. To test if a trajectory can be used to decompose an assembly, we first find the strongly connected components of its precedence graph in $O(n + c)$ time using Tarjan's algorithm[18]. These components are the sets of parts which form cycles and thus cannot be separated from each other with the trajectory $\theta$. Thus, the strongly connected components of the graph in figure 11.3b are $\{A, B, C\}, \{D\}, \{E\}, \{F, G\}$. If the entire graph forms a single strongly connected component, then the trajectory $\theta$ cannot be used, and we try the next trajectory. Otherwise, we collapse each strongly connected component into a single node labeled by the set of parts in the subassembly it represents. This results in an acyclic graph like the one in figure 11.3c. This graph is topologically sorted in $O(n + c)$ time[17] to give an ordered list of parts and subassemblies which can be combined to build the assembly, all using trajectory $\theta$. This leads to a decomposition of the assembly such as the one described by the subassembly tree in figure 11.3d. Thus we can decompose any internal node in the subassembly tree in $O(tn + c)$ time. Since the subassembly tree can have at most $n - 1$ internal nodes, the total time to build a tree will be bounded by $O(n^2t + nc)$. This algorithm will always find a plan if one exists.

Thus, a polynomial algorithm exists to find monotone plans for the problems solved by XAP/1. However, the goal in XAP/1 is to find an optimal plan, and this is a much harder problem. We will see in section 11.3.3 that optimizing the directionality criterion alone is an NP-hard problem. It is thus to be expected that any system to solve such problems will require exponential time.

## 11.3 The XAP/1 Planner

This section will begin by briefly describing the method used by the XAP/1 planner to generate plans, and will then describe the operation of each of XAP/1's major modules.

XAP/1 represents plans by collections of assertions and constraints. Two types of assertions are supported. A *sequencing assertion*, written $P \prec Q$, indicates that part $P$ must be inserted before part $Q$. A *trajectory assertion*, written $P : \Theta$, indicates that part $P$ must be inserted using some trajectory in the set $\Theta$. Both kinds of assertions have the property that they are easily negatable. The negation of $P \prec Q$ is $Q \prec P$, and the negation of $P : \Theta$ is $P : \Theta_P - \Theta$ where $\Theta_P$ is the set of trajectories originally proposed for part $P$.

XAP/1 uses assertion sets to represent sets of plans. Specifically, an assertion set is used to represent the set of all plans which satisfy all the assertions in the set. Initially the set of possible plans would be represented by the set of assertions $\{(P : \Theta_P) \mid P \in \mathbf{P}\}$, where $\mathbf{P}$ is the set of all parts and $\Theta_P$ is the set of trajectories initially proposed for part $P$. Such an initial assertion set is shown as the root of the tree in figure 11.4c. Every plan that can possibly be generated using the proposed trajectories satisfies this assertion set.

Constraints are relations among assertions that must be obeyed by all legal assertion sets. Two kinds of constraints are used in XAP/1. First, assertion sets must obey certain logical constraints. If an assertion set contains assertion $a$, then it may not contain the negation of $a$, and at least one trajectory must be allowed for each part. The sequencing assertions in the set must form a transitive relation. That is, they must obey the rule

$$\forall P, Q, R \in \mathbf{P} \; (P \prec Q \;\wedge\; Q \prec R \;\Rightarrow\; P \prec R).$$

The assertions in the set must also obey the geometrical constraints generated during the constraint generation stage of the preprocessing stage. All of these geometrical constraints take the same form. If moving part $P$ along each trajectory in the set $\Theta$ causes it to collide with part $Q$, then we have the following constraint:

$$P{:}\Theta \;\Rightarrow\; P \prec Q.$$

Thus, for the assembly in figure 11.4a, there would be two geometrical constraints: if part $A$ is inserted from below, it will hit $B$; and if part $B$ is inserted from above, it will hit $A$. These two constraints are shown in figure 11.4b.

XAP/1 generates plans by starting with the initial assertion set, which represents all possible plans, and iteratively subdividing it into subsets until an assertion set that describes just one plan is found. The subdivision is done by selecting any assertion $a$ which has not already been determined to be either true or false and adding $a$ to one child of the set and $\neg a$ to the other. When an assertion is added to a set, any other assertions which are implied by the constraints are also added.

For example, in figure 11.4c, we started by refining the initial assertion set with the assertion $A:\{\uparrow\}$. Note that the assertion $A:\{\uparrow\}$ may be applied because it is not known to be either true or false. It is known that part $A$ uses either trajectory $\uparrow$ or $\downarrow$, but which has not been determined. This implies by the first constraint that $A \prec B$ must be true for all plans in this set. This, in turn, implies by the second constraint that $B:\downarrow$ must be false for all plans in this set. The other child of the initial assertion set is produced by applying the negation of the assertion in the same manner. As shown, this procedure could be repeated until all assertion sets are complete, that is, they propose only one trajectory per part, and they totally order the insertion sequence. This will give a complete enumeration of all plans. Note that different trees can be generated by applying the assertions in different orders, for example as in figure 11.4d, but the same set of plans would always be produced in the leaves.

In normal operation, however, the XAP/1 planner would not expand out the entire tree. Our objective is not to enumerate every possible plan, but to

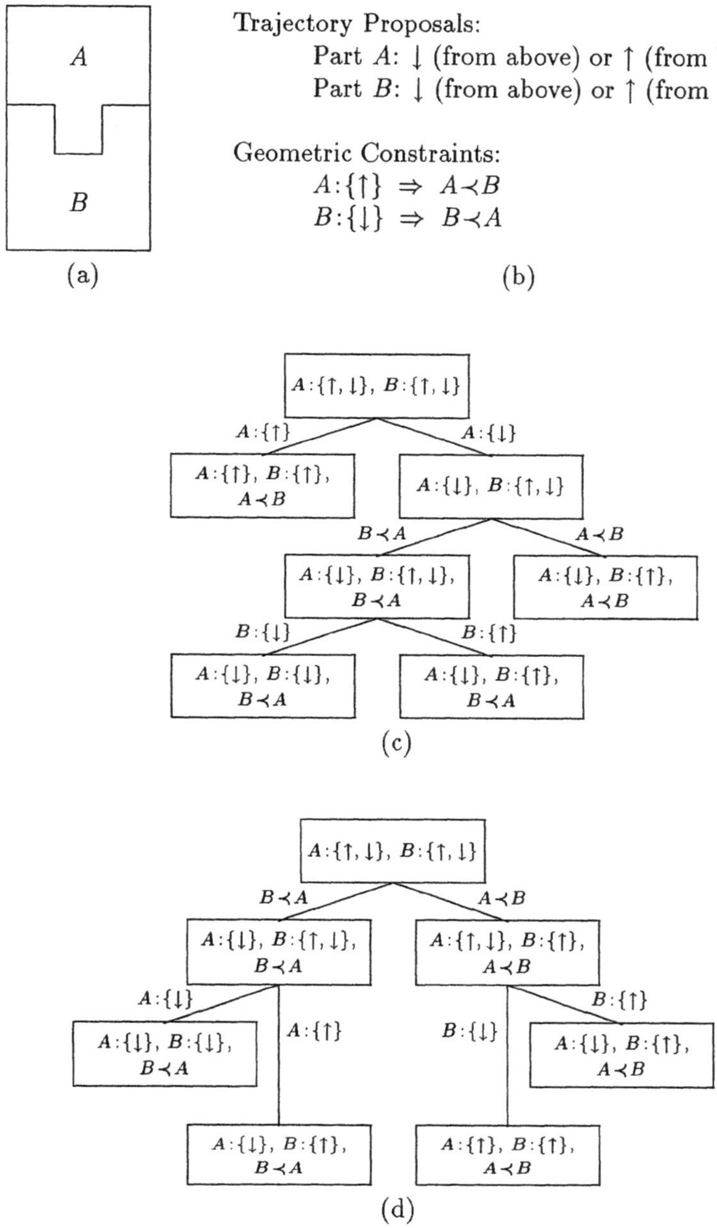


Figure 11.4: The two part assembly (a) might be described by the trajectory proposals and constraints shown in (b). Two possible search trees producing all plans for this assembly are shown in (c) and (d).

identify an optimal plan. XAP/1 uses search heuristics to guide its refinement of sets of plans in two ways: to select which incomplete plan-set to refine next, and to select which assertion to use to refine it. If these decisions are made well, it should be necessary to construct only a small section of the tree.

Note that no possible plan is ever eliminated during the search process. Every plan that was described by the assertion set of a parent node, is described by one of its two children. Because of this, we can guarantee that if there is any possible linear, monotone plan to build the assembly using the proposed trajectories, then XAP/1 will be able to find it.

The XAP/1 system is actually implemented as a collection of distinct modules. The search executive drives the search for an optimal plan by using a heuristic search approach. It is supported by the consequence generator, which applies assertions to assertion sets and finds which other assertions arise as consequences, and the criteria arbitrator, which collects advice from lower-level criterion modules to guide the search. Each criterion is also implemented as a separate module. We will describe the operation of each of these modules in turn.

### 11.3.1 Search Methodology

The goal of the XAP/1 system is not to enumerate all possible plans, as was done in figure 11.4, but to generate one optimal plan. To do this, we must first have a definition of optimality. The system assumes the existence of a rating function $f(\mathcal{P})$ which returns a numeric rating for a plan $\mathcal{P}$. Normally this will be a weighted combination of a number of different heuristic evaluation functions such as the directionality, fixture complexity and manipulability criteria functions. All those possible plans for which $f(\mathcal{P})$ takes a minimal value are optimal plans. It is one of these that the system is designed to find.

The function $f(\mathcal{P})$ can be computed only for complete plans. To be able search effectively, we must be able to estimate ratings for plan-sets that ha not been completely refined. This will allow us to decide if they are worthy further refinement. For this purpose we use a heuristic function $h(\mathcal{S})$ that giv an estimate of the rating of the best plan in the set $\mathcal{S}$. We will normally requi that this function be strictly optimistic—that is, for all complete plans $\mathcal{P}$ in the plan-set $\mathcal{S}$, we must have $f(\mathcal{P}) \geq h(\mathcal{S})$. This is similar to the admissibility criterion for heuristic functions in the A* search.

The XAP/1 planner selects nodes for refinement in a best-first manner based on the nodes' $h(\mathcal{S})$ values. That is, it maintains a list of all leaves of the current search tree on an open list, and in each iteration selects the one with the best rating for further refinement. If there is more than one equally good choice, it prefers assertion sets which are more nearly complete. If the $h$ function is strictly optimistic, this procedure will guarantee that the first complete plan $\mathcal{P}$

selected will be an optimal plan. This is because at that time no other leaf of the search tree may have better $h$ value than $\mathcal{P}$ (otherwise it would have been selected instead of $\mathcal{P}$) so no descendent of another leaf could be better than $P$, since none of them can have ratings better than their ancestors already in the tree.

After an assertion set is selected for refinement the system must choose an assertion to use to split it. This can be any assertion such that neither it nor its negation is already known to be true. The choice of an assertion is a simple type of meta-planning decision—that is, a decision about which decision should be made next. The strategy used by XAP/1 to select assertions is to choose ones that appear likely to lead to two children with very different ratings. If this is done then the child with the worse rating may very well never need to be refined. This should lead to a narrowly directed search which goes quickly to a solution without refining too many alternatives. By doing this the planner is effectively making "obvious" decisions first, an opportunistic strategy commonly used by human planners. Note that XAP/1 would be able to find an optimal plan if one exists even if these decisions were selected at random. However, if they are made well, the time required to find that plan will be greatly reduced.

More details on how assertions are selected are given in the sections on the criterion modules, and some evidence on the effectiveness of this search technique is given in section 11.4.

### 11.3.2 Consequence Generation

Consider the two assertion sets $\{A \prec B, B \prec C\}$ and $\{A \prec B, B \prec C, A \prec C\}$. These are equivalent because the third assertion in the second set is implied by the other two, so any plan for which the first set is true must also satisfy the second set. Clearly there are two different strategies that might be used in representing assertion sets. Either we could maintain a minimal set of assertions that describes the desired set of plans, or we could maintain a complete list of all assertions which are true for the plans in the set.

XAP/1 uses the second approach, because it is easier to find new assertions to apply to fully-specified assertion sets, and because it is easier to detect when such a set is complete. In order to maintain this representation, it must be able to find all new assertions which arise as consequences when an assertion is added to an assertion set.

We define a part $P$ to be *unconstrained* relative to a set of parts $\mathbf{S}$ if it has a proposed trajectory $\theta$ allowed by the assertion set that would let it be removed from $\mathbf{S}$ without violating any sequencing assertion in the assertion set, or any geometric constraint on $\theta$. Using this definition, the following two rules determine if an assertion is a consequence of a set of assertions $\mathcal{S}$:

The sequencing assertion $Q \prec P$ is a consequence of $\mathcal{S}$ if and only if there is a set of parts **S** containing both $P$ and $Q$ such that $Q$ is the only part in **S** that is unconstrained relative to **S**.

The trajectory assertion $P:\Theta$ is a consequence of $\mathcal{S}$ if and only if there is a set of parts **S** containing $P$ such that $P$ is the only part in **S** that is unconstrained relative to **S** and $P$ is constrained to precede some part in **S** for every allowed trajectory not in $\Theta$.

Proofs of these theorems are given in [21].

Unfortunately, these tests are computationally expensive to perform. In practice most (but not all) consequences can be found by simple applications of the transitive property and the geometric constraint rules. If a sequencing assertion $P \prec Q$ is applied, we look for assertions of the form $R \prec P$ or $Q \prec R$ and apply the transitive property if any are found. If we have a constraint $P:\Theta \Rightarrow P \prec Q$ then $P \prec Q$ is a consequence of any assertion $P:\Phi$ where $\Phi \subseteq \Theta$ and $P:(\Theta_P - \Theta)$ is a consequence of any assertion $Q \prec P$.

This method of consequence generation is relatively efficient, but it does not guarantee that all consequences will be found. If some consequence is overlooked, then it is possible that the negation of that assertion will be applied to the plan later. This will result eventually in the generation of consequences whose negations are already in the assertion set. If this kind of contradiction occurs, the assertion set represents an empty set of plans, so it is simply discarded from the search tree. Fortunately, this happens only occasionally, so the time lost is less than what would be spent in generating all consequences directly.

### 11.3.3 Criterion Modules

The criterion modules provide two services to the XAP/1 planner. First, they provide ratings of plan-sets, which are used by the search executive to select one to refine. Second, they select the assertions to apply to those plan-sets in order to refine them. Each of the modules is a specialist in a different criterion. In the current implementation, one module is designed to minimize fixture complexity, another minimizes the number of different insertion directions used, and a third avoids performing difficult operations with parts that are hard to manipulate. A single arbitration module combines the ratings generated by these criterion modules and selects among the assertions they propose.

The criterion modules have only limited information about the rest of the system. Though they function much like the knowledge sources in a blackboard system, there is no shared, global blackboard in XAP/1. Instead, each criterion module maintains its own representations of all the plan-sets on the tree. Whenever an assertion is applied to a plan-set, each criterion module is informed of the fact, and updates its own data structures appropriately. This leads to some redundancy in information storage, but allows each module to arrange its data structures optimally for its own purposes.

It should be noted that only the arbitration module knows what constraint modules are installed. The individual criterion modules do not intercommunicate directly. Similarly, only the consequence generator has any understanding of geometric and logical constraints. When the criterion modules propose assertions, they consider only the assertion itself, not any assertions which might arise as consequences. The exploration of the actual consequences of decisions is the job of the system as a whole.

The following sections will describe each of the modules. First the arbitration module will be briefly described. For the three criterion modules, we will begin by defining the evaluation function $f(\mathcal{P})$ that the criterion attempts to minimize. Then the algorithms used to generate $h(\mathcal{S})$, the optimistic estimate of the rating of the best plan in $\mathcal{S}$, and to propose assertions will be summarized. More detailed descriptions of these criterion modules are given in [21].

## Arbitration Module

The arbitration module is initialized with the list of criterion modules to be used, and a weight for each. Whenever an assertion set has been updated through the consequence generator, the arbitration module is called to compute a new rating for it. It does this by asking each installed criterion module to rate it, and computing the weighted sum of their ratings.

The arbitration module is also called when an assertion needs to be selected to refine an assertion set. To do this, it calls each installed criterion module and asks it to propose an assertion and give a rating for that assertion. The rating of the assertion estimates the difference in the ratings of the two plans that would be produced by applying the assertion and its negation. The arbitration module multiplies the ratings by the weights of the different criterion modules, and uses a simple voting technique to choose the one which seems likely to make the biggest difference.

Note that as the plans near completion, some or all of the criterion modules may make no proposals. This occurs when all the possible plans that satisfy the assertion set appear to be equally good. If no other criterion has any proposal to make, the arbitrator calls a default criterion, which selects assertions that seem likely to have many consequences, and thus will lead rapidly to a complete plan.

## Fixture Complexity Criterion

Fixture complexity measures the number of things which must be held in place during the execution of the assembly plan. Generally insertion operations should be sequenced so that the assembly holds itself together as well as possible during all stages of the assembly process. This is, of course, not the only criterion that effects the cost of fixtures. Directionality, for example, is relevant too.

Different types of fixtures would require different definitions of fixture complexity. For example, fixtures that can dynamically reconfigure themselves during the assembly process have different requirements than static fixtures which cannot. We will consider rigid, static fixtures, which must be able to fully restrain each part at the instant when it is least restrained by other parts of the assembly. For monotone plans, this will always be the moment immediately after insertion. Thus we will define the static fixture complexity of a plan as the sums of the degrees of freedom of the parts immediately after insertion.

To compute static fixture complexity, each part is first classified according to how well it is held in place immediately after insertion. It may be either "attached" such as a nut that has just been placed on a bolt; "lift-only" such as a square peg inserted into a square hole; or "lift-rotate" such as a washer just placed on a cylindrical peg. All other parts are considered to be "free." Parts are classified by using a set of rules which describe for each part what combinations of previously placed parts would put it in each group. These rules would have to be extracted from the CAD model in much the same way that the geometric constraints are. The $f(\mathcal{P})$ function is, thus, computed by classifying each part, counting the remaining degrees of freedom (0, 1, 2, or 6 for attached, lift-only, lift-rotate, and free parts respectively), and adding them up.

Optimistic estimates of the static fixture complexity rating of plan sets are produced by assuming that every part which could have been inserted before a part, is inserted before that part. Thus, in the initial plan, where no sequencing assertions have been made, it assumes that every part is inserted before every other part. Various techniques can be used to tighten up this estimate. For example, every plan must have a first part, which must be free, so XAP/1 always places at least one part into the free class.

To propose assertions with which to refine a plan, we look for an assertion that will harm the estimated rating. In particular, we look for a part $P$ that would be moved to a lower attachment class than we had optimistically placed it in, if we assert that some part $Q$ does not precede it. Then the assertion $P \prec Q$ would be proposed, because it hurts the plan's rating though its negation does not. The change in the part's rating which would be caused is returned as the rating of the proposed assertion.

### Directionality Criterion

The directionality criterion measures the number of different directions from which operations are performed. Generally, plans that work from a single direction are better than plans that require operations to be performed from all sides. They require a less dexterous robot, a simpler fixture, and fewer reorientations of the workpiece.

The directionality criterion counts how many different insertion directions are used in the plan. Sometimes different trajectories may be performed from the same direction. For example, spiraling a screw in from above and inserted a peg straight from above are different trajectories performed from the same direction. The directionality criterion is given a classification of trajectories into directions. The $f(\mathcal{P})$ function it computes is simply the number of directions that are used by at least one part in the plan $\mathcal{P}$.

A good optimistic estimate of $f(\mathcal{P})$ could be produced by finding the smallest set of directions such that every part has at least one proposed trajectory using that direction. However this is the NP-complete HITTING SET problem[6], and solving it for every node in the tree is impractical. Instead, XAP/1 finds all parts with only one direction proposed, and defines $U$ as the set of directions used by those. If every other part has a proposed direction in that set, then the plan-set's rating is $|U|$. Otherwise, if the intersection of the direction sets of the parts with no directions in $U$ is non-empty, the rating is $|U|+1$. Otherwise, the rating is $|U|+2$ is used. This gives a rating that is sometimes overly optimistic, but is still a good guide, especially later in the planning process.

To generate an assertion, the directionality criterion looks for a trajectory proposal which would move us from the $|U|$ case to the $|U|+1$ case, or from the $|U|+1$ case to the $|U|+2$ case. For example, in the first case, we propose that a part that has some trajectories with directions in $U$ and some with directions not in $U$ use one of the latter trajectories. The rating of the trajectory proposed will always be 1, since that is the largest difference a single proposal can make in the plan-sets estimated rating.

### Manipulability Criterion

The manipulability criterion favors doing difficult operations with parts that are easy to handle. For example, when attaching a spark plug to an automobile, it is the spark plug, and not the automobile, that should be rotated. To generate a manipulability rating for a plan, each part is given a manipulability rating, and each trajectory is given a difficulty rating. The product of these gives the rating for an operation. The manipulability of a plan is the sum of the ratings for all the operations.

The manipulability rating of a part tells how hard the part is to handle with a robot relative to how hard it is to hold with a fixture. The mass of a part is a reasonable estimate of this, since heavy parts are much harder to handle than to fixture. Very small parts are hard to handle, but they are also hard to fixture. The difficulty of a trajectory depends on the complexity of the motions and the amount of fine motion control involved. This would be based

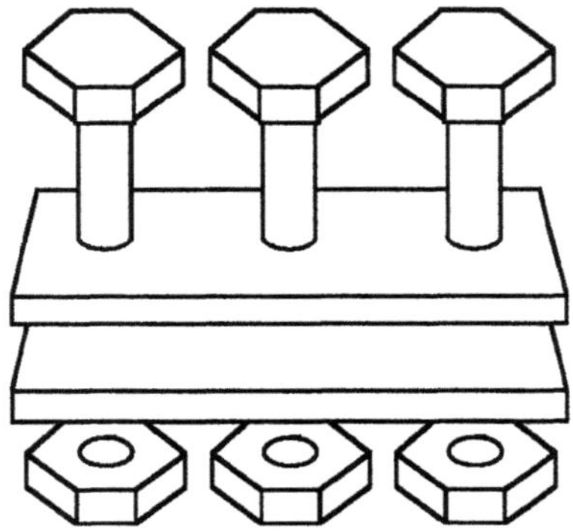

Figure 11.5: Example assembly consisting of 3 bolts holding together 2 plates.

on a classification of trajectories into types. For example, screwing trajectories, and trajectories to insert snap-together parts would be types of trajectories that are more difficult than simple insertions.

XAP/1 estimates manipulability ratings in a manner similar to that used with the static fixture complexity criterion. Each part is evaluated separately, assuming that it uses the simplest trajectory proposed for it. To propose assertions, it looks for a part with trajectories with different difficulties proposed for it. It then proposes the deletion of all but the most difficult trajectories, rating that proposal with the difference in the difficulties times the manipulability.

## 11.4 Performance

The performance of the XAP/1 system varies widely according to the relative weights of the criteria, and the particular characteristics of the plan. Predictably, it works best when one criterion dominates the others, or when the proposed trajectories have a large number of constraints.

As an example, we will consider a class of assemblies similar to the one shown in figure 11.5. These contain two plates which are held together by a set of $b$ nut/bolt pairs. This is a fairly unconstrained assembly, so that, even without subassemblies, there are $4^{b+1}$ possible plans to build an assembly with $b$ bolts. We will use all three criteria, choosing weights that leave them fairly evenly balanced, with a slight edge being given to the static fixture complexity criterion.

Figure 11.7 shows the search tree generated by XAP/1 in the course of finding a plan for an assembly with three bolts. This tree clearly shows how the XAP/1 system makes obvious decisions first. Note that the upper portion of the tree

| number of bolts | nodes expanded | CPU time | tree memory |
|---|---|---|---|
| 1 | 4 | 0.3 sec | 8K |
| 2 | 9 | 0.4 sec | 17K |
| 3 | 31 | 0.7 sec | 64K |
| 4 | 65 | 1.3 sec | 159K |
| 5 | 122 | 2.4 sec | 336K |
| 6 | 206 | 4.6 sec | 492K |
| 7 | 329 | 8.0 sec | 493K |
| 8 | 496 | 13.6 sec | 494K |
| 9 | 730 | 22.0 sec | 495K |
| 10 | 1046 | 36.0 sec | 502K |
| 11 | 1462 | 57.3 sec | 504K |
| 12 | 2298 | 125.9 sec | 506K |

Figure 11.6: Performance of XAP/1 with a 488K soft memory limit on assemblies consisting of $b$ bolts holding together 2 plates.

is very narrow, with very little expansion of side branches, while near the end it bells out into a nearly exhaustive search. In the upper part of the tree, one alternative is usually far superior to the other, so only that one need be explored much further. In the lower part of the search, the system is primarily attempting to decide in which order the bolts and the plates are to be inserted. Since this actually makes little difference to the quality of the plan, the system must inspect many nearly equal alternatives. The search tree would have been much larger if these comparatively trivial decisions had not been deferred to the end. If we had started with one of these decisions, the size of the tree would likely have been nearly doubled, since both subtrees would require almost as much exploration as the entire tree in this example. This demonstrates the importance of being able to make decisions in an opportunistic fashion.

Running XAP/1 on a number of such assemblies with varying numbers of bolts gives rise to the table in figure 11.6. The CPU times here are based on an implementation in C on a Sun 3/50 workstation. As expected, the time grows exponentially with the size of the assembly, approximately doubling with each additional bolt. This kind of behavior is not really satisfactory. Generating a plan for four bolts shouldn't be much harder than generating one for three bolts. This arises largely because the planner currently must consider every possible sequence in which the screws can be inserted separately. Given any valid plan for this assembly, we can generate $n!$ other valid plans by rearranging the labels of the $n$ nut/bolt pairs. All these plans are equivalent as far as the three installed criteria are concerned. Yet the planner must partially generate many of these $n!$ plans to make sure none is better than the other.

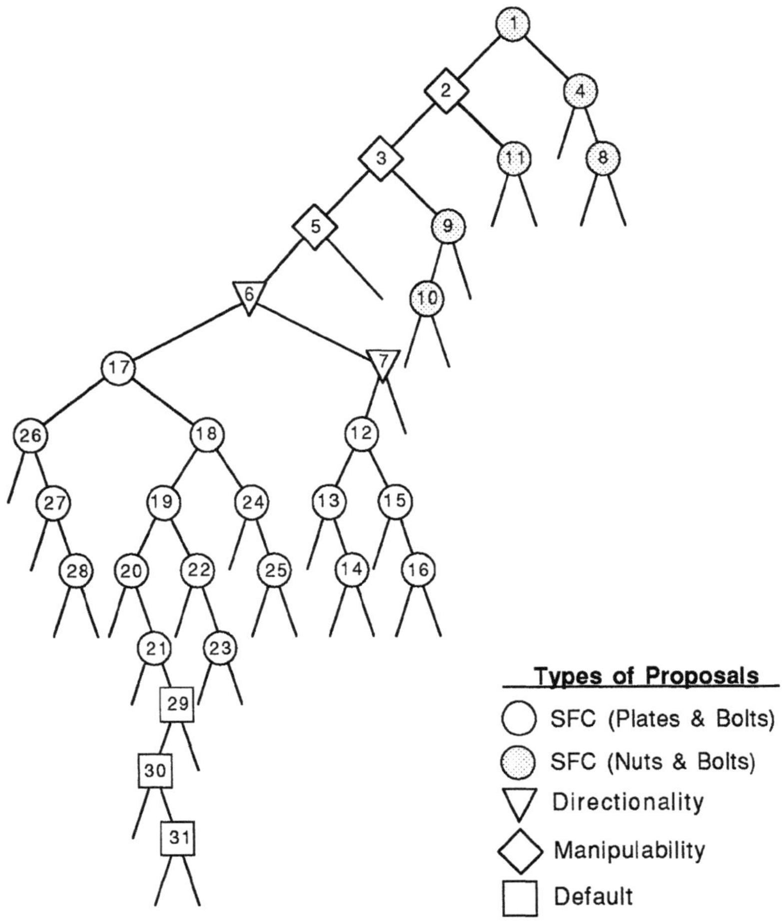


Figure 11.7: Search tree generated in solving the problem in figure 11.5 showing internal nodes numbered in the order in which they were refined and identifying the criterion module that proposed the assertion used in its refinement.

This kind of situation, where there are several similar structures in an assembly, arises very commonly in actual assembly problems. It is clear that future assembly planners should be able to deal more effectively with such cases. A form of internal learning could be used to recognize when an incomplete assertion set is similar to one previously generated. However detecting similar sets can be computationally quite expensive. An alternative approach would be to use unquantified part variables to make plans of the form "insert one screw from above and two from below".

The memory usage in the table includes only the data structures, not the program text. Note that the memory usage grows exponentially until it passes a user-defined limit of 488K. At that point, the XAP/1 system begins discarding the data structures describing poorly rated plan-sets to recover the memory devoted to them. Only the set of assertions that was used to generate the plan-set is kept, so that the other data structures can be rebuilt if necessary. This strategy proves very effective—it is possible to collapse a very large fraction of the nodes without needing to rebuild any—so that even for fairly large problems the increase in memory usage is slight. This is possible because only half of all leaves produced will ever be expanded, and the opportunistic planning methodology ensures that many of those produced will be very poorly rated. Of course, even using this trick, the growth in memory will continue to be exponential, but the constant is small enough to allow large problems to be solved.

The values given in the table in figure 11.6 cannot be taken as typical. The assemblies used are unusually unconstrained and the criteria are more evenly weighted than they would be in realistic applications. Figure 11.8 shows a more realistic assembly which contains 34 parts and for which 26 trajectories were proposed. The 26 trajectories include 6 straight-line insertion trajectories (from above, below, left, right, front and back), 16 spiralling insertion trajectories (two for each screw attachment: one for the male part, one for the female), and a rivet insertion trajectory. Generating the plan shown required about 5.6 seconds of CPU time and 563K of memory (for both program and data structures) on the Sun 3/50. Depending on the weights chosen, the CPU usage may rise to nearly 15 seconds.

Since the set of criteria used is quite small, the plan suffers from a few defects. The insertion of the tray being done in the midst of the foot insertions might require an extra tool change, for example. However, no tool change criterion was installed and it is a good plan under the criteria used. For example, the directionality criterion causes it to avoid doing any operations from below and the manipulability criterion causes it to avoid turning the base onto any of the screws.

It should also be noted that the CPU times reported here do not include the time to generate trajectory proposals, geometric constraints, and similar information from a CAD model. This information is currently generated manually, but work is also in progress on the development of an interface to a geometric modeling system for the XAP/1 system.

The XAP/1 planner, thus, is able to generate reasonable plans for realistic assemblies very quickly. Its major failing is that it cannot produce plans which use subassemblies. The speed with which XAP/1 generates linear plans, however, makes it a promising base for a system to solve more general assembly planning problems. Some initial work has been done in describing an expanded set of three types of assertions that would support the opportunistic generation of plans with subassemblies[23,22]. With these assertions the system can intermix decisions about the sequencing, trajectory assignment, and subassem-

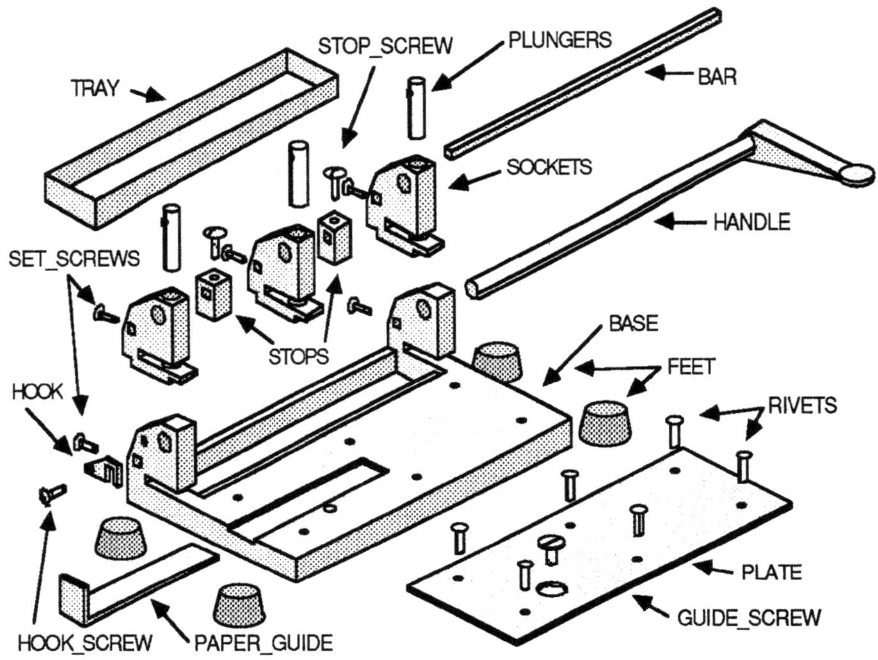


(1) insert FOOT_B from above
(2) insert FOOT_D from above
(3) insert TRAY from above
(4) insert FOOT_A from above
(5) insert FOOT_C from above
(6) snap in BASE from above
(7) insert STOP_A from above
(8) insert STOP_B from above
(9) insert SOCKET_C from above
(10) insert PLUNGER_C from above
(11) insert SOCKET_A from above
(12) insert PLUNGER_A from above
(13) insert SOCKET_B from above
(14) insert PLUNGER_B from above
(15) insert HANDLE from right
(16) insert HOOK from left
(17) screw HOOK_SCREW from left
(18) insert PLATE from above
(19) rivet in RIVET_D
(20) rivet in RIVET_F
(21) rivet in RIVET_B
(22) rivet in RIVET_C
(23) rivet in RIVET_E
(24) rivet in RIVET_A
(25) insert PAPER_GUIDE from left
(26) insert GUIDE_SCREW from above
(27) insert BAR from right
(28) screw STOP_SCREW_B from above
(29) screw SET_SCREW_C from behind
(30) screw SET_SCREW_E from behind
(31) screw SET_SCREW_A from behind
(32) screw STOP_SCREW_A from above
(33) screw SET_SCREW_B from behind
(34) screw SET_SCREW_D from behind

Figure 11.8: Plan produced by XAP/1 for a 34-part hole punch assembly.

bly structure of the plan arbitrarily, leaving it free to make the most obvious decisions first throughout the planning process.

## Acknowledgements

The author would like to thank Dr. Richard Volz and Dr. Tony Woo for their invaluable support and guidance in the work described here. This research was supported by the Air Force Office of Scientific Research under contract number F33615-85-C-5105. The author was funded in part by an IBM Graduate Pre-Doctoral Research Fellowship in Manufacturing Research.

## References

[1] A. Bourjault. *Contribution a une approche méthodologique de l'assemblage automatisé: Elaboration automatique des séquences opératiores.* PhD thesis, L'Université de Franche-Comté, November 1984.

[2] B. K. Choi, M. M. Barash, and D. C. Anderson. Automatic recognition of machined surfaces from a 3D solid model. *Computer-Aided Design*, 16(2):81–86, March 1984.

[3] R. Dawson. On removing a ball without disturbing the others. *Mathematics Magazine*. 57(1):27–30, January 1984.

[4] T. L. De Fazio and D. E. Whitney. Simplified Generation of All Mechanical Assembly Sequences. *IEEE J. of Robotics and Automation*. 3(6):640–658, December 1987.

[5] T. L. De Fazio and D. E. Whitney. Correction to "Simplified Generation of All Mechanical Assembly Sequences." *IEEE J. of Robotics and Automation*. 4(6):705–708, December 1988.

[6] M. R. Garey and D. S. Johnson. *Computers and Intractability: A Guide to the Theory of NP-Completeness*. W. H. Freemand and Co., New York, 1979.

[7] L. Guibas and F. Yao. On translating a set of rectangles. In *12th Annual ACM Symposium on Theory of Computing*, pages 154–160, April 1980.

[8] M. R. Henderson and D. C. Anderson. Computer Recognition and Extraction of Form Features: A CAD/CAM Link. *Computers in Industry*. 5(4):329–339, December 1984.

[9] R. L. Hoffman. Assembly planning for CSG objects. Technical Report 8916, Automation Sciences Lab, Northrop Research and Technology Center, May 1989.

[10] R. L. Hoffman. Automated assembly in a CSG domain. In *IEEE Intl. Conf. on Robotics and Automation*, pages 210–215, May 1989.

[11] L. S. Homem de Mello and A. C. Sanderson. A correct and complete algorithm for the generation of mechanical assembly sequences. In *IEEE Intl. Conf. on Robotics and Automation*, pages 56–61, May 1989.

[12] J. E. Hopcroft, J. T. Schwartz, and M. Sharir. On the Complexity of Motion Planning for Multiple Independent Objects: PSPACE-hardness of the 'Warehouseman's Problem' *Robotics Research*. 3(4):76–88, May 1988.

[13] W. Jentsch and F. Kaden. Automatic generation of assembly sequence. In I. Plander, editor, In *Artificial Intelligence and Information-Control Systems of Robots*, pages 197–200. Elsevier Science Publishers, North-Holland, 1984.

[14] S. Joshi and T. C. Chang. Graph-based heuristics for recognition of machined features from a 3D solid model. *Computer-Aided Design*, 20(2):58–66, March 1988.

[15] H. B. Olsen. A flexible robotic workcell for the assembly of airframe components. In *IEEE Intl. Conf. on Robotics and Automation*, pages 1278–1283, May 1990.

[16] J.-R. Sack and G. T. Toussaint. Separability of pairs of polygons through simple translations. *Robotica*, 5(1):55–63, January–March 1987.

[17] R. Sedgewick. *Algorithms*. Addisson-Wesley, Reading, Massachusetts, second edition, 1988.

[18] R. E. Tarjan. Depth first search and linear graph algorithms. *SIAM Journal of Computing*. 1(2):146–160, 1972.

[19] J.-M. Valade. Geometric reasoning and automatic synthesis of assembly trajectory. In *Proceedings of Int'l Joint Conf on Advanced Robotics*, pages 43–50, 1985.

[20] J. Wolter and P. Chandrasekaran. Knowledge representation for design of mechanical assemblies. Technical Report 90-026, Texas A&M University, Computer Science Dept., October 1990.

[21] J. Wolter. *On the Automatic Generation of Plans for Mechanical Assembly*. PhD thesis, University of Michigan, Dept. of Computer, Information and Control Engineering, September 1988.

[22] J. Wolter. A constraint-based approach to planning with subassemblies. In *IEEE Intl. Conf. on Systems Engineering*, pages 412–415, August 1990.

[23] J. Wolter. Representing subassembly trees by deepest common ancestor relations. Technical Report 90-009, Texas A&M University, Computer Science Dept., May 1990.

# Chapter 12

# A common sense approach to assembly sequence planning

**Richard Hoffman**

Humans are adept at solving assembly and disassembly problems due to the versatility of the hands, sophisticated sensing feedback, and common sense reasoning. Common sense reasoning is most important since it allows educated guesses to be made about reasonable operations to make, and visualization of operations and their consequences. In this chapter, an operation is a translation or rotation of a component or multicomponent subassembly. Educated guesses are a result of taking advantage of previous experience.

Despite this, manual generation of detailed assembly plans can be tedious and time consuming, particularly in environments prone to frequent design changes. Small batch environments typically produce only units or tens of a given product so that the manual assembly plan generation is a substantial portion of the total product cost. Therefore, with a vision of delegating the assembly plan generation to computers, automated assembly planning has become an area of intense research.

In implementing an assembly planner, one could begin with individual components of the assembly and search for a sequence of operations that puts them together. Nevertheless, the correct order in which individual pieces must be merged into the intermediate assembly is not easy to deduce a priori, resulting in a large amount of false starts and, hence, backtracking. On the other hand, the number of movable subassemblies is more tractable if one begins with the assembled product and searches for a sequence of reversible operations that will disassemble the product. Hence the assembly planning problem is normally solved as a disassembly problem.

Automated assembly sequence planners can predict movability of subassemblies in two ways. The first assumes that enough high-level information is included in the problem statement to provide an insertion direction. For example, component liaisons [1] represent contacts and types of contact between components. Systematically breaking these liaisons corresponds to disassembly of the product. Some systems require human interaction to verify directions of insertion [2]. Such approaches only deal with assemblies involving single move insertions of components and subassemblies, because multiple move insertions would require supplying liaison information for intermediate stages of assembly.

The second way to predict subassembly movability is to use geometric data supplied by CAD models of objects. The complexity of this task depends on the complexity of the models and their representations. The simplest case is with polyhedral objects [3,8]. A more complex class of models involves objects produced by combining "nameable" surfaces such as planes and cylindrical surfaces, and solid shape primitives such as boxes, cylinders, and spheres [4,5,10]. The most complex case addressed to date involves objects with parameterized surface patches expressed as bicubic equations [6]. Such models are called boundary representation (B-rep) models.

Bicubic surfaces are common in industrial applications where objects need to have specific aerodynamic properties or for aesthetic reasons. Such surfaces are useful for the design of components that must satify design criteria such as stiffness, mass, strength, moments of inertia, and tools and materials that are available to manufacture components. Therefore it is important that assembly planners be capable of dealing with products with sophisticated representations in order to be useful to industry.

Unfortunately, the computational costs associated with deriving freedom of motion increases with increased model complexity. This cost makes finding shortcuts to freedom determination derivation very attractive. Fortunately, a few simple common sense rules provide substantial computational savings.

This chapter presents the **B-r**ep **A**ssembly **En**gine (BRAEN), a system that generates disassembly sequences of products whose components are represented as B-rep objects with bicubic surfaces. Static workcell objects such as tabletop and walls are also included in the solution process as environment components. Both translational and rotational operations may be involved in the assembly sequence. Also, subassemblies do not need to be removed in a single operation.

We demonstrate how a few common sense rules can be used to reduce the cost of automated assembly sequence planning. For example, rules are presented that allow previous calculations of freedom of motion to be reused for other spatial configurations of components. Also, we demonstrate heuristics to generate promising operations when no moves that break an assembly into two subassemblies are available, and reason about the effect of gravity and stability on assembly to help ensure that disassembly operations are reversible. These results extend previous work reported in [6] and [7].

Section 12.1 presents the overall procedure of BRAEN, Section 12.2 discusses specific applications of common sense knowledge in BRAEN, with experimental results presented in Section 12.3. Section 12.4 discusses the performance and future extensions of the system.

## 12.1 Approach

BRAEN derives an assembly sequence for a product sitting on the flat work area of a workcell such as a tabletop. When an assembly is broken into two subassemblies (namely, the parts that are moved and the parts that do not move during the operation), the system places single components resulting from this operation by the edge of the work area. Multicomponent subassemblies are kept in the central work area.

The system assumes a single robot. Therefore, if an operation results in a subassembly that is suspended in midair — held by the robot — then that subassembly must continue to be moved until it is set down. Only then can an operation can be performed on other components.

The system distinguishes between two classes of components: environment and product. *Product* components are part of the product to be disassembled, whereas *environment* components are fixed (immovable) elements of the workcell environment, such as tabletop or wall. The role of environment components is to act as a support for the assembly (in the presence of gravity) and to define the boundaries of the workcell.

BRAEN assumes that components are rigid. For example, this means that there are no elastic or plastic interactions involved in assembly that could create irreversible operations.

The overall procedure of BRAEN is illustrated in Figure 12.1. An assembly is selected from the center area of the table. If there is more than one assembly to pick from, the selection method is important because it is best to find out as early as possible if a product cannot be disassembled. Therefore, the selected assembly should be more complex than other assemblies. BRAEN uses the number of components in an assembly as a heuristic measure for complexity.

The *Disassembly Module* generates a sequence of operations that break the

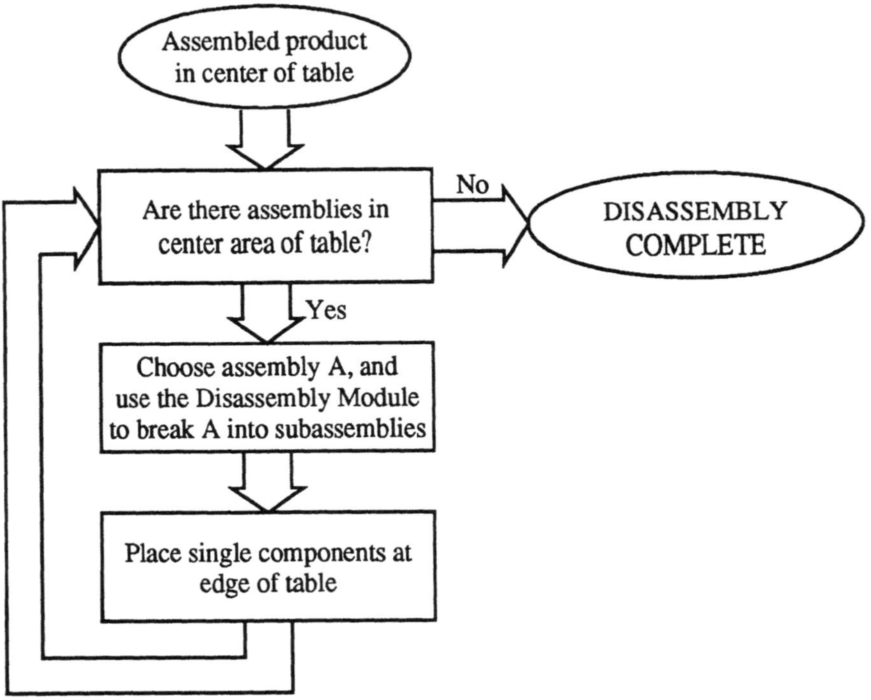


Figure 12.1: Operation of BRAEN

selected assembly into two subassemblies. Each operation is a translation or rotation of a component or multicomponent subassembly. The final operation of the sequence must be a move that physically separates the assembly into two disjoint pieces, called a *separating move.*

The operation of the *Disassembly Module* is shown in Figure 12.2. There are two fundamental processes in this module, *freedom determination* and *search strategy.* Information obtained from these two processes is used to construct a search graph, whose nodes correspond to potential operations on subassemblies.

The function of each process is summarized as follows:

**Freedom determination.** Given a spatial configuration (set of positions) of the components in an assembly, the *freedom determination* routine provides information about which subassemblies can move, directions they can move, and how far they can move before colliding with other components.

**Search strategy.** Given the information provided by *freedom determination,* it is necessary to generate sequences of subassembly motions that will

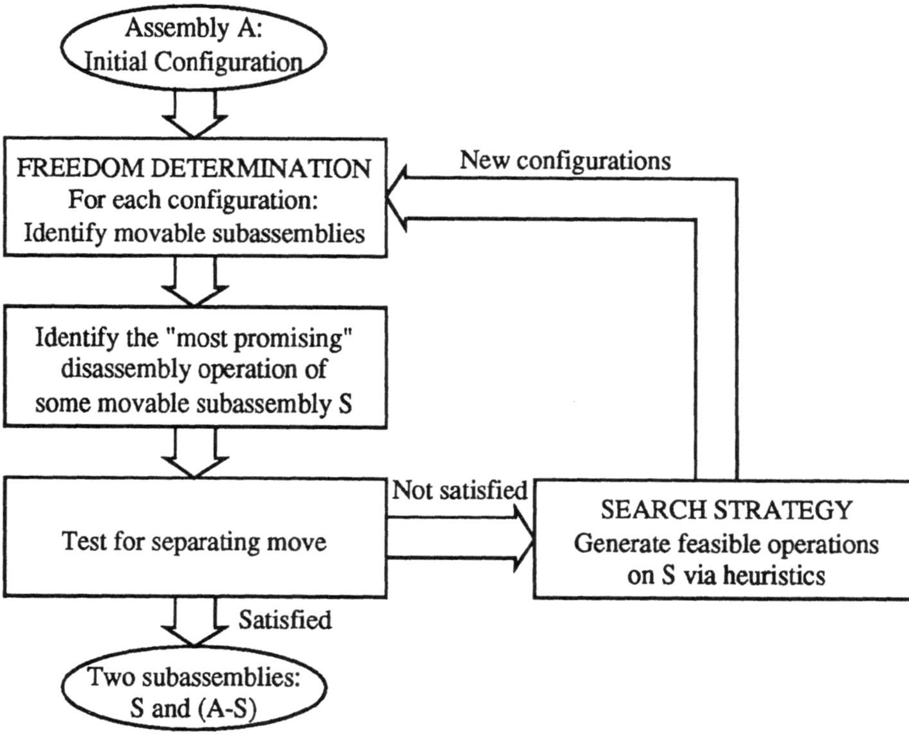


Figure 12.2: BRAEN Disassembly Module

provide disassembly. Since a disassembly sequence will not, in general, consist only of moves that result in collision, it is necessary to determine:

- Which of a number of feasible subassembly moves seems to be the most promising;
- At which of the uncountably many positions along one movement trajectory the motion should be stopped to allow moving a different subassembly or changing the trajectory of movement.

### 12.1.1 Component Representation

All components are represented in BRAEN as sets of oriented bicubic surface patches. The orientation of a patch indicates that side of the patch facing outward from the interior of the component.

In addition, mass, center of mass, and sets of component features are supplied for each component. Component features may be derived by feature extraction routines or supplied explicitly by a feature-based modeler. For each component, there are three types of features:

**Direction** features indicate major axes of the component. For example, "box" structures on a component supply six directions (e.g., $\pm x$, $\pm y$, $\pm z$ directions), and "cylindrical" or "conical" structures supply two directions, along the axis of symmetry. Direction features are unit vectors with tails at the origin.

**Center** features indicate centers of "protrusions" or "holes" of the component. These are useful for postulating moves that line up features of different components. Center features are 3D points.

**Axis** features indicate rotational axes of a component. For example, "cylindrical" or "conical" structures each provide one axis feature. If a component displays rotational symmetry about an axis, then that axis is not an axis feature. Axis features are unit vectors.

Figure 12.3 shows the representation for the *key* object shown in Figure 12.4. This object belongs to the *capsule* assembly shown in Figure 12.5, along with a *roof* component which has tabs that slide in tracks in the *base* component. The first part of the representation gives 40 bicubic surface patches; the second part supplies the component features, mass, and center of mass (COM).

### 12.1.2 Freedom of Motion Determination

We call the set of positions of all components the *spatial configuration*, denoted by $\Pi$. The position of each component is given by a $4 \times 4$ transformation matrix and reports the results of any translational and rotational operations performed on the component. The freedom of motion of a subassembly depends on the relative positions of other components, that is, on $\Pi$.

Two types of freedom of motion are obtained, translational and rotational. Given an assembly $A$ (in configuration $\Pi$) to be decomposed by the *Disassembly Module* into two subassemblies, the possible directions of translation are supplied by the **direction** features of components in $A$, and the possible axes of rotation are supplied by the **axis** features of components in $A$. These direction and axis features are, of course, modified to reflect the current position of $A$.

To determine the translational freedom of a subassembly $S$ in a direction $d$, all patches of components of $S$ are translated in direction $d$ until they collide with patches not in $S$. The smallest motion of any patch in $S$ indicates the maximal collision-free translation of subassembly $S$ in direction $d$, denoted as $\Delta^d(\Pi, S)$. To illustrate, Figure 12.6 shows a 2D example of this process. Here, "patches" are 2D curves. The freedom of translation of subassembly $S$ is being evaluated for the direction $d$ (translation to the right in the figure). There

```
(
  ;;; Bezier patch definitions
  ;;; X, Y, and Z matrices of control points
  ((( 0.5000000  0.5000000  0.5000000  0.5000000 )
    (  0.5000000  0.5000000  0.5000000  0.5000000 )
    (  0.5000000  0.5000000  0.5000000  0.5000000 )
    (  0.5000000  0.5000000  0.5000000  0.5000000 ))
   (( -0.2000000 -0.0666667  0.0666667  0.2000000 )
    ( -0.2000000 -0.0666667  0.0666667  0.2000000 )
    ( -0.2000000 -0.0666667  0.0666667  0.2000000 )
    ( -0.2000000 -0.0666667  0.0666667  0.2000000 ))
   ((  0.0000000  0.0000000  0.0000000  0.0000000 )
    (  0.7500000  0.7500000  0.7500000  0.7500000 )
    (  1.5000000  1.5000000  1.5000000  1.5000000 )
    (  2.2500000  2.2500000  2.2500000  2.2500000 )))
                        .
                        .
      plus 39 other patches not shown
                        .
                        .
  ) ;;; end of patch definitions
(
  ;;; component features
  (:DIRECTION (0.0 0.0 1.0) (0.0 0.0 -1.0))
  (:CENTER (PROTRUSION (0.0 0.0 1.125)))
  (:AXIS ((0.0 0.0 0.0) (0.0 0.0 1.0)))
      ;axis from cylindrical bottom of key
      ;rotational axis passes through (0,0,0)
      ;in z (0,0,1) direction.
  (:MASS 1.3062)
  (:COM (0.01375 6.8e-4 0.3309))
  ) ;;; end of component features
```

Figure 12.3: Key object representation

Figure 12.4: Key object

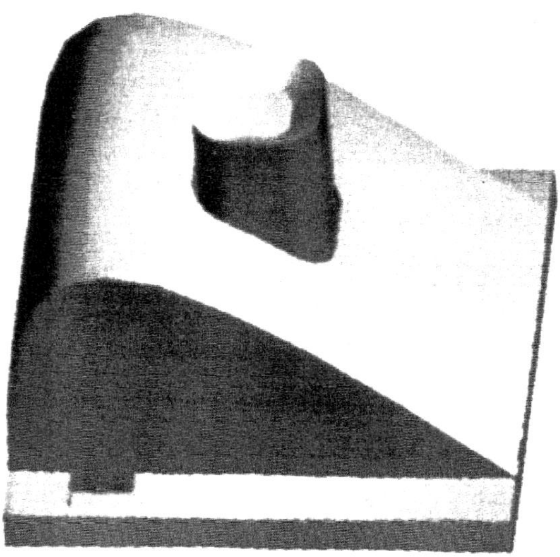

Figure 12.5: Capsule assembly

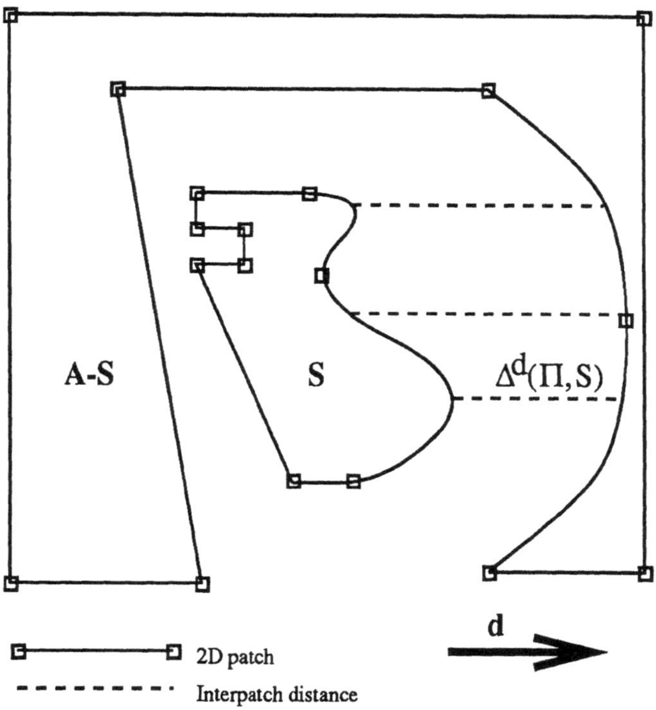


Figure 12.6: Derivation of $\Delta^d(\Pi, S)$

are three interpatch distances involved in this evaluation; the linear patches of $S$ are either parallel to $d$ or face away from the direction $d$ and thus do not contribute to the result. The shortest interpatch distance shown indicates the value $\Delta^d(\Pi, S)$.

The distance that a patch $P_1$ may translate in direction $d$ before colliding with a patch $P_2$, denoted by $\delta^d(P_1, P_2)$, is computed recursively. Patches $P_1$ and $P_2$ are each subdivided into subpatches, and an approximate distance of translatability in direction $d$ is derived of every subpatch of $P_1$ to every subpatch of $P_2$. A polyhedral approximation of each subpatch reduces these calculations to distances between triangular facets. The distance is evaluated for those pairs of subpatches with smaller distances, until the desired level of accuracy is attained.

To determine the rotational freedom of a subassembly $S$ about an axis $a$, all patches of components of $S$ are rotated about axis $a$ until they collide with patches not in $S$. As with translation, the smallest motion of any patch in $S$ indicates the maximal collision-free rotation of $S$ about axis $a$. This is denoted as $\Theta^a(\Pi, S)$, and reports the angle of rotation available in both clockwise and counterclockwise directions about axis $a$. The angle through which patch $P_1$ may rotate about axis $a$ before colliding with patch $P_2$, denoted by $\theta^a(P_1, P_2)$, is computed analogously to $\delta^d$.

From a given spatial configuration $\Pi$ and assembly $A$, the *freedom determination* module determines a set of potential operation triples $\mathcal{M}(\Pi, A)$,

$$\mathcal{M}(\Pi, A) = \{\langle S_i, t_i, f_i\rangle\}$$

where

$S_i$ specifies the subassembly to be moved in the $i$th potential operation;

$t_i$ indicates the trajectory of the $i$th potential operation, either a direction of translation or an axis of rotation of $S_i$;

$f_i$ is the freedom of motion of $S_i$ for trajectory $t_i$: if $t_i$ is a translational direction $d$, then $f_i = \Delta^d(\Pi, S_i)$, and if $t_i$ is a rotational axis $a$, then $f_i = \Theta^a(\Pi, S_i)$.

Generation of $\mathcal{M}(\Pi, A)$ is accomplished as follows:

### Identify translatable components

For each component $C$ in $A$, evaluate the freedom of translational motion of $C$ along directions indicated by the **direction** features of components in $A$. A potential operation triple $\langle C, d, f\rangle$ is supplied when a component $C$ can translate a non-zero distance $f$ in a direction $d$.

### Identify movable subassemblies

For each component $C$ and direction $d$ (from the set of directions indicated by the **direction** features of components in $A$) for which $\Delta^d(\Pi, C) = 0$, determine the subassembly $S$ containing $C$ which must move when $C$ moves an incremental distance in direction $d$. A potential operation triple

$$\langle S, d, \Delta^d(\Pi, S)\rangle$$

is supplied if $S$ is a proper subset of $A$. This disallows operations that move all of $A$ or move any environment components.

### Identify rotatable components

For each component $C$ in $A$, evaluate the freedom of rotational motion of $C$ about those axes indicated by the **axis** features of $C$. A potential operation triple is supplied when a component can rotate through a clockwise or counterclockwise non-zero angle about an axis.

The current implementation only considers rotations of single components, primarily because applying a torque to one component $C$ of a subassembly does not necessarily lead to rotation about the same axis of other components that are pushed by $C$ as a result of the rotation. If $C$ sufficiently constrains the freedom of the other components then they will rotate as a unit; we need to determine these conditions of sufficient constraint before allowing rotating subassemblies.

As an example, let $A$ be the capsule assembly, with its initial configuration $\Pi_0$ as shown in Figure 12.5, then $\mathcal{M}(\Pi_0, A)$ contains two potential operation triples:

- $S_1 = key$, $t_1 =$ translation $+z$ direction, and $f_1 = 0.832$ units.
- $S_2 = base\ \&\ roof$, $t_2 =$ translation $+z$ direction, and $f_2 = 0.1$ units.

### 12.1.3 Search Strategy

The function of the *Disassembly Module* is to search for a sequence of operations that will break an assembly $A$ into two subassemblies. The $A^*$ algorithm sans heuristic estimator [11] is used for a best-first search over a graph whose nodes correspond to feasible operations on specific configurations.

Given a potential operation triple $\langle S, t, f\rangle$, the operation $\langle S, t, f^*\rangle$ is an $\langle S, t, f\rangle$-*feasible operation* if $f^*$ does not exceed the freedom of motion indicated by $f$. For example, if a subassembly $S$ can translate at most 2 units in the $t$ direction, then translations of $S$ from 0 to 2 units in the $t$ direction are $\langle S, t, 2\rangle$-feasible. The techniques used to identify a finite number of $\langle S, t, f\rangle$-feasible operations from a potential operation $\langle S, t, f\rangle$ are discussed in Section 12.2.2.

When the *Disassembly Module* begins, the freedom of motion of an assembly $A$ within an initial configuration $\Pi_0$ is evaluated to obtain the set of potential operations $\mathcal{M}(\Pi_0, A)$. For each potential operation triple $\langle S, t, f\rangle$ in $\mathcal{M}(\Pi_0, A)$, a number of $\langle S, t, f\rangle$-feasible operations $\langle S, t, f^*\rangle$ are identified, each of which is paired with the configuration $\Pi_0$ to form a node $(\Pi_0, \langle S, t, f^*\rangle)$.

A node is a goal node if its operation is a separating move. At any stage of search in which no goal node is found, it is necessary to expand the node having maximal merit. A node $(\Pi, \langle S, t, f\rangle)$ is expanded as follows:

- Apply the operation $\langle S, t, f\rangle$ to configuration $\Pi$, obtaining a new configuration $\Pi'$. If there already exists nodes with configuration $\Pi'$, then the new configuration has already been encountered, so no new nodes are generated.

- Otherwise, derive the set of potential operation triples $\mathcal{M}(\Pi', A)$.
- For each potential operation $\langle S', t', f' \rangle$ in $\mathcal{M}(\Pi', A)$, identify $\langle S', t', f' \rangle$-feasible operations; for each feasible operation $\langle S', t', f^* \rangle$ create a new node $(\Pi', \langle S', t', f^* \rangle)$.

The merit of a node $(\Pi, \langle S, t, f \rangle)$ is based on favoring the following node characteristics:

- Large magnitude of motion.
- Large $|S|$; that is, a large number of parts.
- Trajectories that involve robot approach from above the assembly, and therefore provide easier (uncluttered) access to the assembly. This is indicated by a large $z$-component to the trajectory.
- Small total path length to get to configuration $\Pi$ from the initial configuration $\Pi_0$.

In particular, the associated merit value is infinite if and only if the potential operation provides a separating move, guaranteeing that if a separating move is available the search will terminate.

To illustrate, Figure 12.7 shows the search graph generated by the Disassembly Module to break the capsule assembly in its initial configuration into two subassemblies. All operations are translations for this example, therefore each node specifies a configuration, subassembly, translation trajectory, and translation magnitude. Configuration $\Pi_0$ is the initial configuration (Figure 12.5), and in general the configuration of a node corresponds to application of the operations specified by the node's ancestors. The shaded node indicates the separating move for the assembly.

## 12.2 Common Sense

Brute force manipulation of geometric data can require prodigious amounts of computation. Computation of freedom of translational and rotational motion is expensive because the number of evaluations of $\delta$ and $\theta$ is on the order of the square of the number of patches in all components. The search strategy is problematic because, given a potential operation of magnitude $m$ along a trajectory $t$, it is impossible to examine all feasible operations along the trajectory. This section examines the intuitively simple techniques used by BRAEN for freedom of motion determination and search strategy.

In addition, some operations that are geometrically feasible cannot be performed with physical objects. For example, if one subassembly is sitting on

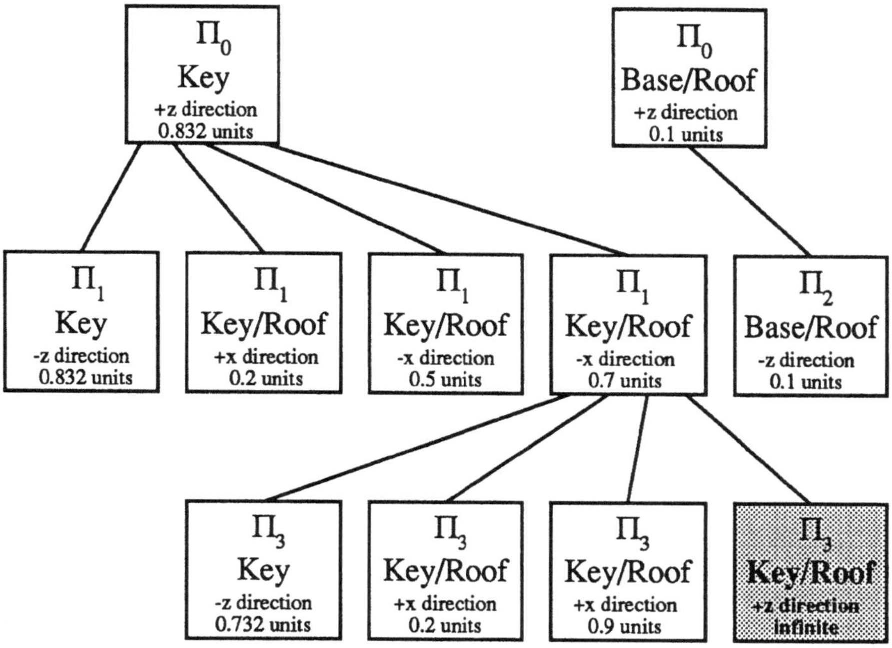


Figure 12.7: Search graph for capsule breakup

top of another subassembly, then moving the bottom subassembly will cause the top subassembly to move as well. We show how simple physics is used to model gravity and stability issues in planning the assembly sequence.

### 12.2.1 Freedom of Motion

There are situations in which the freedom of $S$ in configuration $\Pi$ can be used to obtain the freedom of a subassembly $S'$ in a configuration $\Pi'$. These situations — essentially indicated by common sense rules on how $S'$ may differ from $S$ and how $\Pi'$ may differ from $\Pi$ and still allow the same freedom result (or a slight variant) to hold — allow significant reductions in the number of evaluations of $\Delta$ and $\Theta$.

The rules for translational freedom are best visualized by considering a train on an infinitely long straight rail. Without loss of generality, assume the track runs east-west. The train on the track may be impeded to the east by an east barricade and to the west by a west barricade. The track accessible to the train is the swept area of the train as it passes from its west-most position to its east-most position.

Three observations may be made:

1. Objects that do not intersect the accessible track may be moved anywhere except on the accessible track and still not modify the freedom of the train.

2. If there is an east barricade, then at the train's extreme east-most position some components of the train will collide with components of the barricade. As long as at least one pair of colliding components remains, other components of the train and east barricade can be moved off the track without modifying the train's eastward freedom. The same argument holds for other freedoms.

3. Moving the train (legitimately) $r$ units to the east reduces its eastward freedom by $r$ and increases its westward freedom by $r$. Moving the east barricade due west by $r$ units decreases the train's eastward freedom by $r$ but leaves its westward freedom unchanged.

To map back to the disassembly problem, let the train be a subassembly $S$. The accessible track corresponds to the volume swept out by $S$ as it passes between the extreme points of its motion along a translational trajectory. Then the observations above indicate how to recognize new situations with the subassembly $S$ in altered configurations for which the freedom of motion is easily deduced.

### 12.2.2 Search Strategy

Given a potential operation $\langle S, t, f\rangle$, heuristics are needed to identify a finite number of $\langle S, t, f\rangle$-feasible operations that are likely to be useful for discovering a successful disassembly sequence. We use two heuristics, a collision heuristic and a line-up heuristic.

For the collision heuristic, the largest possible motion is made, subject to the freedom. Mattikalli and Khosla [9] only use this heuristic in their search strategy for disassembly. When finding a disassembly sequence using this heuristic alone, a sequence of operations is obtained for disassembling a product that can be guided by force sensing conditions. That is, a trajectory, once initiated, is maintained until a collision occurs as flagged by a sensed force condition. Nevertheless, the reverse of the disassembly sequence will not necessarily have the same property.

The line-up heuristic uses the idea that moving a subassembly $S$ until a component feature of $S$ lines up with a component feature of $A - S$ can provide a useful alignment of components. For example, lining up a tab on $S$ with a slot on $A - S$ may allow the tab to pass through the slot in a future operation. **Center** and **direction** features are useful to this heuristic. This is a powerful heuristic that is not used in any other known assembly sequence planners.

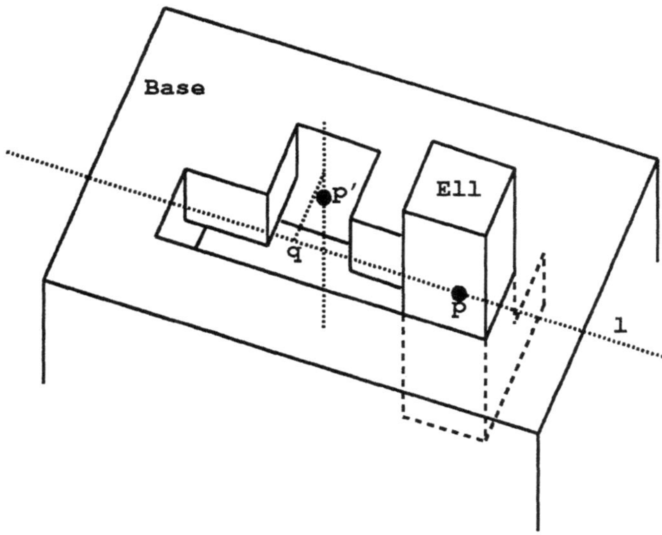


Figure 12.8: Translational Line-Up

An example of lining up **center** features is shown in Figure 12.8. The "L"-shaped component labeled Ell has been found to be able to translate in the slot of the component labeled Base, in direction indicated by line l. Translating **center** feature p of Ell to point q lines up Ell with the notch feature in the Base having **center** feature $p'$, allowing vertical translation of Ell in the next move.

Figure 12.9 illustrates how **direction** features may be useful for alignment. Vectors $v0$ and $v1$ correspond to **direction** features of the components labeled Ell and Base, respectively. The angle $\beta$ between $v0$ and $v1$ is the angle of rotation of Ell about axis $a$ that will line up the lower arm of Ell with the slot in Base, allowing vertical translation of Ell in the next move.

### 12.2.3 Simple Physics

Given a potential operation triple $\langle S, t, f \rangle$, gravity and stability effects are used to either modify the operation or reject it. Gravity effects may cause some components in $A - S$ to move when $S$ moves. Stability effects may cause subassemblies $S$ or $A - S$ to topple over if $S$ is removed from $A$. By detecting and accounting for the side effects of moving a subassembly $S$, irreversible disassembly operations are avoided.

Detecting gravitational effects involves determining if any components in $A - S$ are sitting on top of subassembly $S$. Components sitting on $S$ should move

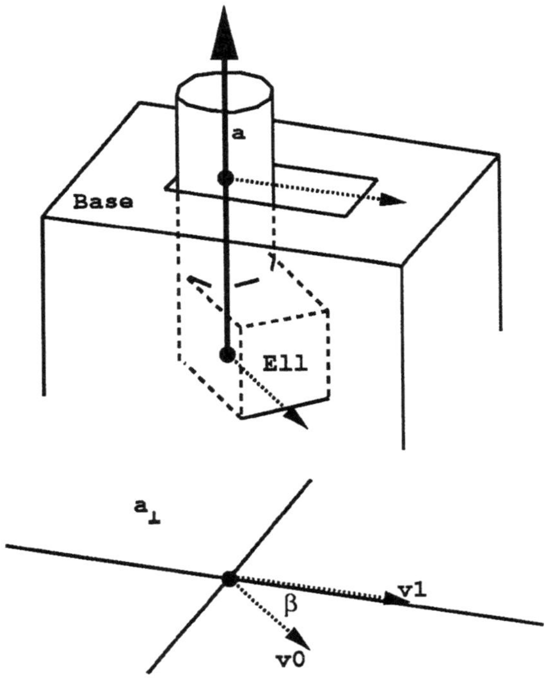


Figure 12.9: Rotational Line-Up

when $S$ is moved. The union of $S$ and components sitting on $S$ form an augmented subassembly $S^g$. Components sitting on $S$ are identified by (hypothetically) removing $S$ from $A$ and observing whether any subassembly of $A - S$ can then translate in the $-z$ direction. If $S^g$ is equal to $A$, then an operation on $S^g$ is meaningless and the potential operation is rejected. Otherwise, the augmented subassembly $S^g$ may have a different freedom of motion along trajectory $t$, say $f^g$. If $f^g$ is non-zero, then $\langle S^g, t, f^g \rangle$ replaces $\langle S, t, f \rangle$.

Instability of a subassembly $S$ is determined as follows. First, the footprint is derived for $S$ when it is set down on the tabletop. If the projection of the center of mass of $S$ in the $-z$ direction onto the tabletop falls outside of the footprint, then $S$ is deemed unstable.

Evaluation of stability occurs if a potential operation $\langle S, t, f \rangle$ is a *separating move*. The gravitational stability of the resulting subassemblies having more than one component is evaluated, and $\langle S, t, f \rangle$ is rejected if instability results.

BRAEN currently does not try to find a reorientation strategy that will remove an instability condition. For example, if only $S$ is unstable, then after $S$ is separated from $A$ it could be rotated sufficiently so that it is stable when set down on the tabletop. If $A - S$ is unstable, then an appropriate reorientation

Table 12.1: Characteristics of disassembly examples

| Statistic | pipe | latch | capsule | safe | record | industrial | puzzle |
|---|---|---|---|---|---|---|---|
| Product components | 2 | 2 | 3 | 4 | 4 | 10 | 12 |
| Bezier patches | 90 | 44 | 162 | 104 | 79 | 376 | 230 |
| Nodes generated | 16 | 7 | 14 | 17 | 58 | 32 | 52 |
| Calls to freedom | 6 | 3 | 6 | 7 | 6 | 9 | 18 |
| Calls to search | 5 | 2 | 5 | 3 | 1 | 0 | 8 |
| Moves in disassembly | 4 | 3 | 5 | 6 | 4 | 9 | 14 |
| $\Delta^d$ calls w/o reuse | 162 | 124 | 260 | 684 | 1350 | 1540 | 6882 |
| $\Delta^d$ calls w/reuse | 66 | 58 | 102 | 158 | 234 | 182 | 652 |
| Time w/o reuse (min) | 18 | 26 | 81 | 54 | 163 | 1033 | 500 |
| Time w/reuse (min) | 9 | 14 | 32 | 28 | 78 | 60 | 74 |

of $A$ would be needed before removing $S$.

## 12.3 Experimental Results

BRAEN is implemented on a Symbolics 3640, with computation of $\Delta^d$ and $\Theta^a$ delegated to an HP 9000 Model 835. Component models were generated with the PATRAN® solid modeler package. Table 12.1 reports various characteristics for a number of disassembly runs. Each example included a tabletop environment component. For each example the elapsed time and number of evaluations of $\Delta^d$ are reported when the reuse rules are not used (*w/o reuse*) and when they are used (*w/reuse*). Note that the improvement in performance provided by the reuse rules becomes more dramatic as the number of components increases. Figure 12.10 shows each product in its initial configuration, except the industrial example for which a cross section is supplied.

In the pipe example, the pipe component was suspended in midair in most configurations during the disassembly process, preventing the base component in the assembly from slipping around underneath the pipe (since that would require two robots to implement). The complete sequence of operations generated for capsule disassembly is given in Figure 12.11. Each step indicates the subassembly to be moved and the operation to be performed: the type of motion (translation or rotation), the direction (or axis) of motion, and the magnitude of motion. This sequence is illustrated in Figures 12.12 and 12.13; for brevity, those operations that transport components to the edge of the tabletop have not been shown. Figure 12.12 shows the sequence of operations that break up the initial assembly. This sequence corresponds to the search graph shown in Figure 12.7. Figure 12.13 shows how the subassembly *key & roof* is broken up. Initially, the *key* component is being held up and must be

®PDA Engineering, Costa Mesa, CA.

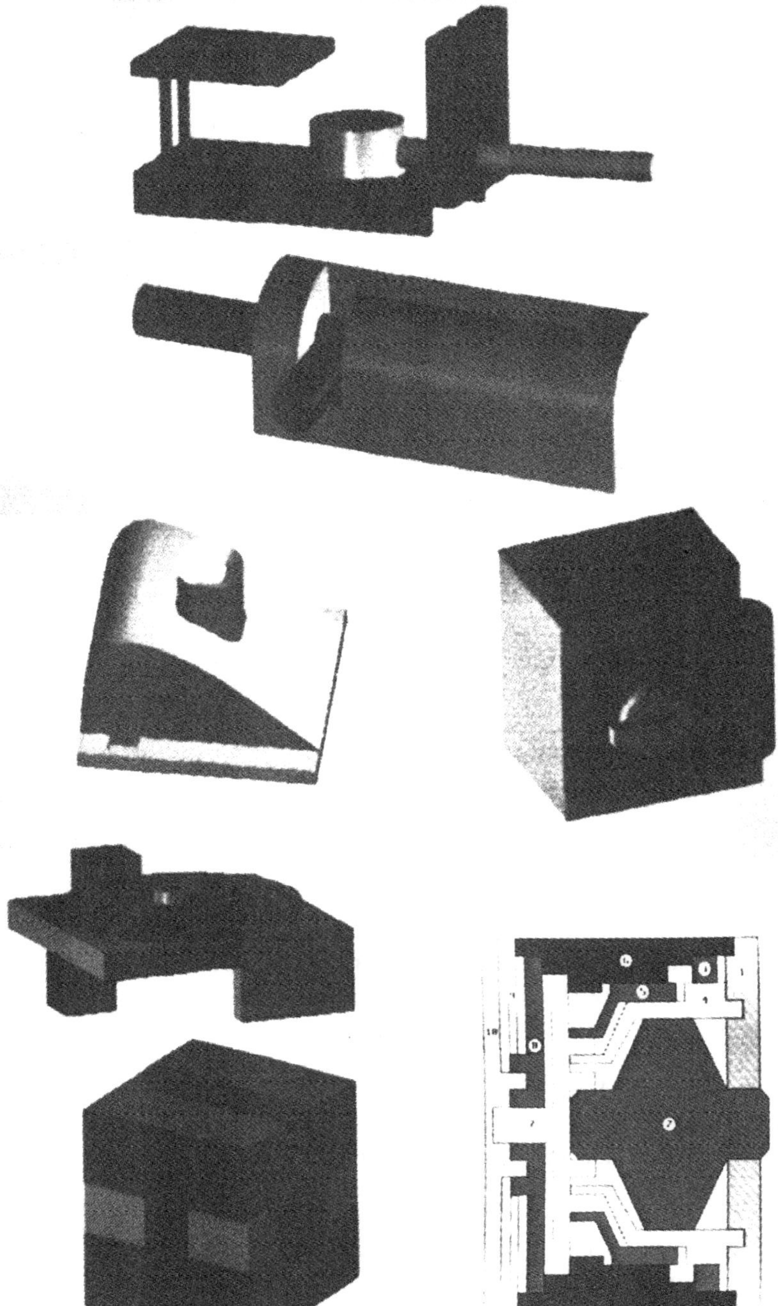

Figure 12.10: Example assemblies

```
Moving Components:  (key)
     Operation:  (TRANSLATION (0.0 0.0 1.0) 0.832)
Moving Components:  (roof key)
     Operation:  (TRANSLATION (-1.0 0.0 0.0) 0.7)
Moving Components:  (roof key)
     Operation:  (TRANSLATION (0.0 0.0 1.0) 3.2)
Moving Components:  (roof key)
     Operation:  (TRANSLATION (1.0 0.0 0.0) 10.7)
Moving Components:  (roof key)
     Operation:  (TRANSLATION (0.0 0.0 -1.0) 3.3)
Moving Components:  (key)
     Operation:  (TRANSLATION (0.0 0.0 -1.0) 0.732)
Moving Components:  (base)
     Operation:  (TRANSLATION (0.0 0.0 1.0) 3.3)
Moving Components:  (base)
     Operation:  (TRANSLATION (-0.393175 -0.919463 0.0) 10.8759)
Moving Components:  (base)
     Operation:  (TRANSLATION (0.0 0.0 -1.0) 3.3)
Moving Components:  (roof)
     Operation:  (TRANSLATION (0.0 0.0 1.0) 3.3)
Moving Components:  (roof)
     Operation:  (TRANSLATION (-0.680074 -0.733143 0.0) 13.639896)
Moving Components:  (roof)
     Operation:  (TRANSLATION (0.0 0.0 -1.0) 3.3)
Moving Components:  (key)
     Operation:  (TRANSLATION (0.0 0.0 1.0) 3.3)
Moving Components:  (key)
     Operation:  (TRANSLATION (-0.481919 -0.876216 0.0) 11.412712)
Moving Components:  (key)
     Operation:  (TRANSLATION (0.0 0.0 -1.0) 3.3)
```

Figure 12.11: Capsule disassembly sequence

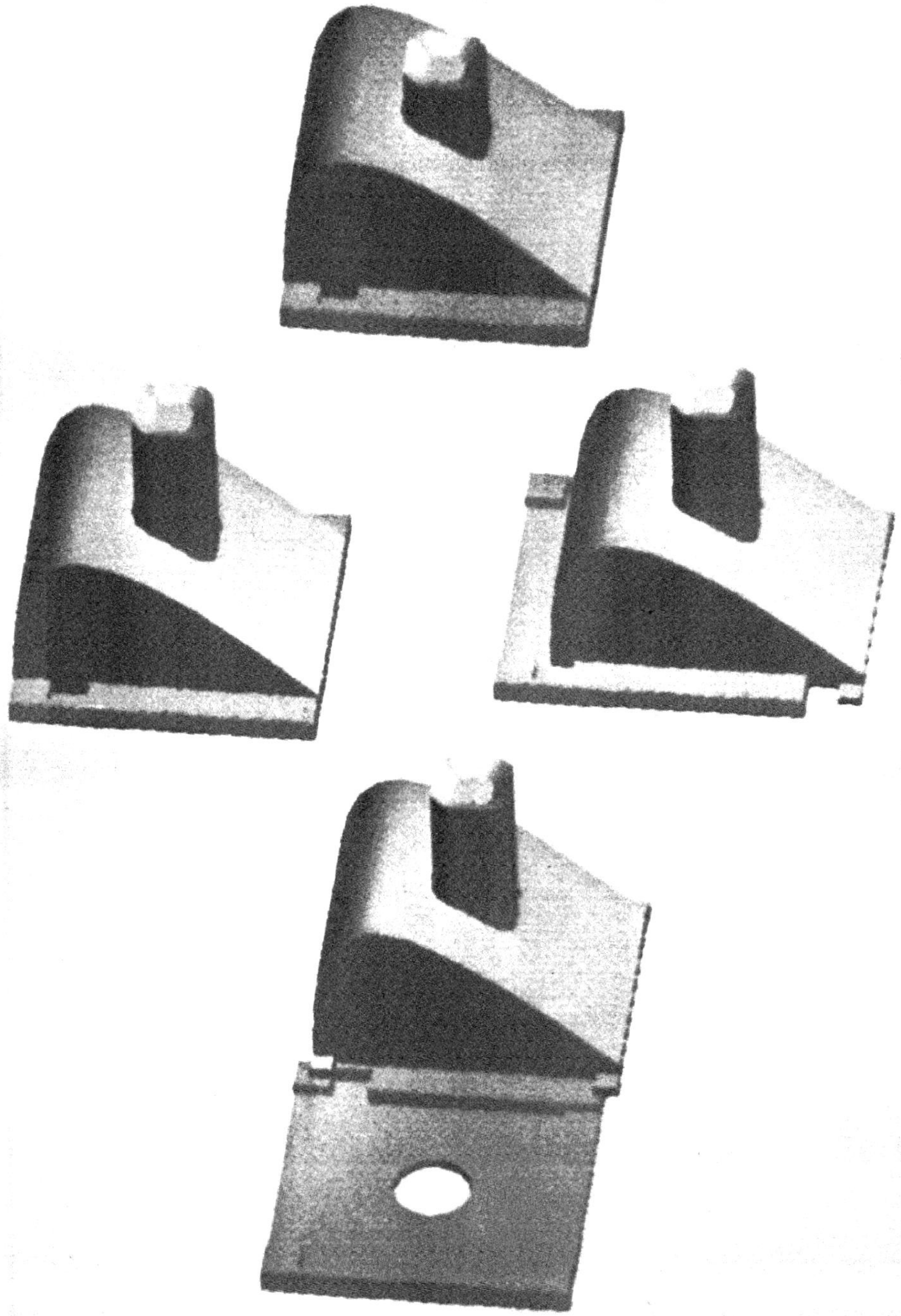

Figure 12.12: Part 1 of capsule disassembly

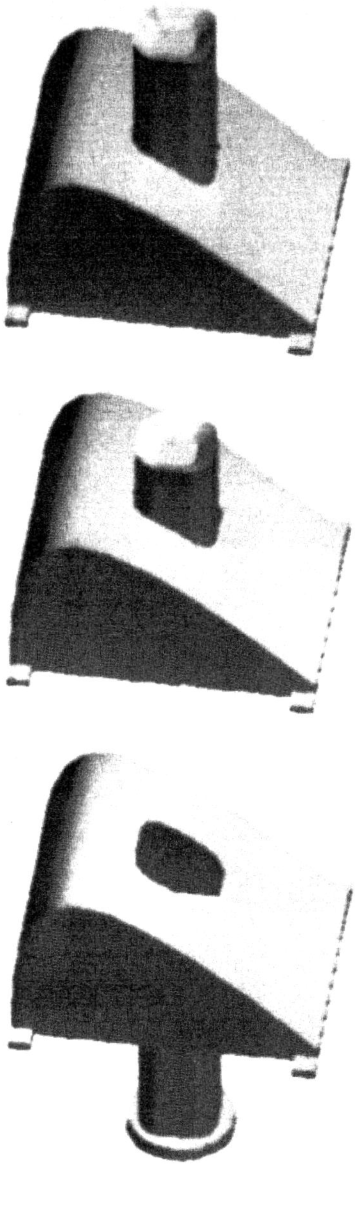

Figure 12.13: Part 2 of capsule disassembly

set down. Then, the *roof* component may be lifted from the subassembly.

The record player disassembly in Figure 12.14 demonstrates the operation of the stability maintenance mechanism. There are four components; a disk that has a weight affixed to one side, a peg that holds the disk down, a foot that props the whole assembly up on one end (especially when the disk is rotated so the weight is on the same side as the foot) and also fastens the peg, and the base. The *freedom determination* module initially finds that the foot can be removed from the assembly, but rejects that move when it discovers the remaining three components are unstable. Nevertheless, the disk can rotate freely, and it is found that rotating the disk by 180° shifts the center of mass of the disk-peg-base subassembly enough so that the foot may be removed. Disassembly of the disk-peg-base subassembly then follows.

The industrial example was adapted from [2]. For this example, the derived disassembly sequence was to remove components in the order {10, 9, 8, 7, 1, 2, 4, 3, 5, 6}. Note that after components {10, 9, 8, 7} were removed, component 6 was able to be separated from the remaining components. The stability test discarded this potential move, however, and the safer operation of removing component 1 was used instead. After components {10, 9, 8, 7, 1} were removed, although component 6 could still be removed on initial analysis, gravity would cause all other remaining components to move along with it, and thus component 6 could not be profitably moved until it was the sole remaining component. Note that the number of calls to *search strategy* is 0; this is because *freedom determination* always succeeded in discovering a *separating move*.

## 12.4 Discussion

We have demonstrated an approach for generating an assembly sequence of a product expressed in B-rep format with bicubic surfaces. This treats components as part of a real workcell environment by allocating sections of the table to subassemblies and disassembled parts. We have demonstrated a number of common sense techniques for enhancing the system performance.

We are currently developing the capability for BRAEN to focus on specific disassembly goals rather than disassembling until all components are separated. For example, this capability is useful when:

- Several components are linked together to form an articulated object and cannot be taken apart.
- A maintenance sequence for a particular component is desired. Movements of other components should be minimized.

One modification made to BRAEN involves choosing the assembly $A$ in the work area based on the specified goal rather than on number of components in

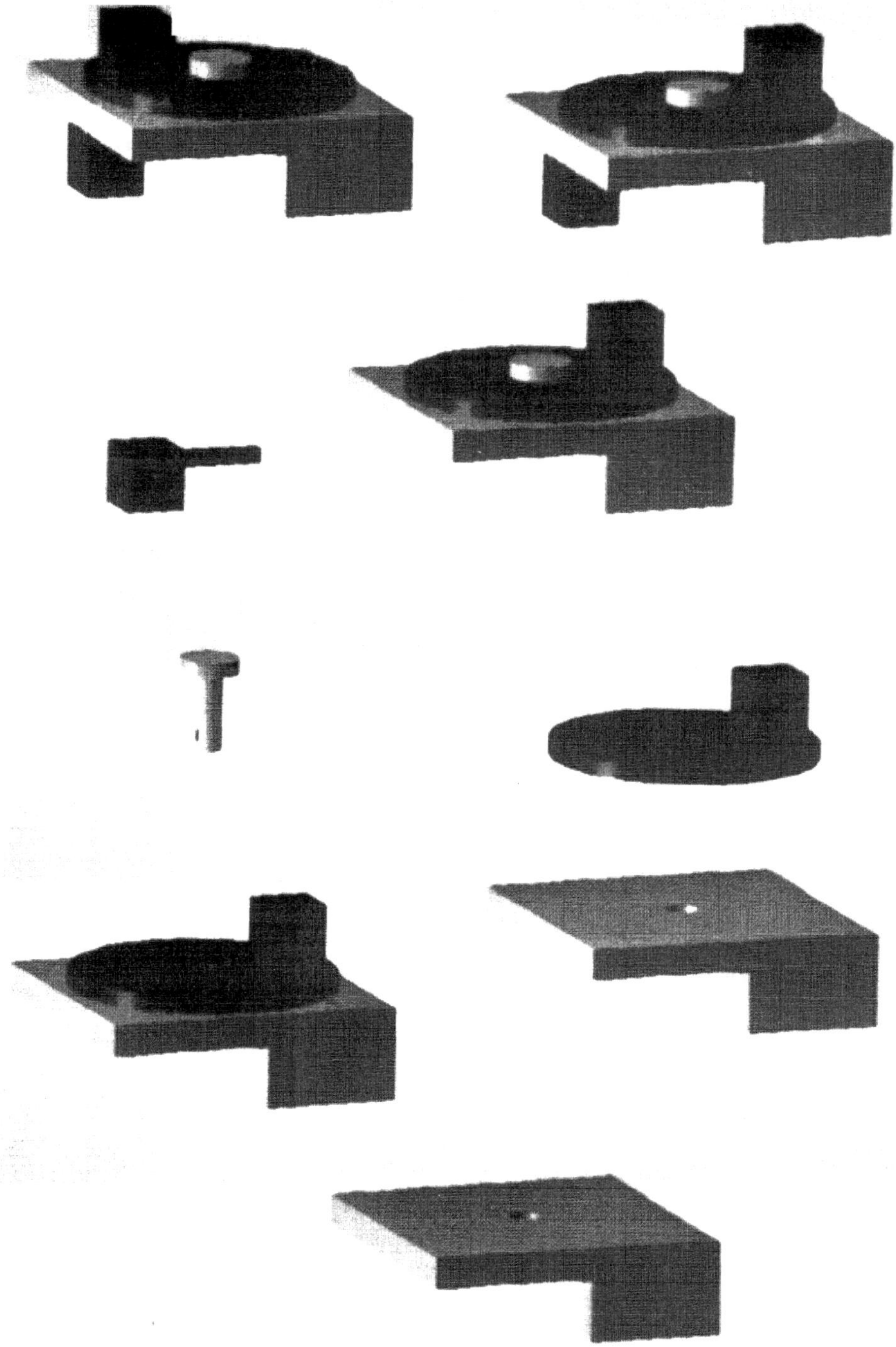

Figure 12.14: Record player disassembly sequence

*A*. Other issues for successful focused disassembly, such as predicting the utility of a given operation toward a given goal configuration, are being addressed.

Other directions for future work are:

- Implement reuse rules for rotational motions; this should be fairly similar in flavor to the translational reuse rules, except that swept volumes will be cylindrical.
- Improve capabilities for detecting instability. In particular, some consideration of friction is needed.
- Introduce a tool component class, which can be used to determine accessibility of a component to be moved as well as determine grasp points.
- Use *fixtures* as environment components.
- Relax assumptions about operations for assembly, such as allowing simultaneous translation and rotation operations. For example, it is possible for one component to rotate as another component translates, as with a doorknob and door latch.

A number of further issues must be addressed before these results can be put to practical use. For example, the operations supplied are relatively high level, and must be converted to robot-specific command sequences. In addition, robot path planning is required to provide a collision-free path for the robot arm and tool during operation setup and execution. If an operation involves a location that cannot be reached by the robot or a singularity point, then it may be necessary to modify the assembly sequence. Therefore, a practical system will either perform robot simulation concurrently with assembly sequence planning, or do it after assembly sequence planning and provide a feedback mechanism for replanning when needed.

We have shown how a few simple ideas for reusing freedom of motion results has provided substantial reduction of cost of automated assembly sequence planning. This bodes well for the development of more sophisticated rules to provide further savings.

# Acknowledgement

Many thanks to Jeff Barnett and Dan Geiger for valuable suggestions.

# References


[1] A. Bourjault, *Contribution a une Approche Méthodologique de L'Assemblage Automatisé: Elaboration Automatique des Séquences Opératoires,* Thèse d'état, Université de Franche-Comté, Besançon, France, November 1984.

[2] T.L. De Fazio and D.E. Whitney, "Simplified generation of all mechanical assembly systems", *IEEE Journal of Robotics and Automation,* Vol. RA-3, Number 6, pp. 640–658, December 1987.

[3] B.R. Donald, "A search algorithm for motion planning with six degrees of freedom", *Artificial Intelligence,* Vol. 31, pp.295–353, 1987.

[4] B. Faverjon, "Object level programming of industrial robots", *Proc. IEEE International Conference on Robotics and Automation,* San Francisco, pp.1406–1412, 1986.

[5] R.L. Hoffman, "Disassembly in a CSG domain", *1989 IEEE International Conference on Robotics and Automation,* Scottsdale Arizona, pp.210–215, May 1989.

[6] R.L. Hoffman, "Assembly planning for B-rep objects", 2nd International Conference on Computer Integrated Manufacturing, pp. 314–321, Troy, New York, May 1990.

[7] R.L. Hoffman, "Automated assembly planning for B-rep products", IEEE International Conference on Systems Engineering, pp. 391–394, Pittsburgh, August 1990.

[8] C. Laugier and P. Theveneau, "Planning sensor-based motions for part-mating using geometric reasoning techniques", *Proc. 7th European Conference on Artificial Intelligence,* pp. 494-506, Brighton England, July 1986.

[9] R.S. Mattikalli, P.K. Khosla, Y. Xu, "Subassembly identification and motion generation for assembly: a geometric approach", IEEE International Conference on Systems Engineering, pp. 399–403, Pittsburgh, August 1990.

[10] J.M. Miller and R.L. Hoffman, "Automatic assembly planning with fasteners", *1989 IEEE International Conference on Robotics and Automation,* Scottsdale Arizona, pp.69–74, May 1989.

[11] E. Rich, "Artificial Intelligence", *McGraw-Hill series in artificial intelligence,* McGraw-Hill, New York, 1983.

# Chapter 13

# Assembly Coplanner : Cooperative Assembly Planner based on Subassembly Extraction

**Sukhan Lee and Yeong Gil Shin**

The use of multiple assembly workstations enables assembly operations to be done in parallel. The routing of parts and the capabilities of robotic systems provide flexibility in assembly. To maximize performance of a system of multiple robotic workstations, an assembly plan that provides proper parallelism and flexibility is required. The problem of assembly planning can be stated formally as follows:

> Given the description of parts, $\mathcal{P} = \{p_1, p_2, \ldots, p_n\}$, and the geometric and topological relations on $\mathcal{P}$, $\mathcal{R} = \{r_1, r_2, \ldots, r_m\}$, how does one determine sequences of assembly operations that satisfy $\mathcal{R}$ subject to (1) minimizing cost, (2) maximizing parallelism, and (3) satisfying feasibility conditions.

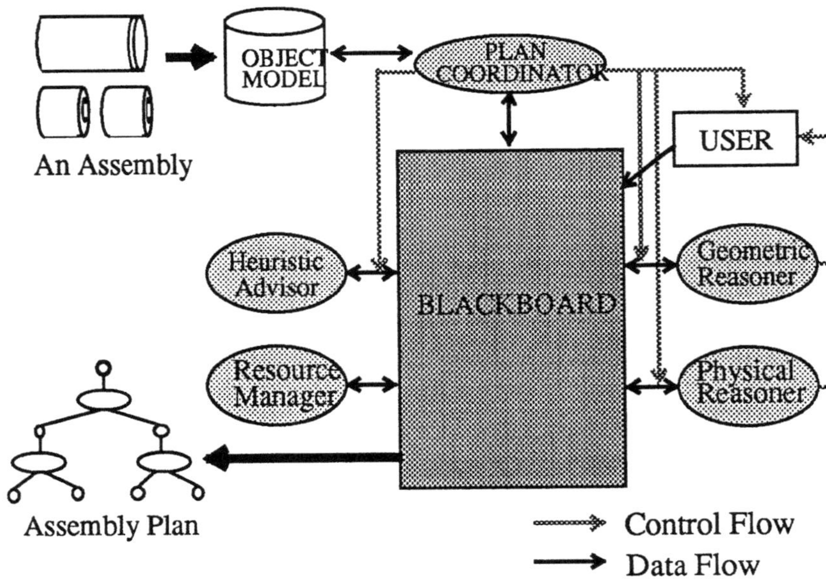


Figure 13.1: The Block Diagram of COPLANNER

It is assumed here that: 1) an assembly operation joins two parts or subassemblies, 2) the order of assembly is the reverse of the order of disassembly, and 3) the geometric relationship between individual parts remains fixed after they are assembled; we deal with nonlinear, nonsequential but monotonic assembly plans[14]. It is also assumed that assembly planning is supported by a proper path planning algorithm[2] to detect geometric interference in part assembly.

## 13.1 COPLANNER: A Cooperative Assembly Planning System

As an assembly planning module of the flexible assembly system, we developed COPLANNER. COPLANNER has been implemented in Common Lisp and C on a Sun260 workstation. Lisp functions are used for the implementation of reasoning processes and C codes are incorporated for the mathematical computation and the communication protocols. COPLANNER is organized under the "Cooperative Problem Solving(CPS)" paradigm. The cooperative problem solving system which is shown in figure13.1 consists of the following models:

- **The plan coordinator** : The plan coordinator coordinates the cooperation of the knowledge sources by controlling their access to a common message buffer, the blackboard.
- **The heuristic advisor** : The heuristic advisor extracts subassemblies based on the embedded heuristics. It evaluates the difficulty and cost of each assembly operation required for each primary liaison and assigns the weight to each edge of the liaison graph.
- **The geometric reasoner** : The geometric reasoner checks the feasibility of an assembly operation by calculating the directional freedom of motion, the manipulability, interference with the neighboring parts, and accessibility of a part.
- **The physical reasoner** : The physical reasoner reasons about stability of a part, the weight of a subassembly, and the connection type of each liaison.
- **The resource manager** : The resource manager keeps the information of currently available resources and decides the feasibility of an assembly.
- **The blackboard** : This is a common working memory and communication protocol among advisors. The blackboard contains all the relevant information for planning, such as part descriptions, liaisons, and any data structure constructed during planning. The access to the blackboard can be controlled by the slot ACCESSIBLE of the blackboard schema. Advisors who can access the content of the blackboard are specified in the slot ACCESSIBLE.

In the traditional blackboard systems[5, 6, 10], only one agent can read or write on the blackboard at any given time. Under this single access mode, the most important problem is how to schedule access to the blackboard since performance of the system is affected by the available knowledge in the blackboard. Most blackboard systems have a scheduler which controls the access to the blackboard by evaluating the degree of contributions of each knowledge source. In COPLANNER, several advisors are allowed to read the blackboard but only one advisor can modify it at any given time, with the higher priority given to the advisors who intend to write.

## 13.2 Attributed Liaison Graph

In this chapter, an assembly is represented by an attributed liaison graph. An attributed liaison graph is a connected graph, $G = (N, E)$, with $N$ representing a set of nodes, and $E$ representing a set of edges. A node $n$, $n \in N$, is assigned to each part of the assembly, and an edge $e$, $e \in E$, is assigned to each liaison.

Table 13.1: The Relative Stability of an Edge as a Function of Interconnection Type

| Interconnection Type | Mating Type | | |
|---|---|---|---|
| | Insert | Semi-Insert | Place-On |
| Attach | 0.3 | 0.2 | 0.1 |
| Sticky | 0.4 | 0.3 | 0.2 |
| Force-Fit | 0.5 | 0.4 | 0.3 |
| Push & Twist | 0.7 | 0.6 | - |
| Screw | 0.8 | 0.7 | - |
| Connectors | 0.9 | 0.8 | 0.7 |
| Weld | 1.0 | 1.0 | 0.9 |

A liaison is said to exist between a pair of parts if one part constrains the freedom of motion of the other either by a direct contact or by a near contact.[1]

A part frame is attached to each node to describe attributes associated with a part. The attributes of a part frame contains 1) the part geometry which specifies its shape and volume, 2) the mating volumes and the contact subfaces as part features, and 3) the physical properties of the part such as weight.

A liaison frame is attached to each liaison to describe attributes associated with the liaison. The attributes of a liaison consist of 1) the mating features, 2) the mating type such as insertion, semi-insertion, and place-on, 3) the interconnection mechanism, 4) the relative stability of a liaison after the corresponding interconnection is completed (refer to Table13.1), and 5) the functional and physical dependency among liaisons. The functional support of a liaison is a list of the liaisons which functionally assist the achievement of the given liaison. The stability support of a liaison is a list of the liaisons required for the stabilization of the given liaison which is otherwise unstable after the interconnection is done. Figure 13.3 illustrates the attributed liaison graph representation of the flashlight shown in figure 13.2.

## 13.3 Geometric Reasoning

Geometric reasoning is needed to decide the feasibility of an assembly operation. To be a feasible assembly operation, it is necessary that there is no interference in part motion. The local constraints in part motion can be deduced based on the geometry of each part and constraints embedded in the associated liaisons.

[1]A *near contact* is defined between two contact surfaces having distance smaller than the prespecified threshold.

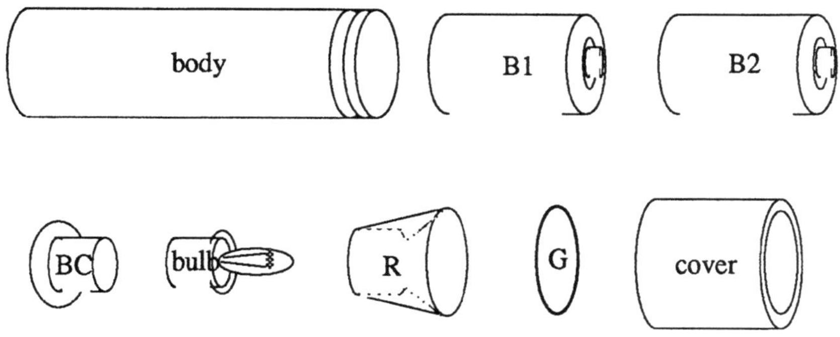


Figure 13.2: The Flashlight Assembly

**Definition 13.3.1 (Freedom of Separation (FS))** *Let us denote the freedom of separation of a part or a cluster of parts, $P_1$, against another part or another cluster of parts, $P_2$, by $FS(P_1/P_2)$ or simply by $FS(P_1)$ if $P_2$ is clear from the context. $FS(P_1/P_2)$ is represented by a tuple composed of the principal axes of motion in which $P_1$ can be separated from $P_2$ and the directional tolerances along the principal axes of motion. Formally, $FS(P_1/P_2) = \{(d_{+x}, d_{-x}, d_{+y}, d_{-y}, d_{+z}, d_{-z}), (\alpha_{+x}, \alpha_{-x}, \alpha_{+y}, \alpha_{-y}, \alpha_{+z}, \alpha_{-z})\}$, where $d_a = 1$, if $FS(P_1/P_2)$ has the freedom of separation in the direction of $a$, otherwise $d_a = 0$; $\alpha_b$ is the directional tolerance along the $b$ directional freedom of separation, such that $\alpha_b$ = (the maximum angle of directional errors that the $b$ direction of separation can tolerate)/90.*

For instance, in figure13.4, $FS(B/A \cup B) = \{(1\ 0\ 1\ 1\ 0\ 0),(15/90\ 0\ 0\ 0\ 0\ 0)\}$ indicating that B has the freedom of separation in the direction of ($+x$, $+y$, $-y$) and the tolerance of 15/90 along $+x$.

**Definition 13.3.2 (Degrees of Freedom of Separation (DFS))** *The degrees of freedom of separation of $P_1$ with respect to $P_2$, $DFS(P_1/P_2)$, is obtained from $FS(P_1/P_2)$, $FS(P_1/P_2) = \{(d_{+x}, d_{-x}, d_{+y}, d_{-y}, d_{+z}, d_{-z}), (\alpha_{+x}, \alpha_{-x}, \alpha_{+y}, \alpha_{-y}, \alpha_{+z}, \alpha_{-z})\}$, by*

$$DFS(P_1/P_2) \triangleq \sum_a (d_a + \alpha_a),\ a \in \{+x,\ -x,\ +y,\ -y,\ +z,\ ,-z\}$$

Table 13.2 illustrates the FSs and the DFSs calculated for all the part clusters of the subassembly {A, B, C} in Fig.13.4:

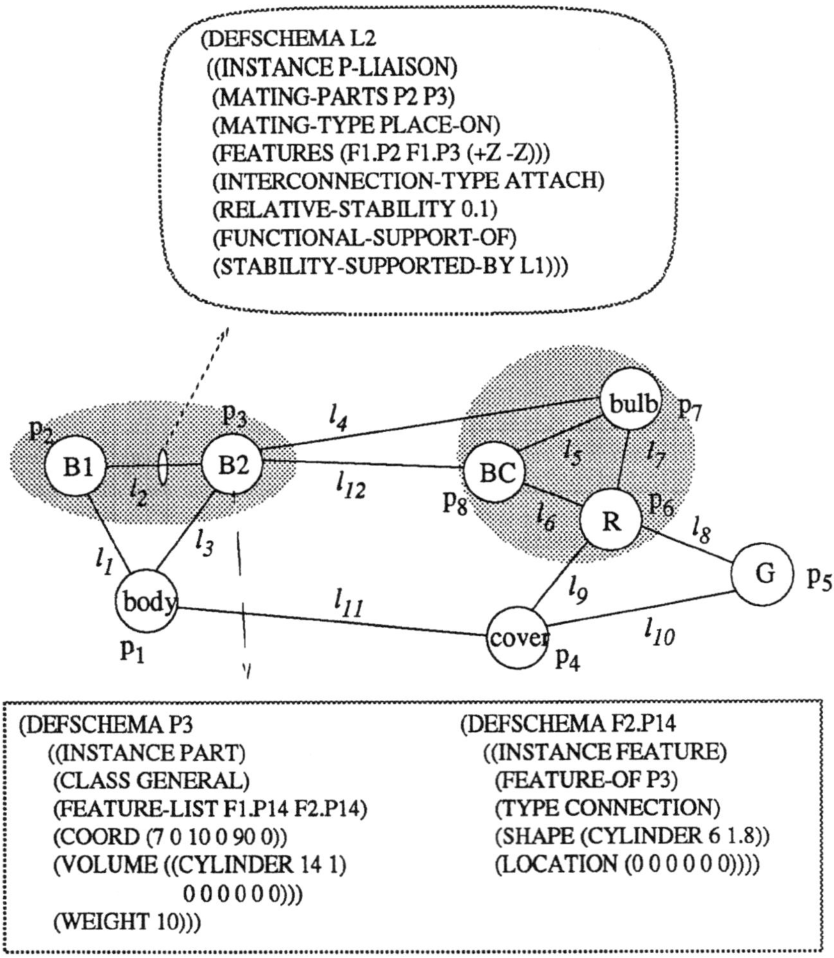


Figure 13.3: The Attributed Liaison Graph for the Flashlight Assembly

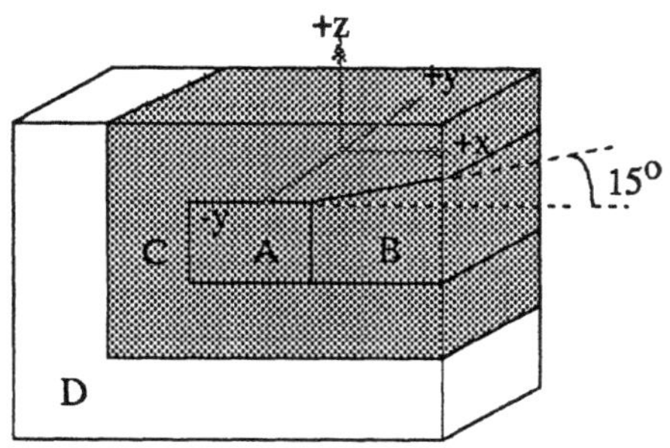


(a) S = A U B U C

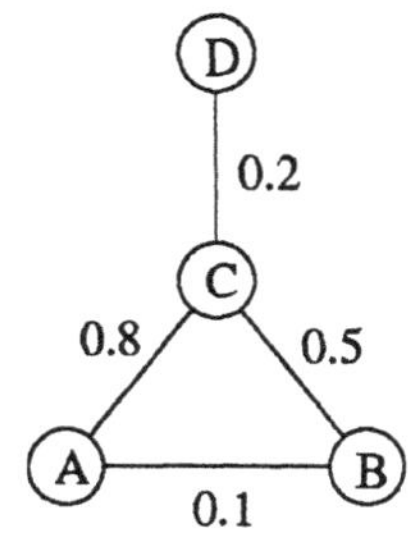


(b) The numbers indicate the relative stability of interconnection for individual edges

Figure 13.4: An Example to illustrate the Degree of Freedom of Separation(DFS)

Table 13.2: FSs and DFSs for the subassembly $S$ shown in Fig.3

| Parts and Clusters | Principal Direction of separation | Tolerance | DFS |
|---|---|---|---|
| {A} | (0 0 1 1 0 0) | (0 0 0 0 0 0) | 2 |
| {B} | (1 0 1 1 0 0) | (15/90 0 0 0 0 0) | $3\frac{1}{6}$ |
| {C} | (0 1 1 1 0 0) | (0 0 0 0 0 0) | 3 |
| {A B} | (1 0 1 1 0 0) | (0 0 0 0 0 0) | 3 |
| {A C} | (0 1 1 1 0 0) | (0 0 0 0 0 0) | 3 |
| {B C} | (0 0 1 1 0 0) | (0 0 0 0 0 0) | 2 |

Note that FS{A,B} and FS{C} are equivalent. So are FS{A,C} and FS{B}, and FS{B,C} and FS{A}. This is because the $+x$ directional freedom of separation of {A,B} represents the $-x$ directional freedom of separation of the rest of the cluster (C), and vice versa. Therefore, $FS\{A\} \equiv FS\{\overline{A}\}$, where $A \cup \overline{A}$ represents the whole subassembly.

## 13.4 Construction of an Abstract Liaison Graph

The construction of an Abstract Liaison Graph(ALG) is based on merging a set of mutually inseparable nodes(as defined later in this section), into a single node called a *supernode*. Let us first introduce the following definitions:

**Definition 13.4.1 (Manipulable Node(M-node))** *A part is said to be manipulable if it is accessible and manipulable by a tool for disassembly. A node is accessible and manipulable if any of the parts forming the node is accessible and manipulable. Shaded nodes in Fig.13.5(b) show manipulable nodes.*

**Definition 13.4.2 (Accessible Path(A-path))** *An accessible path to a node $n$ is a simple path starting from an M-node and ending with the node $n$ without having any other M-node on the path. A set of A-paths to the node $n$ is called independent if they share no common node except the node $n$.*

**Definition 13.4.3 (Satellite Node)** *A node $n_1$, which is not an M-node, is said to be a satellite of a master node $n_2$, if all the A-paths to the node $n_1$ pass through the node $n_2$. In Fig.13.5, Part E is a satellite of part D. A satellite node of the node $n$ is not independently separable from the master node $n$, since it is not possible to deliver the force required for breaking the liaison between the satellite and the master nodes.*

**Definition 13.4.4 (Floating Liaison)** *A liaison between two nodes is said to be floating if each node has its own independent A-paths and the force required for breaking the liaison is not deliverable to that liaison.*

*Let us define $D_p(l_i; n_{i_1}/n_{i_2})$ as a set of principal directions of the motion and force involved in the separation of the liaison $l_i$ between $n_{i_1}$ and $n_{i_2}$ under the condition that $n_{i_2}$ is fixed. The direction of separation of a liaison $l_i$ is assumed arbitrary, which may not be coincident with a principal axis of an assembly, but may have components on a set of principal axes.*

*Since there exist different ways of separating $n_{i_1}$ from $n_{i_2}$, a liaison $l_i$ may have a collection of such $D_p(l_i; n_{i_1}/n_{i_2})$, represented by $D(l_i; n_{i_1}/n_{i_2})$, $D(l_i; n_{i_1}/n_{i_2}) \triangleq \{D_p(l_i; n_{i_1}/n_{i_2}), p \in P, P =$ an index set$\}$.*

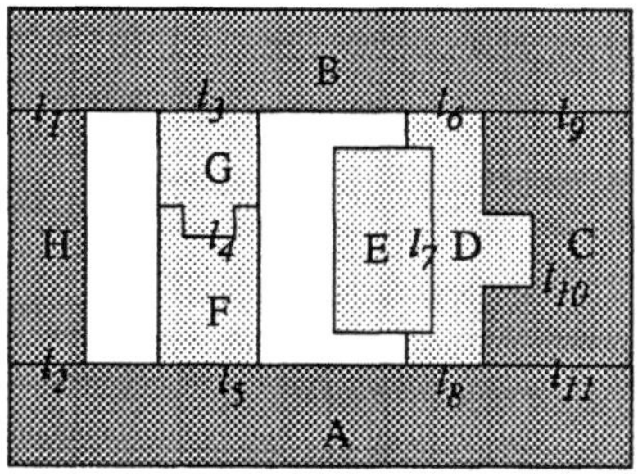


(a) An Example of an Assembly

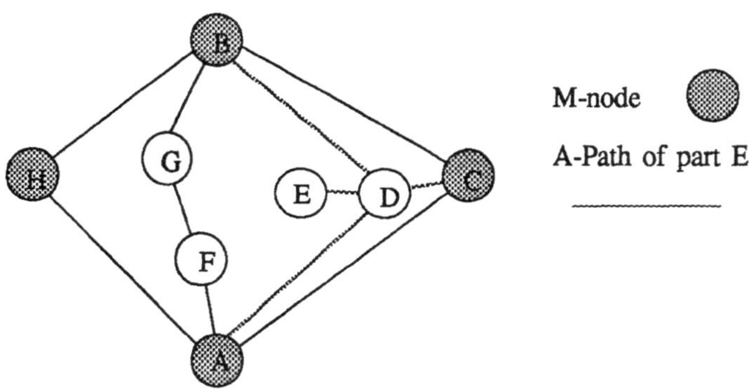


(b) An Example of a Satellite Node

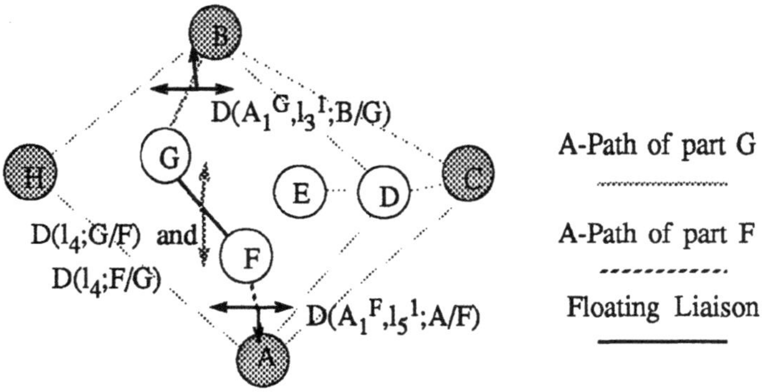


(c) The Existence of a Floating Liaison

Figure 13.5: Examples of Mutually Inseparable Nodes

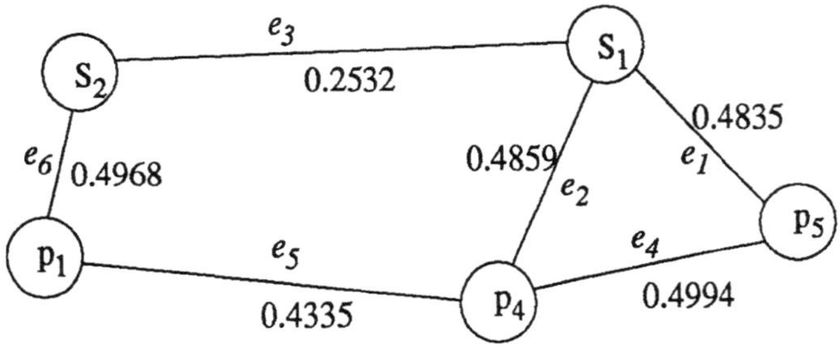


Figure 13.6: The Weighted Abstract Liaison Graph for the Flashlight Assembly. The nodes marked with S represent supernodes.

M1: The node $n_1$ is a satellite of the node $n_2$, or vice versa.

M2: The liaison between the two nodes $n_1$ and $n_2$ is floating.

M3: The liaison between the two nodes $n_1$ and $n_2$ includes at least one unremovable connector/retainer. (A connector/retainer which is neither accessible nor manipulable is considered unremovable.)

M4: The liaison between the two nodes $n_1$ and $n_2$ is the precondition for another liaison, due to its role as a functional support or a stability support.

M5: The node $n_1$ is immobilized by a supernode $n_2$, or vice versa.

An ALG is constructed by merging those nodes which are mutually inseparable until all the nodes of the ALG are free from any of the above merging conditions. From the liaison graph of the flashlight(see Figs 13.2 and 13.3) two supernodes are generated: $S_1$ by merging $P_6$, $P_7$, and $P_8$ which are connected by floating liaisons $l_5$, $l_6$, and $l_7$; and $S_2$ results from merging $P_2$ and $P_3$ which are connected by a floating liaison $l_2$.

## 13.5 Construction of a Weighted ALG

A weighted ALG (WALG) is an ALG with a weight assigned to each of its edges. The weight of an edge is determined by the total strength of the edge in terms of physical stability and structural connectivity, and the cost involved in disconnecting or connecting the edge.

## The Total Strength of an Edge

**Definition 13.5.1 (The Net Strength of an Edge)** *The net strength of an edge $e_i$, $S_N(e_i)$, is defined by the relative stability of $e_i$, $X_s(e_i)$, and the directional constraints of a motion during part separation, $X_d(e_i)$, as follows:*

$$S_N(e_i) \triangleq \alpha X_s(e_i) + (1-\alpha)X_d(e_i)$$

*where $\alpha$ is the weighting coefficient, $0 \leq \alpha \leq 1$.*

$X_s(e_i)$, $0 < X_s(e_i) \leq 1$, is specified in the liaison frame based on the interconnection type of the edge. $X_d(e_i)$, $0 < X_d(e_i) \leq 1$, represents how the motion of the node $n_{i1}$ is restricted by the node $n_{i2}$ during the separation of $e_i$, or vise versa. $X_d(e_i)$ is obtained by $X_d(e_i) = 1 - DFS(n_{i1}/n_{i2})/6$, where $DFS(n_{i1}/n_{i2})$ represents the degree of freedom of separation between the two nodes, $n_{i1}$ and $n_{i2}$, connected by $e_i$. Refer to the Section 3 for the details of how to compute $DFS(n_{i1}/n_{i2})$.

**Definition 13.5.2 (The Total Strength of an Edge)** *The total strength of an edge $e_i$, $S_T(e_i)$, is determined by reinforcing $S_N(e_i)$ with the structural connectivity and the structural preference of the edge $e_i$. The structural connectivity and the structural preference evaluate the strength of an edge in the context of its surrounding structure. $S_T(e_i)$ can be computed by the following two steps:*

**Step 1.** The Reinforcement by Structural Connectivity

The structural connectivity between two nodes, $n_{i1}$ and $n_{i2}$, of the edge $e_i$ is due to the indirect as well as the direct paths between $n_{i1}$ and $n_{i2}$. A direct path between $n_{i1}$ and $n_{i2}$ represents a path of a single edge($e_i$), whereas an indirect path between $n_{i1}$ and $n_{i2}$ represents a path of multiple edges. An indirect path gives an additional strength to $S_N(e_i)$, *e.g.*, the net strength of an edge between $P_3$ and $P_{11}$ in Fig.13.6 is reinforced by an indirect path through nodes $P_3, P_{12}$, and $P_{11}$. The modification of $S_N(e_i)$ between $n_{i1}$ and $n_{i2}$ by the structural connectivity is accomplished by the following algorithm:

**(a)** Find all the indirect paths between $n_{i1}$ and $n_{i2}$ of $e_i$, $\{P_j^i | j = 1, \ldots, n\}$, sharing no common edges. Assume that $P_j^i$ consists of a set of edges $\{e_{jl} | l = 1, \ldots, m_j\}$ for $j = 1, \ldots, n$.

**(b)** Define the net strength of a path $P_j^i$, $S_N(P_j^i)$, by

$$S_N(P_j^i) = \Pi_{l=1}^{m_j} S_N(e_{jl}), \text{ for } j = 1, \ldots, n,$$

where $\Pi$ is the multiplication operator.

(c) Reinforce $S_N(e_i)$ by adding $c_s \sum_{j=1}^{n} S_N^2(P_j^i)$ to $S_N(e_i)$, where $c_s$, $0 < c_s < 1$, is a weighting constant. Note that the reinforcement $S_N(P_j^i)$ attenuates geometrically according to its magnitude.

(d) Normalize the result of (c) by a sigmoid function to obtain the total strength of an edge $e_i$, $S'_T(e_i)$ reinforced by the structural connectivity:

$$S'_T(e_i) = 1/[1 + exp\{-(S' - \theta)/T\}]$$

where $S' = S_N(e_i) + c_s \sum_{j=1}^{n} S_N^2(P_j^i)$, and $\theta$ and $T$ are the parameters (the threshold and the temperature, respectively) of the sigmoid function.

**Step 2.** The Reinforcement by Structural Preference

The structural preference of an edge $e_i$ is measured by the attractiveness between the associated nodes $n_{i1}$ and $n_{i2}$. The attractiveness between the two nodes, Attr($n_{i1}$, $n_{i2}$), is defined by Attr($n_{i1}$, $n_{i2}$) = $|D(n_{i1}) - D(n_{i2})|$, where $D(n)$ represents the degree of $n$, the number of edges incident on $n$. A higher attractiveness implies a stronger tendency of associated parts to be merged into a supernode. Thus, in Fig. 3, $P_1$ is more likely to be merged with $P_7$ as opposed to $S_1$ inasmuch as the attractiveness between $P_7$ and $P_1$ is higher than that between $P_1$ and $S_1$.

The total strength of an edge, $S_T(e_i)$, is now obtained by modifying $S'_T(e_i)$ with the reinforcement from the structural preference. This can be done by shifting $S'_T(e_i)$ to the left proportional to the attractiveness:

$$S_T(e_i) = 1/[1 + exp\{-(S' - \theta + \Delta\theta)/T\}]$$

where $\Delta\theta = c_a Attr(n_{i1}, n_{i2})$, and $c_a$ is the weighting constant.

## The Assembly Cost of an Edge

The cost of assembly of an edge depends on such factors as the difficulty of aligning and positioning parts, the resistance during insertion, the difficulty of part handling, the need to hold down a part after assembly, etc.[1] To be more specific, we consider that the cost of assembly of an edge is a function of the interconnection type of the edge, the relative stability of the edge after the interconnection is completed, the degrees of freedom of separation(DFS) of the two parts associated with the edge, the mating tolerance, the number of mating volumes involved in an edge, the number of connectors/retainers, and the weight of the mating part.

Table 13.3: The Relative Cost of an Edge as a Function of Interconnection Type

| Type | Attach-ment | Stick-ing | Force-Fit | Push & Twist | Screw | Conne-ctors | Weld-ing |
|---|---|---|---|---|---|---|---|
| Relative Cost | 0.1 | 0.3 | 0.4 | 0.5 | 0.6 | 0.7 | 1.0 |

The relative cost of assembly of an edge $e_i$, $C_r(e_i)$, $0 < C_r(e_i) \leq 1$, is then determined by:

$$C_r(e_i) = \sum_{i=1}^{5} \alpha_i X_i$$

where

$X_1$ = the relative cost as a function of interconnection type (see Table 13.3),
$X_2$ = 1 − the relative stability associated with an interconnection type,
$X_3 = 1 - DFS(e_i)/6$,
$X_4 = 1/[1+exp\{-$(the number of mating volumes$-1)/$normalization factor$\}]$,
$X_5 = min(1, 0.2\times$ (the number of connectors $-1)$),

and $\alpha_i$, $0 \leq \alpha_i \leq 1$, $i = 1, \ldots, 5$, are asembly coefficients, and $\sum \alpha_i \leq 1$. The values of $\alpha_i$'s and normalization factor are dependent upon the actual assembly environment and also on whether it is manual assembly, robot assembly, or hard automation assembly. The value of normalization factor is set to 1 based on the cost analysis of robot assembly by Boothroyd and Dewhurt[1]. The normalization factor 1 gives 1.5 times of the cost of single insertion to the cost of a multiple insertion since multiple insertion increases the difficulty of alignment.

**Weight Assignment**

The weight of an edge $e_i$, $W(e_i)$, is determined by the linear combination of the total strength of the edge, $S_T(e_i)$, and the assembly cost of the edge, $C_r(e_i)$, as follows:

$$W(e_i) = \beta S_T(e_i) + (1-\beta)C_r(e_i),$$

where $\beta$ represents the weighting coefficient.

It is noted that the assignment of weights relies upon various heuristic functions having a number of parameters associated with them. The heuristic functions, though not from analytic results, can be tuned to particular assembly operations by optimizing their parameter values based on experimentation. Fig.13.6 illustrates a WALG for the assembly of the flashlight using the following values of assembly coefficients: $\alpha = 0.5$, $\beta = 0.5$, $c_s = 0.5$, $c_a = 0.5$, and $\alpha_i = 0.1$, $(i = 1, \ldots, 5)$.

## 13.6 Decomposition of a Weighted ALG

A set of tentative subassemblies is generated by successive merging of nodes in the WALG: 1) The nodes connected by edges having weights greater than or equal to a threshold are merged into supernodes. 2) The disassemblability[2] of each supernode is checked. 3) The disassemblable supernodes become tentative subassemblies. 4) By adjusting the threshold, different sets of tentative subassemblies can be obtained.

Fig.13.7 illustrates the tentative subassemblies generated by applying the decomposition process to the flashlight assembly. At first, among the individual nodes and supernodes of the WALG, $P_1$ and $P_4$ are found disassemblable. Since the disassembly of $P_1$ makes $P_2$ and $P_3$(embedded in $S_2$) unstable and the disassembly of $P_4$ makes $P_5$ unstable, both subassemblies become cut-supernodes. The subsequent reduction of threshold up to 0.4968 and the test of disassemblability results in three tentative subassemblies $[S_2, P_1]$, $[P_4, P_5]$, and $[S_1, P_4, P_5]$.

There can be at most $2n-1$ tentative subassemblies for a WALG with $n$ nodes. This is because the upper bound of the number of generated supernodes occurs when the merging process successively combines a pair of nodes, resulting in a binary tree: The total number of nodes in a binary tree with $n$ terminal nodes is $2n-1$. Table 13.4 shows a set of tentative subassemblies generated for the flashlight with several different threshold settings.

[2] A node(part) is said to be disassemblable from the rest of an assembly if there is a path along which the part can be taken out by a single motion without colliding with the rest of the assembly. A node is said to be separable from the other node if the edge between the two is breakable, and a node is said to be decomposable if it is separable from the rest of the graph. A decomposable node is not necessarily disassemblable, whereas a disassemblable node must be decomposable

Table 13.4: A set of Tentative Subassemblies Generated for Flashlight Assembly

| threshold | tentative subassemblies |
|---|---|
| 0.4994 | $[P_4, P_5]$ |
| 0.4968 | $[S_2, P_1]$ |
| 0.4859 | $[S_1, P_4, P_5]$ |

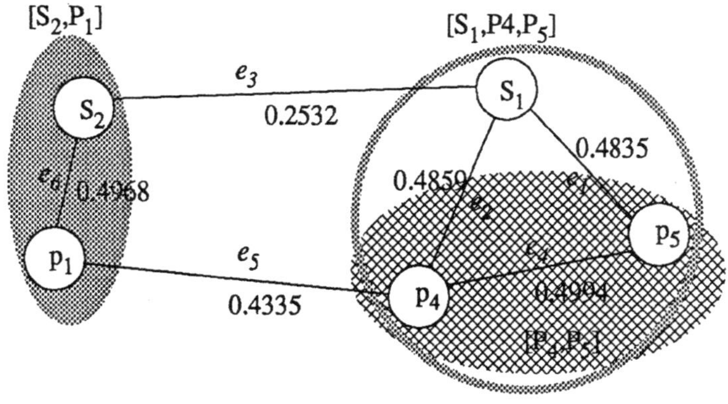


Figure 13.7: The Tentative Subassemblies Generated from the Decomposition of the ALG of the Flashlight

## 13.7 Selection of Preferred Subassemblies

The preferred subassemblies can be selected by evaluating tentative subassemblies based on subassembly selection indices(SSIs). The SSIs evaluate a cluster of parts as a subassembly based on the following criteria: 1) the stability of a cluster of parts during assembly, 2) the difficulty of disconnecting a cluster of parts from the rest of an assembly, and 3) the structural preference of a cluster of parts as a subassembly. Note that these criteria are closely related to the cost of assembly since they determine the required assembly set-ups, such as fixtures, jigs, special tools, robots, etc.

SSIs are composed of the stability index, SI, and the structural preference index, SPI, which are measured by the intra- and inter-cluster mobilities and the intra- and inter-cluster structural complexities, respectively.

### 13.7.1 Stability Index

**Definition 13.7.1 (Intra-cluster mobility)** *The intra-cluster mobility, $M_\phi(S)$, of a subassembly $S$ represents how easily the subassembly can be broken into two or more pieces. $M_\phi(S)$ is determined by the freedom of separation and the relative stability of individual cutsets of the subassembly $S$. Suppose that $c_1, c_2, \ldots, c_n$ are valid cutsets[3] of $S$, and the maximum relative stability of $c_i$ is $r(c_i)$, where the maximum relative stability of $c_i$, $r(c_i)$, is the highest value among the relative stabilities of the liaisons involved in $c_i$. assembly. The intra-cluster mobility of $S$ is calculated by*

$$M_\phi(S) \triangleq \sum_{d_a} max\{DFS_{d_a}(c_i) \times (1 - r(c_i)) | c_i \text{ is a cutset in } S,\ i = 1, \ldots, n\}$$

*where $d_a$ = x, y, and z-axis, and $DFS_{d_a}(c_i)$ is calculated by the free reference axes and the tolerance of $c_i$ along the direction of the free reference axes: $DFS_{d_a}(c_i)$ = (the number of free reference axes in the axis of $\pm d_a$) + (the tolerance of the free reference axes).*

Fig.13.8 illustrates the intra-cluster mobility of various subassemblies. In the flashlight assembly, the $M_\phi(S)$ of $[S_2, P_1]$ is relatively lower than other subassemblies since liaisons in the subassembly are not rigid.

**Definition 13.7.2 (Inter-cluster mobility)** *The inter-cluster mobility, $M_\pi(S)$, of a subassembly $S$ represents how easily the subassembly $S$ can be connected to or separated from the rest of the assembly.*

[3] For the generation of separable cutsets from a subassembly, the rigid liaisons are considered to be non-separable edges. Therefore, a set of cutsets is extracted from a liaison graph in which nodes that are connected by rigid liaisons are merged.

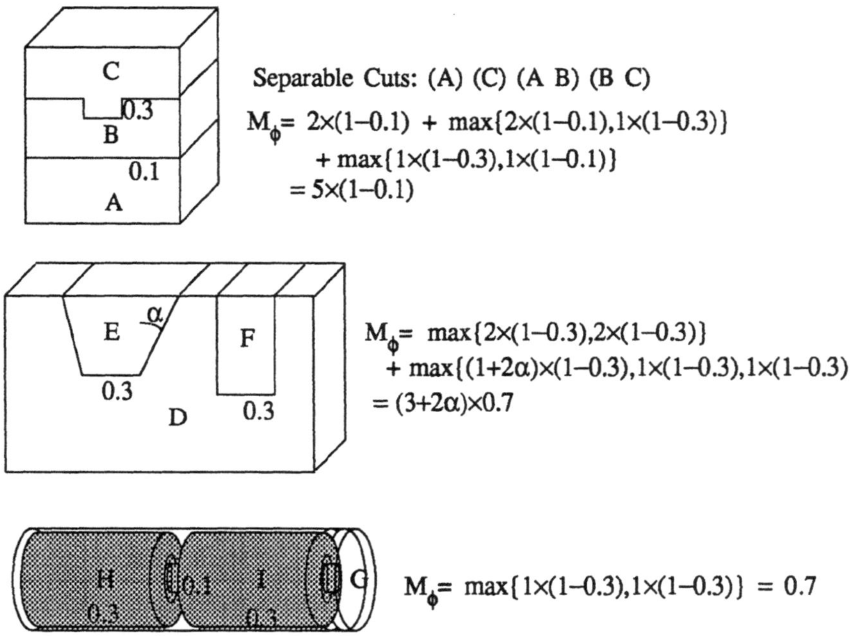


Figure 13.8: Intra-cluster Mobility of Various Subassemblies. The number in each subassembly represents the relative stability of a liaison.

$$M_\pi(S) \triangleq DFS(S_i/\overline{S}_i) \times (1- \textit{the maximum relative stability of the interconnections between } S \textit{ and } \overline{S})$$

*where $S \cup \overline{S}$ represents the original assembly.*

For instance, in Fig.13.8, $M_\pi(B \cup C) = DFS(B \cup C) \times (1 - 0.1) = 5 \times 0.9 = 4.5$, and $M_\pi(C) = DFS(C) \times (1 - 0.3) = 3 \times 0.7 = 2.1$.

**Definition 13.7.3 (Stability Index)** *The stability index, SI(S), of a subassembly S is measured by the intra-cluster and inter-cluster mobility of the subassembly S: a high stability index is assigned to a subassembly with a lower intra-cluster mobility and a higher inter-cluster mobility.*

$$SI(S) = exp\{-[\omega_d M_\phi(S) + \omega_s(6 - M_\pi(S))]\}$$

*where $M_\phi(S)$ and $M_\pi(S)$ represent the intra-cluster mobility, and the inter-cluster mobility of the subassembly S; and $\omega_d$ and $\omega_s$, $0 \leq \omega_d, \omega_s \leq 1$, are assembly coefficients.*

### 13.7.2 Structural Preference Index

**Definition 13.7.4 (Intra-cluster structural complexity)** *The intra-cluster structural complexity, $C_\phi(S)$, of a subassembly S is represented by a tuple $(\overline{d}_w, \eta_w)$. $\overline{d}_w$ is the average of the weighted degrees of individual nodes in the subassembly S. $\eta_w$ is the weighted connectivity of the subassembly S: $\overline{d}_w = \sum_{i=1}^{n} d_w(n_i)/n$, where $d_w(n_i)$, the weighted degree of a node $n_i$, is the sum of the weights of the edges incident upon $n_i$ in the weighted ALG of the subassembly S; and n is the total number of nodes in the weighted ALG of S. $\eta_w$ = the sum of the weights of the edges which belong to the minimal cut-set of the weighted ALG of the subassembly S.*

**Definition 13.7.5 (Inter-cluster structural complexity)** *The inter-cluster structural complexity, $C_\pi(S)$, of a subassembly S represents the complexity of connection between the subassembly S and the rest of the assembly. $C_\pi(S)$ is the sum of the weights of the edges connecting the subassembly with the rest of the assembly.*

**Definition 13.7.6 (Structural Preference Index)** *The structural preference index, SPI(S), of a subassembly S is measured by the intra-cluster and inter-cluster structural complexities: a higher structural preference index is assigned to a subassembly with a higher intra-cluster structural complexity and a lower inter-cluster structural complexity. SPI(S) is computed by:*

$$SPI(S) = exp\{-[(1 - \eta_w(S)/n) + \gamma_1(1 - \overline{d}_w(S)/n) + \gamma_2 C_\pi(S)/n]\}$$

*where*

| | |
|---|---|
| $n$ : | *the number of nodes in the weighted ALG representing a cluster of parts,* |
| $\eta_w(S)$, $\overline{d}_w(S)$ : | *the intra-cluster structural complexity of a subassembly S,* |
| $C_\pi(S)$ : | *the inter-cluster structural complexity of a subassembly S,* |
| $\gamma_1, \gamma_2$ : | *the assembly coefficients.* |

Note that $\frac{\eta_w}{n}$ and $\frac{\overline{d}_w}{n}$ are less than 1 since $\eta_w$ and $\overline{d}_w$ are bounded by $n - 1$.[4]

Table 13.5 shows SIs and SPIs of the tentative subassemblies of the flashlight. We assume that $\omega_d = 0.2$, $\omega_s = 0.2$, $\gamma_1 = 0.5$, and $\gamma_2 = 0.5$. The SI of $[S_2, P_1]$ is relatively low since the liaisons between parts in $[S_2, P_1]$ are free.

[4] The edge connectivity of a graph with $n$ nodes and $e$ edges is bounded by $\lfloor 2e/n \rfloor$, where $\lfloor \rfloor$ indicates the maximum integer value not exceeding $2e/n$. The maximum number of edges in a graph with $n$ nodes is $n(n-1)/2$.

Table 13.5: Stability Indices(SIs) and Structural Preference Indices(SPIs) calculated for the Tentative Subassemblies

| subassembly | SI | SPI |
|---|---|---|
| $[P_4, P_5]$ | 0.2466 | 0.2285 |
| $[S_2, P_1]$ | 0.2369 | 0.2728 |
| $[S_1, P_4, P_5]$ | 0.2466 | 0.3236 |

### 13.7.3 Selection Process

The selection of subassemblies is based on the stability and structural preference of each tentative subassembly as measured by the SI and SPI. The system calculates the subassembly selection index, SSI(S), for each tentative subassembly $S$ by the linear combination of SI(S) and SPI(S). The weights can be determined based upon the relative significance of SI(S) and SPI(S) on the overall assembly cost, if such a measure is available, or can be subject to designer's preference. In the flashlight example, considering SSI with the assignment of equal weights to SI and SPI, $[S_1, P_4, P_5]$ is selected as the best subassembly(see Fig.13.7 and Table 13.5).

Some additional considerations are given in the following: (1) Through SI and SPI, the system prefers to select a stable and cost-effective subassembly which can be easily connected to the rest of assembly; (2) there is a possibility that the extraction of a subassembly may cause the rest of assembly to be unstable. However, such a possibility is obviated in this system by the use of the stability support in the construction of an ALG; (3) Additional constraints due to the limitations of assembly environment, such as the maximum allowable weight or volume of a subassembly, can be incorporated in the selection process. For instance, if we limit the maximum size of a subassembly by at most three parts, $[S_1, P_4, P_5]$ cannot be selected even though it shows the best SSI.

### 13.7.4 Assembly Instruction

The selected subassemblies define the cut sets along which an assembly is decomposed. Each cut set is to be assembled after the associated subassemblies are assembled. An *assembly instruction* is generated for each cut set to guide the assembly of two subassemblies decomposed by the cut set. An assembly instruction script contains the following attributes:

- PRINCIPAL-PART – the part or subassembly which is moving. A subassembly which has a higher SI becomes the principle part since a moving subassembly is more likely to be unstable.

- SECONDARY-PART – the part or subassembly which is fixed during assembly operation.
- PRINCIPAL-DIRECTION – the suggested direction of motion of the principle part in assembly. The direction with more tolerance is selected from the free axes.
- ALTERNATIVE-DIRECTION – the alternative direction of motion of the principle part in assembly.
- INTERCONNECTION-TYPE – the type of the interconnection which requires the highest cost.
- WEIGHT – the weight of the subassembly.
- TOOLS – tools required for assembly.
- PART-TO-FIX – parts which require fixtures.
- STABLE-ORIENTATION – the orientation of the parts in assembly to secure all the parts not to be separated. A stable orientation of an assembly is represented by tuples: $(\theta_x, \theta_y, \theta_z)$, in which each tuple represents the counterclockwise rotation along the x-, y-, and z-axis of the reference coordinate.

An assembly instruction in Fig.13.9 shows an example of a script for mating subassemblies $s_4$ and $s_6$. The script implies that $s_4$ is assembled with $s_6$ by a screw type connection, and $s_4$ is moved into $s_6$ in the direction of $-X$-axis of the reference coordinate.

## 13.8 Generation of a Assembly Plan

### 13.8.1 Hierarchical Partial-Order Graph

The recursive application of decomposition process results in a disassembly plan in which temporal and spatial parallelism is embedded. For example, the selection of $[S_1, P_4, P_5]$ as a subassembly decomposes the original assembly into two: $[S_1, P_4, P_5]$ and $[S_2, P_1]$, which defines a temporal relationship between the liaisons connecting the two subassemblies and the liaisons of individual subassemblies: $\{l_4, l_{11}, l_{12}\} \prec \{l_1, l_2, l_3, l_5, l_6, l_7, l_8, l_9, l_{10}\}$. The disassembly plan also presents spatial parallelism among assembly operations, since individual subassemblies, such as $[P_1, P_2, P_3]$ and $[P_4, P_5, P_6, P_7, P_8]$, can be assembled in different workstations. The temporal and spatial relationships among liaison resulting from the recursive application of decomposition process can be orga nized into a hierarchical partial order graph(HPOG) as shown in Fig.13.9 fo the assembly plan of the flashlight shown in figure 13.2.

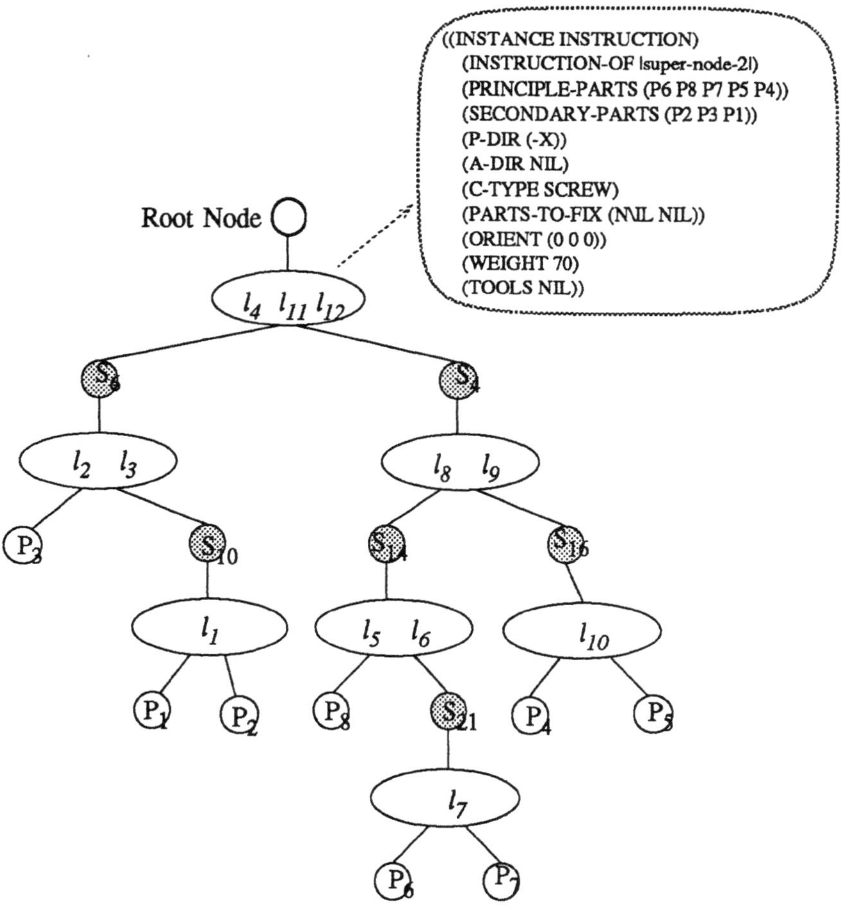


Figure 13.9: A Hierarchical Partial Order Graph generated by the Assembly Planning System for the Assembly of the Flashlight

A HPOG is an acyclic digraph, $G = (V, E)$, which consists of a finite nonempty set of vertices $V$ and a set of edges $E$ connecting vertices. The set of vertices $V$ is composed of two subsets $V = (N, S)$ with $N$ representing a set of nodes and $S$ representing a set of supernodes. A vertex of a node is represented by a circle vertex, and a vertex of a supernode is represented by an oval vertex. A node $n$, $n \in N$, represents either a part or a subassembly. A node which corresponds to a completely assembled product is called a *rootnode*. For each node $n$ of a subassembly, there is a corresponding supernode $s$ which describes how the node $n$ is assembled. $S$ consists of such supernode, and all the supernodes are reachable[5] from the rootnode. A supernode $s$ of a node $n$ contains a set of liaisons of which connections result in $n$ and the pointer to an assembly instruction that contains the detailed information for the assembly of $n$. An edge $e$, $e \in E$, is a relation on the set of vertices. A set of edges which is identified with an ordered pair $(s_i, s_j)$, (i.e. there is a path from $s_i$ to $s_j$ in a HPOG), defines temporal relationship in assembly of $s_i$ and $s_j$, such that the completion of the assembly of $s_j$ is required for starting the assembly of $s_i$. For example, in Fig.13.9, the assembly of $\{l_4, l_{11}, l_{12}\}$ must be started after $\{l_2, l_3\}$ and $\{l_8, l_9\}$ are completely assembled. Note that each edge in a HPOG is a directed edge which connects a node with a node in lower level.

A HPOG shows temporal and spatial parallelism in assembly by explicitly representing subassemblies. A set of supernodes which is reachable from a node $n$ shows all the liaisons in the subassembly $n$. In other words, all the nodes reachable from $n$ correspond to subassemblies which are extracted from $n$. A pair of subassemblies which are not reachable each other have no temporal relationship, thus both subassemblies can be done in arbitrary order. For example, in Fig.13.9, $s_4$ and $s_6$ show such a loosely parallel operation mode in which a set of subassemblies can be done in different workstations or in an arbitrary order. On the other hand, a set of liaisons in a supernode, such as $l_4$, $l_{11}$, and $l_{12}$ in Fig.13.9, shows a tightly parallel mode in which assembly operations are required to be done at the same workstation simultaneously.

### 13.8.2 Procedure of Generating HPOG

A HPOG is generated by recursively decomposing a liaison graph into a set of subgraphs. The following algorithm gives the details of generating a HPOG.

**Step1:** *Select a w-node for the decomposition.* A leaf node which is newly generated by the previous decomposition is selected. If there is no such a node then the decomposition process stops.

**Step2:** *Decide the type of the node.* Check the disassemblability and manipulability of each part in the w-node to decide whether liaisons in the

[5] In a digraph a node $n$ is said to be *reachable* from node $m$ if there is a directed path from $m$ to $n$.

w-node have any temporal and spatial relationship.

**Step3:** *Decompose the node.* The w-node of which parts have any temporal and spatial relationships each other is decomposed into a set of subassemblies.

**Step4:** *Modify HPOG.* The w-node is modified based upon the result of Step3. The extracted subassemblies become new nodes and liaisons between the new nodes locate at the supernode of the w-node.

**Step5:** *Generate an assembly instruction.* For the supernode of the w-node, an assembly instruction is generated.

Fig.13.9 shows a HPOG for the assembly of the flash light shown in figure 13.2. To generate a HPOG for the flashlight assembly, the following values of assembly coefficients are used in the construction of a weighted ALG: $\alpha = 0.5$, $\beta = 0.5$, $c_s = 0.5$, $c_a = 0.5$, and $\alpha_i = 0.1$ $(i = 1, \ldots, 6)$. However, the adjustment of various parameters involved in the WALG and SSIs and the use of different criteria for selecting preferred subassemblies result in a different set of subassemblies which leads to a different assembly partial order. There may exist a correlation between the optimal values of these adjustable parameters and the particular assembly tasks.

## 13.9 Conclusion

This chapter presents a method for the automatic determination of assembly partial orders from a liaison graph representation of an assembly through the extraction of preferred subassemblies. The resulting assembly partial order is inherently cost-effective in the sense that effects of all extraneous factors such as instability, difficulty in handling and manipulation of subassemblies, extra fixturing requirements, as well as the concepts of temporal and spatial parallelism which have a direct consequence on the implementation cost of an assembly plan have been carefully incorporated in the method suggested here. The procedure is performed in three stages: 1) Selecting a set of tentative subassemblies by decomposing a liaison graph into a set of subgraphs based on feasibility and difficulty of disassembly, 2) evaluating each of the tentative subassemblies based on the subassembly selection indices, and 3) constructing a Hierarchical Partial Order Graph by the recursive extraction of subassemblies. For each selected decomposition, the assembly instruction is generated for the lower level plan refinement. A HPOG is a unified representation scheme for temporal and spatial relationship among assembly operations.

## Acknowledgements

This research was supported in part by the National Science Foundation under grants CDR-87-17322, and in part by the industrial members of the Institute for Manufacturing and Automation Research (IMAR).


## References


[1] G. Boothroyd and P. Dewhurt, "Design for Assembly," Penton/IPC, Inc, 1984.

[2] John W. Boyes, "Interference Detection Among Solids and Surfaces," *Comm. of the ACM*, Vol. 22, No.1, pp 3 – 9, 1979.

[3] L. DeFloriani and G. Nagy "A Graph Model for Face-to-Face Assembly," *Proceedings of IEEE Conference on Robotics and Automation*, pages 75–79, 1989.

[4] T.L. De Fazio and D.E. Whitney, "Simplified Generation of All Mechanical Assembly Sequences," *IEEE Tr. on Robotics and Automation,RA-3(6)*, pages 640–658, Dec, 1987.

[5] L.D., Erman, F. Hayes-Roth, V.R. Lesser, and R.D. Reddy, "The Hearsay-II Speech-Understanding System: Interacting Knowledge to Resolve Uncertainty," *ACM Computing Survey*, 12(2), June 1980.

[6] B. Hayes-Roth, "A Blackboard Architecture for Control," *Artificial Intelligence*, 26:251–332, 1985.

[7] Andrew Kusiak, "Modelling and Design of Flexible Manufacturing Systems," Elsevier Pub., 1986.

[8] L.S. Homem de Mello and A.C. Sanderson, "Automatic Generation of Mechanical Assembly Sequences," *The Robotics Institute, Carnegie-Mellon Univ.*, Technical Report 1988.

[9] M.E. Mortenson, "Geometric Modelling," *John Wiley & Sons.*, 1985.

[10] H. Penny Nii, "Blackboard Systems: The Blackboard Model of Problem Solving and the Evolution of Blackboard Architectures," *The AI Magazine*, pages 38–53, Summer 1986.

[11] A.A.G. Requicha, "Representations for Rigid Solids: Theory, Methods, and Systems," *Computing Surveys*, 12(4), pages 437–464, Dec. 1980.

[12] A.C. Sanderson and L.S. Homem de Mello, "Task Planning and Control Synthesis for Flexible Assembly Systems," NATO ASI series, Vol, F33, *Machine Intelligence and Knowledge Engineering for Robotic Applications,* edited by A.K.C. Wong and A. Pugh, 1987.

[13] H.B. Voelcker and A.A.G. Requicha "Geometric Modelling of Mechanical Parts and Processes," *IEEE Computer, Vol.10,* pages 48–57, Dec. 1977.

[14] J.D. Wolter, "On the Automatic Planning with Fasteners," *Proceedings of IEEE Conference on Robotics and Automation*, pages 62–68, 1989.

# Chapter 14

# Backward Assembly Planning with DFA Analysis

**Sukhan Lee**

Most of the automatic assembly planners developed up to date are concerned mainly about generating assembly sequences or assembly partial orders based on reasoning of geometric interference among parts and subassemblies during assembly. Although geometric reasoning should be a fundamental mechanism for automatic assembly planning, and a considerable advancement has been made in this regard[11,12,13,14], there still remains a number of important problems to be resolved, in order to bring automatic assembly planning closer to reality.

First, the assembly of a product often requires special processes such as testing, cleaning, painting, labeling, etc, to be intermixed with assembly (mating) operations. Since special processes not only require particular sets of parts to be processed, but also have precedence relationships, special processes play a role as additional constraints on the feasible assembly sequences and should be incorporated into assembly planning.

Second, due to a large number of feasible assembly sequences or partial orders, it is desirable to select a few best sequences or partial orders which incur minimum assembly cost. However, this has been hampered by the difficulty of selecting proper performance criteria[7,8,9,10] and relating them directly to assembly cost. In addition, it requires to deal with the combinatorially explosive search space[2,3,4,5], should a globally optimal plan be searched for.

Third, one of the roles of automatic assembly planning in computer integrated manufacturing automation is to analyze a product in terms of the assemblability and assembly cost that can be fed back to the designer for design improvement. An automatic assembly planner with the capability of "design for assembly (DFA[1])" analysis will be an extremely valuable tool for automating design evaluation and modification cycles based on concurrent engineering. This problem has yet to receive proper attention.

This chapter presents an algorithm for backward assembly planning which takes the above three issues into consideration. Backward assembly planning recursively identifies decomposable subassemblies and can handle the case where an assembly sequence is not necessarily the reverse of a disassembly sequence.

First, the special processes involved in a product are represented by a "special process forest", and are incorporated into the backward assembly planning based on the "grouping principle". The grouping principle identifies, based on the given special process forest, those parts that should be grouped together in a subassembly at the current stage of backward assembly planning, in order for the special processes to be carried out properly. The grouping principle together with the merging principle (which merges those parts that are not decomposable at the current stage of backward assembly planning due to the infeasibility of interconnection) helps reduce the complexity of search space.

Second, as a criteria for selecting best subassemblies in backward assembly planning, stability, directionality, assembly pose, manipulability, process planning and parallelism are introduced and quantified. Most significantly, the above criteria are evaluated with a direct connection to assembly cost based on 1) the identification of the number of holding devices to stabilize assembly operations, 2) the derivation of the number of reorientations required during mating operations, 3) the determination of the best assembly poses for individual subassemblies generated during planning, and 4) the estimation of the effect of part/subassembly manipulability on mating cost.

Third, the search for a globally optimal plan is performed based on the AO* algorithm[15] with a cost function and a heuristic function defined in terms of the above criteria. In the process of searching for an optimal assembly plan, DFA analysis is performed for each assembly operation based on the detailed evaluation of the above criteria. The result is summarized into a DFA analysis table.

[1]DFA is a term representing design guidelines for easy assembly[1,6].

# 14.1 Representation of an Assembly

This section provides a definition of an assembly and presents a method of representing an assembly based on an attributed liaison graph, a special process forest, and an assembly constraint forest.

## 14.1.1 Definition of an Assembly

**Definition:** An *Assembly*, $A$

An assembly $A$ is a cluster of parts assembled together by a certain assembly sequence, which maintains a particular geometric relationship among parts. More specifically, a necessary and sufficient condition that a cluster of parts, $P$, becomes an assembly is as follows:

1) Every part in $P$ has at least one geometric constraint imposed on itself in relation with other parts in $P$.
2) There exists an assembly sequence which brings all the parts in $P$ together to satisfy the geometric constraints imposed on themselves.
3) $P$ is stable in a sense that it can maintain its geometric constraints either by itself or through the aid of holding devices.

A single part is an assembly by definition. An assembly is said to be connected, if there exists a path (through part connections) between every pair of parts in the assembly. An assembly is said to be self-stable, if it can maintain stability by itself without assistance of external devices.

The definition of an assembly presented above is quite general, since it includes two or more connected assemblies assembled with certain geometric constraints but without any physical contact. A product is defined as a connected and self-stable assembly. However, a disconnected and non-self-stable assembly may be generated during the assembly of a product.

## 14.1.2 Representation of an Assembly

Formally, an assembly $A$ is represented by 4-tuples: $A = \{P(A), G_L(A), G_P(A), \prod(A)\}$, where $P(A)$ represents the set of parts constituting $A$, $G_L(A)$ the attributed liaison graph representation of $A$, $G_P(A)$ the special process and constraint forest associated with $A$, and $\prod(A)$ the set of all the feasible assembly sequences for $A$.

### Attributed Liaison Graph

**Definition:** A *Liaison*

A liaison is said to exist between a pair of parts if one part constrains the freedom of motion of the other by a direct contact.

**Definition:** A *Liaison Graph*

A liaison graph is a graph, $G$, $G = (N, E)$, with $N$ representing a set of nodes, and $E$ representing a set of edges. A node $n$, $n \in N$, is assigned to each part of an assembly, and an edge $e$, $e \in E$, is assigned to a liaison between a pair of parts.

**Definition:** An *Attributed Liaison Graph*, $G_L(A)$

An attributed liaison graph is a liaison graph with frames attached to individual nodes and edges of the liaison graph to describe the attribute associated with a node or an edge. A part frame attached to each node describes the attributes associated with a part, including 1) the part geometry, 2) the mating volumes and the contact surfaces as part features, and 3) the physical properties of a part such as weight. An edge frame attached to each liaison describes attributes associated with the liaison. The attributes of a liaison consist of 1) the mating parts, 2) the mating features, and 3) the interconnection type such as Attachment, Force-fit, Connectors/Retainers, Push-and-Twist, Screw, Glue, or Welding. Thus, $G_L(A)$ contains information on the topology of part configurations, the geometry and relative pose of parts in $A$, the interconnection mechanisms of part connections, and the local freedom of motion in part mating.

### Special Process and Constraint Forest

The assembly of a product involves not only the interconnection of parts to form required liaisons but also the execution of special processes such as testing, adjusting, surface treatment, painting and packaging during assembly, while observing certain assembly constraints. This implies that assembly planning should consider generating feasible assembly sequences not only by reasoning about geometric and physical interference in part matings, but also by reasoning about the accomplishment of special processes and the satisfaction of assembly constraints.

Special processes may be subject to a certain precedence relationship in case 1) several processes share a common part, or 2) there is a need to prevent electrostatic, electromagnetic, and thermal interference, as well as mechanical vibrations and chemical pollutions during processing. The latter may also incur constraints on assembly sequences, and impose precedence relationships among some assembly operations.

Special processes and assembly constraints associated with $A$ are represented by a special process forest, $G_s(A)$, and an assembly constraint forest, $G_c(A)$. The collection of $G_s(A)$ and $G_c(A)$ is referred to here as a special process and constraint forest, $G_P(A)$.

**Definition:** Special Process Forest, $G_s(A)$

$G_s(A)$ consists of a set of trees having the following properties:

1) A node $n_i^s$ of $G_s(A)$ represents a special process, $S_i$. $n_i^s$ is associated with a tuple $(P_i^s, \sum_i^s)$, $n_i^s \sim (P_i^s, \sum_i^s)$, where $P_i^s$ represents a set of parts involved in $S_i$, and $\sum_i^s$ represents the union of the parts involved in $S_i$ and the special processes corresponding to all the offsprings of $n_i^s$: $\sum_i^s = P_i^s \bigcup\{\bigcup_j \sum_j^s\}, \forall n_j^s : n_j^s \sim (P_j^s, \sum_j^s)$ and $n_j^s$ is a child of $n_i^s$.
2) A branch $b_i^s$ connecting $n_i^s$ and $n_j^s$ represents the precedence relationship between $S_i$ and $S_j$: $S_j < S_i$ if $n_j^s$ is a child of $n_i^s$. Special processes corresponding to the sibling nodes of a tree or to the nodes of different trees have no precedence relationship.

**Definition:** Assembly Constraint Forest, $G_c(A)$

$G_c(A)$ consists of a set of trees having the following properties:

1) A node $n_i^c$ of $G_c(A)$ represents a liaison $l_i$ of $G_L(A)$. $n_i^c$ is associated with a tuple $(P_i^c, \sum_i^c)$, $n_i^c \sim (P_i^c, \sum_i^c)$, where $P_i^c$ represents a pair of parts involved in $l_i$, $\sum_i^c$ represents the union of $P_i^c$ and the parts involved in all the offsprings of $n_i^c$: $\sum_i^c = P_i^c \bigcup\{\bigcup_j \sum_j^c\}, \forall n_j^c$, $n_j^c \sim (P_j^c, \sum_j^c)$ and $n_j^c$ is a child of $n_i^c$.
2) A branch $b_i^c$ connecting $n_i^c$ and $n_j^c$ represents a precedence relationship between $l_i$ and $l_j$: $l_j < l_i$ if $n_j^c$ is a child of $n_i^c$. Liaisons corresponding to the sibling nodes of a tree or to the nodes of different trees have no precedence relationship.

The existing assembly planners to date focus on generating feasible assembly sequences based mainly on geometric reasoning on path interference. Since special processes as well as assembly constraints impose constraints on assembly order, the generation of an assembly sequence should consider effective and efficient accomplishment of special processes under the satisfaction of assembly constraints.

### 14.1.3 Example: Raster Output Scanner (ROS) Optical Assembly

The raster output scanner (ROS) optical assembly (a component of a Xerox printer), as shown in Figure 14.1, is used for the illustration of the concepts and algorithms developed throughout the chapter.

The ROS optical assembly consists of a base (B), lenses (L1,L2), mirrors (M1, M2 assembly), an I/O test unit (IOU), a motor assembly, and the PWB guard assembly. The M2 assembly is composed of a mirror (M2) and a mirror bracket

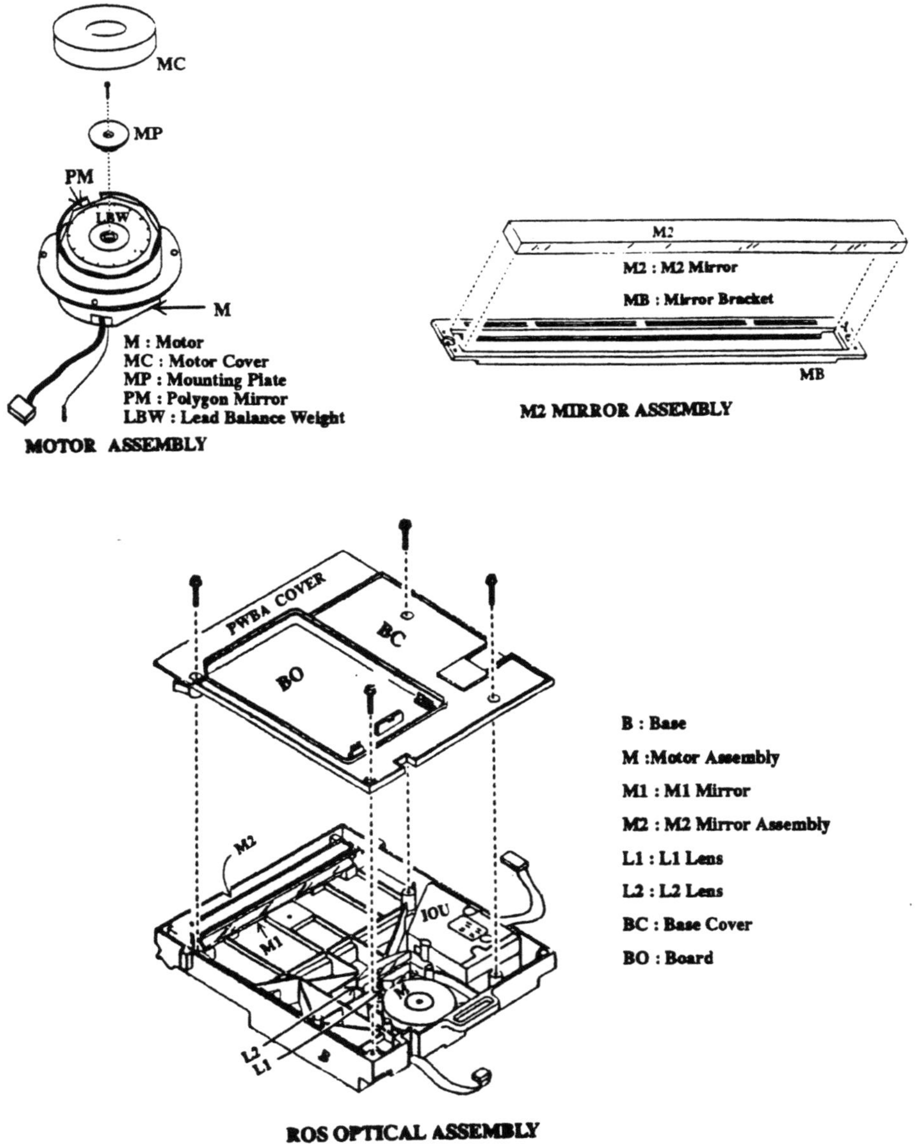


Figure 14.1: An Exploded View of the Raster Output Scanner (ROS) Optical Assembly (courtesy of Xerox, El Segundo)

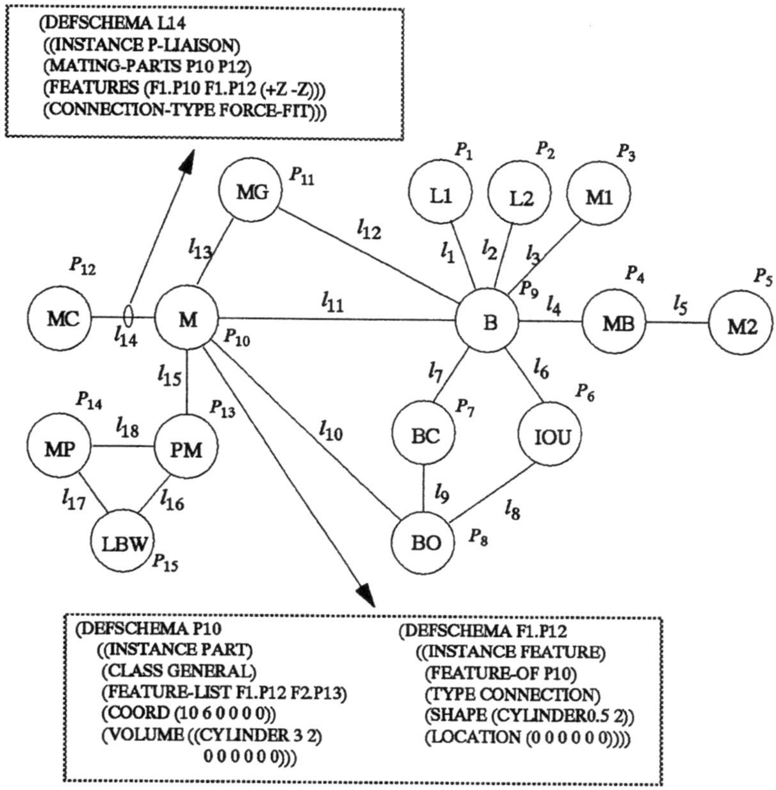


Figure 14.2: The Attributed Liaison Graph Representation of ROS Optical Assembly

(MB). A motor assembly consists of a motor (M), a motor cover (MC), a polygon mirror (PM), a mounting plate (MP), and a lead balance weight (LBW), while PWB guard assembly consists of a board (B), and a base cover (BC). A motor guard (MG) which protects a motor assembly under the base is not shown in Figure 14.1.

Figure 14.2 shows the attributed liaison graph of the ROS optical assembly, where a typical form of a node frame and an edge frame are shown in the boxes.

Figure 14.3 illustrates the special process forest associated with the ROS optical assembly. It shows that the *balancing and labeling* process of a motor assembly, the *adjusting* process of a base and M2 assembly(M2, MB), and the *cleaning* process for a lens, L1, should be done prior to the *testing* process.

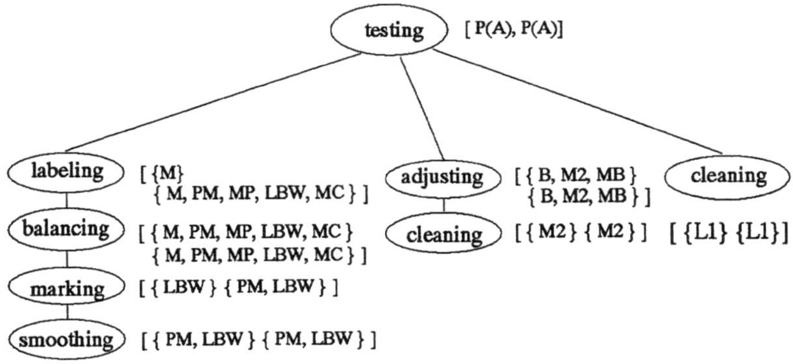


Figure 14.3: Special Process Forest for ROS Optical Assembly

# 14.2 Backward Assembly Planning

## 14.2.1 Definition of Backward Assembly Planning

**Definition:** A *Subassembly of an Assembly* $A$, $S_i|A$

A subassembly of an assembly $A$, $S_i|A$, is an assembly. $S_i|A$ represents a nonempty, proper subset of $A$, i.e., $S_i|A \neq \emptyset$, $S_i|A \subset A$, which can be generated in one of the feasible assembly sequences for $A$.

**Definition:** *A Direct Subassembly of an Assembly* $A$, $S_i^d|A$

A direct subassembly of an assembly $A$, $S_i^d|A$, is a subassembly of $A$, which can be directly assembled with $A - S_i^d|A$ at the last step of assembly in one of feasible assembly sequences for $A$.

**Definition:** *Assembly Planning of an Assembly* $A$, $AP(A)$

The assembly planning, $AP(A)$, of an assembly $A$, $A = \{P(A), G_L(A), G_P(A), \prod(A)\}$, is the process of generating a set of assembly sequences $\prod_a(A) \subseteq \prod(A)$ based on the given $P(A)$, $G_L(A)$ and $G_P(A)$, and the criteria for selecting desirable assembly sequences.

**Definition:** *Backward Assembly Planning of an Assembly* $A$, $BAP(A)$

$BAP(A)$ is a particular method for achieving $AP(A)$, based on the recursive identification and selection of desirable direct subassemblies. $BAP(A)$ first identifies and selects a direct subassembly of $A$, $S_i^d|A$(or a set of direct subassemblies of $A$, $\{S_i^d|A, i = 1, \cdots, m\}$) and decompose $A$ into $S_i^d|A$ and $A - S_i^d|A$ (or $\{S_i^d|A$ and $A - S_i^d|A, i = 1, \cdots, m\}$). Then, it recursively applies the process of decomposition to the subassemblies generated by the previous decomposition until no further decomposition can be applied (i.e., all the generated subassemblies consist of a single part).

**Algorithm:** $BAP(A)$

**Step 1:** If $A =$ a single part, then return.

**Step 2:** Select $S_i^d|A$ and $(A - S_i^d|A)$.

**Step 3:** Generate $G_L(S_i^d|A)$, $G_P(S_i^d|A)$, $G_L(A - S_i^d|A)$, and $G_P(A - S_i^d|A)$.

**Step 4:** Call $BAP(S_i^d|A)$ and $BAP(A - S_i^d|A)$.

### 14.2.2 Backward Assembly Planning vs. Disassembly planning

$BAP(A)$ differs from disassembly planning in that $BAP(A)$ can handle the case where an assembly sequence can not be obtained from the reverse of a disassembly sequence. For instance, a sequence of operations to disconnect a liaison of force-fit interconnection type is often quite different from the reverse of a sequence of operations to interconnect the liaison. As shown in Figure 14.4 (a), the snaps of $A$ should be widened in order to disassemble part B from part A, which may require a new tool. In Figure 14.4 (b), we can have the following disassembly sequence: C-A-D-B, since Screw C can be disassembled before Cover A is removed. However, the reverse of the above disassembly sequence, B-D-A-C, can not be an assembly sequence, since the inability of holding part D during assembly of part C prohibits such an assembly sequence. $BAP(A)$ identifies this problem and generates a feasible assembly sequence, B-D-C-A.

### 14.2.3 Identification of $S_i^d|A$

The necessary and sufficient condition that a cluster of parts $P|A$, $P|A \subset A$, can be a direct subassembly of $A$ is as follows:

1) Accessibility Condition : $P|A$ is accessible.

2) Stability Condition : $P|A$ and $A - P|A$ are stable.

3) Local Mating Motion Condition : All liaisons between $P|A$ and $A - P|A$ have at least one common axis of separation.

4) Path Existence Condition : $P|A$ can be brought to $A - P|A$ from the free space for mating.

5) Interconnection Feasibility Condition : $P|A$ can be interconnected to $A - P|A$ by applying the interconnection operations defined for the liaisons between $P|A$ and $A - P|A$.

6) Process Constraint Condition : $P|A$ meets the constraints defined by the special process and constraint forest of $A$, $G_P(A)$.

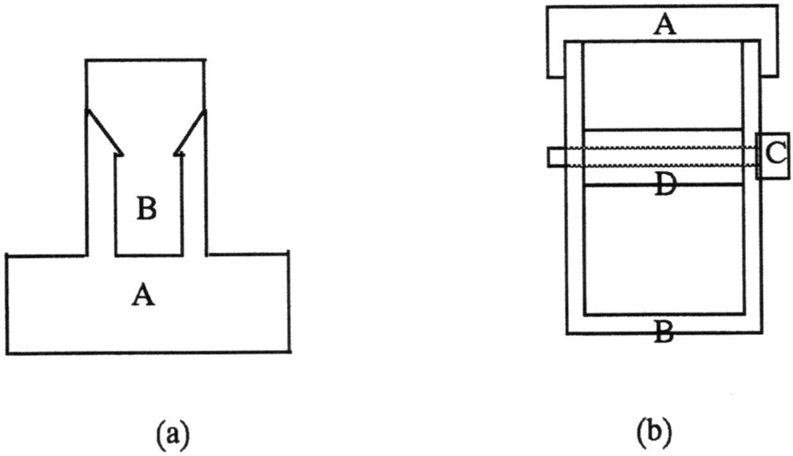


Figure 14.4: Two Typical Assemblies which illustrate the Situation where an Assembly Sequence is different from the Reverse of a Disassembly Sequence.

## 14.3 Abstract Liaison Graph by Part Clustering

The identification of $S_i^d|A$ can be accomplished by checking the above conditions for individual $P|A$'s obtained from all the cut-sets of $G_L(A)$. Since the number of cut-sets of $G_L(A)$ is often large and grows exponentially with respect to the number of nodes in $G_L(A)$, and the cost of testing the path existence condition for a cut-set is very high, the consideration of all the cut-sets of $G_L(A)$ results in inefficiency in assembly planning. This problem hampers the applicability of assembly planning to a product with a large number of parts.

The number of the cut-sets subject to the test of path existence condition may be reduced considerably if we first select those cut-sets that pass the interconnection feasibility condition, the process constraint condition, and the stability condition, prior to the test of the path existence condition. This is well justified by the fact that the cost of testing the interconnection feasibility condition and the process constraint condition is minimal due to the locality involved in the test of these conditions. The process of selecting those cut-sets that pass the test of the interconnection feasibility condition and the process constraint condition is equivalent to finding cut-sets in a simpler form of liaison graph called an abstract liaison graph, $\hat{G}_L(A)$. $\hat{G}_L(A)$ is obtained by transforming $G_L(A)$ based on:

1) Merging those parts of $A$ that cannot be interconnected to form $A$, although it is assumed that those parts are already brought into their mating position. This process identifies the liaisons of $G_L(A)$ that violate the interconnection feasibility condition for $S_i^d|A$, based on reasoning whether the preconditions for a particular interconnection mechanism associated with a liaison can be satisfied.

2) Grouping those parts that should belong to the same subassembly to meet the requirement of special processes.

The parts merged together become a super node, whereas the parts grouped together become a group node in $\hat{G}_L(A)$. A super node and a group node are considered as a single node in finding cut-sets, resulting in a considerable reduction of the number of cut-sets in search space.

Finally, the accessibility and stability conditions for $S_i^d|A$ will be tested for each cluster of parts defined by the cut-sets of $\hat{G}_L(A)$ before the decision on valid cut-sets is made.

### 14.3.1 Interconnection Feasibility and Part Merging

A liaison $l_i$ between $P|A$ and $A-P|A$ cannot be completed, in spite of the fact that $P|A$ can be brought into its mating position without path interference, if the following conditions are not satisfied:

**Condition 1:** The feasibility of applying tools and connectors required for the interconnection of $l_i$.

**Condition 2:** The feasibility of applying a force, while maintaining stability, required for the interconnection of $l_i$.

This implies that a liaison of $G_L(A)$ that violates one of the above conditions cannot be included in a cut-set (or $P|A$) to be tested for $S_i^d|A$. Pruning out those cut-sets of $G_L(A)$ that include the liaisons violating the above conditions is equivalent to defining cut-sets in a simplified $G_L(A)$ obtained by merging those nodes associated with the liaison.

The test of Condition 1 requires the verification of accessibility of the part (associated with a liaison $l_i$) by the tools and connectors required for the interconnection of $l_i$. This can be done by checking the existence of an open channel to the designated part locations, through which tools and connectors can be operated without geometric interference[11]. The test of Condition 2 requires reasoning on the force delivery to a liaison $l_i$ through intermediate liaisons.

To be more specific regarding the testing of interconnection feasibility, let us first categorize a liaison into one of the following three classes:

**Definition:** A *floating liaison*

A liaison is said to be floating if there exists no physical force holding the parts (associated with the liaison) together. For instance, a liaison with the interconnection type of "attachment" is a floating liaison. A floating liaison may or may not be stable, depending on the geometric constraints imposed on the parts associated with the liaison.

**Definition:** A *rigid liaison*

A liaison is said to be rigid, if there exists physical force holding the associated parts together, by which the liaison becomes self-stable even under the presence of external force. For instance, a liaison with the interconnection type of "force-fit", "welding", or "connectors" may be classified as a rigid liaison.

**Definition:** A *firm liaison*

A liaison is said to be firm if there exists a physical force holding the associated parts together, by which the liaison becomes self-stable, although it may exhibit a deformation or a freedom of motion under the presence of external force. For instance, a liaison with the interconnection type of "glue", "push & twist", or "screw" may be classified as a firm liaison.

A liaison is associated with its local freedom of motion:

**Definition:** The *Local Freedom of Motion of a Liaison* $l_i$, $LFM(l_i)$

The local freedom of motion of a liaison $l_i$ connecting the two parts, $P_1$ and $P_2$, $l_i \sim (P_1, P_2)$, is represented by the freedom of motion of $P_1$ against $P_2$, $LFM(l_i; P_1|P_2)$ or the freedom of motion of $P_2$ against $P_1$, $LFM(l_i; P_2|P_1)$, where $LFM(l_i; P_1|P_2)$ or $LFM(l_i; P_2|P_1)$ is represented in terms of the coordinates of the assembly to which $P_1$ and $P_2$ belong, $\{\pm x, \pm y, \pm z, \pm\psi, \pm\theta, \pm\phi\}$. Note that $LFM(l_i; P_1|P_2)$ and $LFM(l_i; P_2|P_1)$ are symmetric about individual coordinates, e.g., if $LFM(l_i; P_1|P_2) = \{+x, -z, +\phi\}$, then $LFM(l_i; P_2|P_1) = \{-x, +z, -\phi\}$.

A local freedom of motion of a liaison may be changed after the interconnection is completed: A floating liaison does not change its local freedom of motion after the interconnection is done. A rigid liaison of the interconnection type, "welding", "force-fit", or "connector", however, completely loses its local freedom of motion after the interconnection is completed. On the other hand, a firm liaison of interconnection type, "screw" or "push & twist" may show a local freedom of motion, when a certain amount of force is applied to the liaison. To distinguish the local freedom of motion of $l_i$ after the interconnection from that before the interconnection, $LFM_+(l_i)$ is used to represent $LFM(l_i)$ after the interconnection. Note that, due to the orthogonality between motion space and force space, a liaison cannot deliver force to the direction where a local freedom of motion exists.

**Definition:** An *Accessible Node*, A-node

A part is accessible by a tool if it is reachable and graspable by a tool for an assembly operation. A node are accessible if any of the parts forming the node is accessible.

**Definition:** An *Access Path to a node $n$*, A-path

An access path to a node $n$, A-path, is a path starting from an A-node and ending with the node $n$ without having any other A-node in the middle of the path. By definition, an A-node has an A-path to itself. An access path is represented by an ordered set of liaisons or parts on the path. A-node may have one or more A-paths. A pair of A-paths, A-path($n_1$) and A-path($n_2$), $n_1 \neq n_2$, are said to be independent each other, if they share no common part.

**Definition:** The *Internal Motion Space,* $\mathcal{M}$[A-path($n$)], *and the Static Force Space,* $\mathcal{F}$[A-path($n$)], *associated with* A-path($n$)

The internal motion space, $\mathcal{M}$[A-path($n$)], of A-path($n$) is defined by the union of the local freedom of motion of individual liaisons in A-path($n$), and represents the flexibility that the configuration of parts along A-path($n$) can be deformed by an external force, with the first part (corresponding to A-node) and the last part (corresponding to the node $n$) fixed in space. Note that $\mathcal{M}$[A-path($n$)] is a function of the external force applied to A-path($n$) when A-path($n$) includes firm liaisons, since the external force determines which firm liaisons can be broken. As will be explained shortly, the external force to be applied to A-path($n$) is given as the force required for the interconnection of the liaison which $n$ is associated with, and is subject to the evaluation of force-deliverability.

Assume that A-path($n$) is represented by an ordered set of liaisons, $\{l_1, l_2, \cdots, l_r\}$, with $l_i$ formed by a pair of nodes, $(n_{i_1}, n_{i_2})$, and that $(n_{i_1}, n_{i_2})$ is ordered along A-path($n$) in the direction toward $n$. Then, $\mathcal{M}$[A-path($n$)] $=\bigcup_{i=1}^{r} LFM(l_i; n_{i_1}|n_{i_2})$.

The static force space, $\mathcal{F}$[A-path($n$)], of A-path($n$) defines the static force that can be delivered to the node $n$ through A-path($n$). $\mathcal{F}$[A-path($n$)] is represented by the orthogonal complement of $\mathcal{M}$[A-path($n$)], i.e., $\mathcal{F}$[A-path($n$)] $=\{\pm x, \pm y, \pm z, \pm\psi, \pm\theta, \pm\phi\} - \bigcup_{i=1}^{r} LFM(l_i; n_{i_1}|n_{i_2})$.

**Definition:** A *Force-Deliverable* A-path

A-path($n_{i_1}$) is said to be force-deliverable to $n_{i_1}$ for the liaison $l_i$, $l_i \sim (n_{i_1}, n_{i_2})$, if $\mathcal{F}$[A-path($n_{i_1}$)] includes the force required for the interconnection of $n_{i_1}$ to $n_{i_2}$.

The test of Condition 2 for a liaison, $l_i \sim (n_{i_1}, n_{i_2})$, is now transformed to the verification of the existence of an independent force-deliverable A-path for $n_{i_1}$ and $n_{i_2}$.

The force-deliverability of an A-path to $n_{i_1}$ or $n_{i_2}$ depends on the force required for the interconnection of $l_i$. That is, $\mathcal{M}$[A-path($n_{i_1}$)] or $\mathcal{M}$[A-path($n_{i_2}$)] is determined by the freedom of motion of individual liaisons along the A-path under the presence of an external force equivalent to the force required for the interconnection of $l_i$. For instance, in case $l_i$ is floating, the amount of force required by $l_i$ is negligible, and the decision of the internal motion space of the A-path is based solely on the floating liaisons along the path. Note that, in this case, the force-deliverability of $l_i$ becomes equivalent to the feasibility of maintaining stability during interconnection. As a summary, we present the following merging principle:

**Merging Principle** :

A liaison $l_i$, $l_i \sim (n_{i_1}, n_{i_2})$, can not be listed as a cut-set, and consequently $n_{i_1}$ and $n_{i_2}$ should be merged together, if one of the following conditions is true:

1) It is not feasible for the tools and connectors (required for the interconnection of $l_i$) to access the designated locations.

2) Either $n_{i_1}$ or $n_{i_2}$ has no independent force-deliverable A-path, including the case where either $n_{i_1}$ or $n_{i_2}$ has no A-path at all.

By applying the merging principle described above to an assembly $A$, we can transform $G_L(A)$ into $\hat{G}_L(A)$ in which those parts of a liaison that can not be separable for $S_i^d|A$ are clustered together, as described by the following merging process:

**Step 1:** Put all the liaisons of $G_L(A)$ in Open set. Identify A-nodes of $G_L(A)$.

**Step 2:** If Open set is empty, then stop.

**Step 3:** Select liaison, $l_i \sim (n_{i_1}, n_{i_2})$, from Open set in an increasing order from an A-node, and remove $l_i$ from Open set.

**Step 4:** Check whether $l_i$ requires tools or connectors to complete the interconnection. If not, go to Step 6.

**Step 5:** Check the accessibility of tools or connectors to the designated locations. If not, merge $n_{i_1}$ and $n_{i_2}$ and go to Step 2.

**Step 6:** Check whether $n_{i_1}$ and $n_{i_2}$ have independent, force-deliverable A-paths. If not, merge $n_{i_1}$ and $n_{i_2}$ and go to Step 2.

**Example :** Part Merging of the ROS Optical Assembly

A-nodes of the ROS optical assembly are identified as B, BC, and MG. First, it is identified that $l_1$, $l_2$, $l_3$, $l_4$, $l_5$, $l_{14}$, $l_{15}$, $l_{16}$, $l_{17}$, and $l_{18}$ can be merged, since each of them does not have an independent force-deliverable $A$-path for one of the associated parts. The analysis of part merging for some other liaisons is given in the following:

1) A liaison $l_6$, $l_6 \sim$ (B,IOU)

A-paths of B : {B}
A-paths of IOU : {B,IOU}, {BC,BO,IOU}, and {MG,M,BO, IOU}.
$l_6$ does not require a tool or a connector to complete the interconnection. A part, IOU, has independent force-deliverable paths which can deliver a force needed to interconnect the floating liaison, $l_6$. (It is unnecessary to check whether the part, B, has a force-deliverable A-path because B is an A-node.) Therefore, $l_6$ is not merged.

2) A liaison $l_9$, $l_9 \sim$ (BC, BO)

A-paths of BC : {BC}
A-paths of BO : {BC,BO}, {B,IOU,BO}, and {MG,M,BO}.
$l_9$ does not require a tool or a connector to complete the interconnection. The part, $BO$, has no independent force-deliverable paths which can deliver a force needed to interconnect $l_9$. Therefore, $l_9$ is merged.

The result of applying the merging process to the ROS optical assembly is shown in Figure 14.5.

### 14.3.2 Special Process Constraints and Part Grouping

A cluster of parts required for a special process should be grouped together and included in a subassembly for processing. A subassembly may be associated with one or more special processes that operate on the parts included in the subassembly.

The recursive determination of $S_i^d|A$ in backward assembly planning should support the execution of special processes in an order specified by $G_s(A)$ and satisfy the assembly constraints specified by $G_c(A)$.

This requires 1) the determination of those parts of $A$ that should be grouped together and should not be separated into $S_i^d|A$ and $A - S_i^d|A$, based on the special process forest $G_s(A)$ and the assembly constraint forest $G_c(A)$, and 2) the generation of a special process forest and an assembly constraints forest for $S_i^d|A$ and $A - S_i^d|A$, i.e., $G_s(S_i^d|A)$, $G_c(S_i^d|A)$, $G_s(A - S_i^d|A)$, and $G_c(A - S_i^d|A)$.

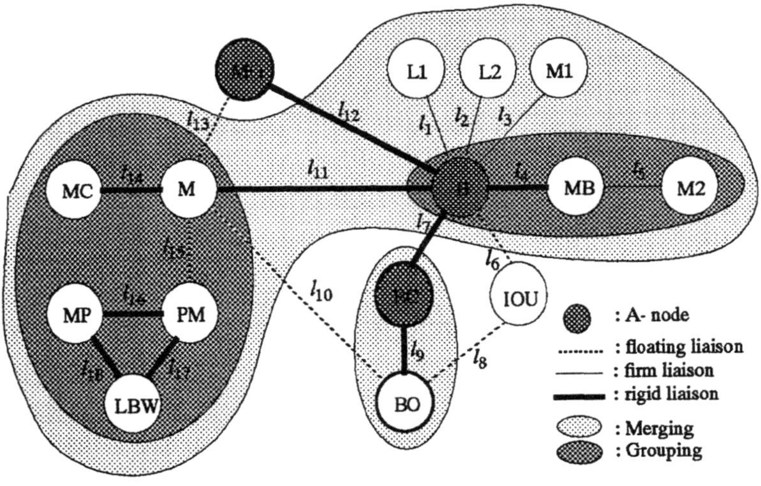


Figure 14.5: The Clustering Process for ROS Optical Assembly

The determination of parts for grouping can be done by the following process:

**1)** Given $G_s(A)$, we first determine which special processes should be accomplished with $A$, by selecting one or more special processes from $G_s(A)$ in a top-down order from the root nodes of $G_s(A)$. We then remove out those nodes corresponding to the selected special processes from $G_s(A)$ and transform $G_s(A)$ into $\widehat{G}_s(A)$. Since the special processes remaining in $\widehat{G}_s(A)$ need to be accomplished in the later stage of backward assembly planning, we should group those parts in the accumulated part list of a root node of $\widehat{G}_s(A)$, i.e., we group those parts in $\sum_{0_i}^{s}$, $n_{0_i} \sim (P_{0i}^{s}, \sum_{0_i}^{s})$ with $n_{0_i}^{s}$ a root node of $\widehat{G}_s(A)$.

**2)** In order to preserve the precedence relationship defined by $G_c(A)$, we should group those parts in the accumulated part list of a child of a root node of $G_c(A)$, i.e., we group those parts in $\sum_{1_i}^{c}$, $n_{1_i}^{c} \sim (P_{1_i}^{c}, \sum_{1_i}^{c})$ with $n_{1_i}^{c}$ a child of a root node of $G_c(A)$. A liaison represented by a root node of $G_c(A)$ is eligible for decomposition at the current stage of backward assembly planning, and is exempt from grouping.

In summary, we present the following grouping principle:

**Grouping Principle** :

For a given $A$, $G_L(A)$ and $G_P(A)$ where $G_P(A) = G_s(A) \bigcup G_c(A)$, we group those nodes of $G_L(A)$ that belong to either of the following lists:

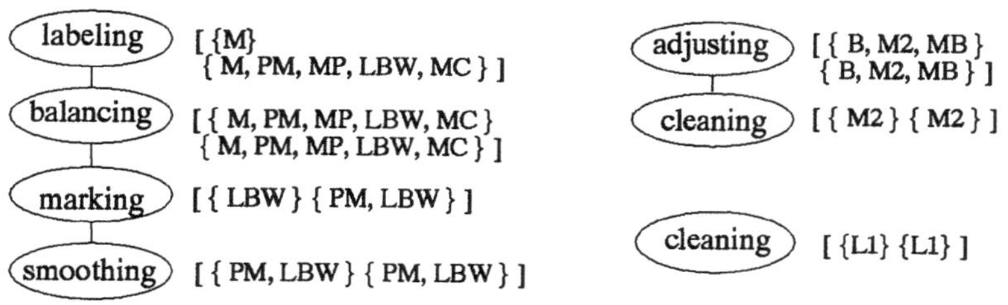


Figure 14.6: Special Process Forest, $\widehat{G}_s(A)$, after removing the Root Node from $G_s(A)$, where the Process associated with the Root Node of $G_S(A)$ is assigned to $A$

1) $\sum_{0_i}^{s}$ associated with a root node, $n_{0_i}^{s}$, of a $\widehat{G}_s(A)$, where $\widehat{G}_s(A)$ is obtained by removing out from $G_s(A)$ those nodes that are to be processed with $A$.
2) $\sum_{1_i}^{c}$ associated with a child, $n_{1_i}^{c}$, of a root node of $G_c(A)$.

The application of grouping principle to the ROS optical assembly is shown in the following example:

**Example:** Part Grouping of the ROS Optical Assembly

First, $\widehat{G}_s(A)$ is obtained in Figure 14.6 by removing the root node of $G_S(A)$ shown in Figure 14.3, since the testing of the whole assembly should be done with $A$.

Then, the grouping of the nodes in $G_L(A)$ is performed with the accumulated part lists of individual root nodes of $\widehat{G}_s(A)$: {M,PM,MP,LBW,MC} becomes a group and {B,M2,MB} becomes another group, as shown in Figure 14.5.

### 14.3.3 Abstract Liaison Graph

Merging and grouping operations transform the original liaison graph, $G_L(A)$, into the simplified liaison graph, called the abstract liaison graph, $\widehat{G}_L(A)$, as shown in Figure 14.7. $\widehat{G}_L(A)$ is composed of super nodes and group nodes obtained from part merging and part grouping, as well as simple nodes remaining from $G_L(A)$.

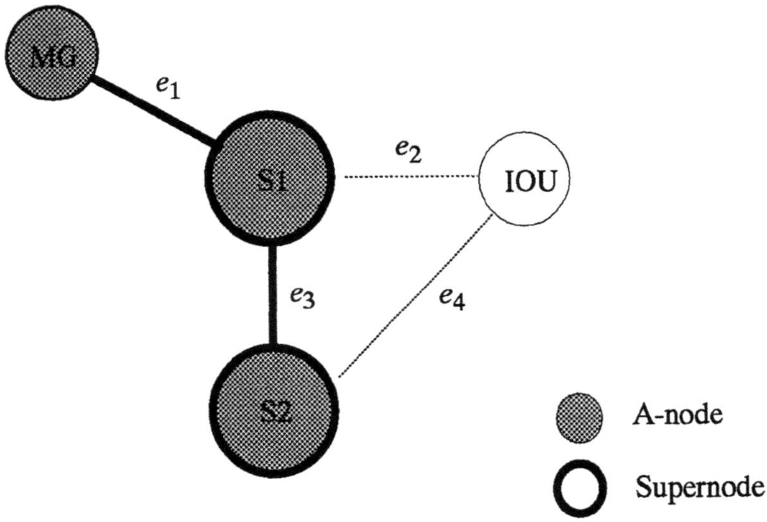


Figure 14.7: The Abstract Liaison Graph of the ROS Optical Assembly

## 14.4 Identification of Direct Subassemblies

The problem of finding direct subassemblies is now transformed into the problem of finding valid cut-sets of the abstract liaison graph, $\widehat{G}_L(A)$. A cut-set,$\gamma_i$, decomposes $\widehat{G}_L(A)$ into disjoint subgraphs, $\widehat{G}_L^{S_1}(A,\gamma_i)$ and $\widehat{G}_L^{S_2}(A,\gamma_i)$, where $\widehat{G}_L^{S_1}(A,\gamma_i)$ or $\widehat{G}_L^{S_2}(A,\gamma_i)$ may not be a connected graph, but may be a collection of multiple connected subgraphs, $\widehat{G}_L^{S_1}(A,\gamma_i) = \{\widehat{G}_L^{S_{1_j}}(A,\gamma_i),\ j = 1,2,\cdots,l_1\}$ or $\widehat{G}_L^{S_2}(A,\gamma_i)=\{\widehat{G}_L^{S_{2_j}}(A,\gamma_i),\ j = 1,2,\cdots,l_2\}$. A cut-set, $\gamma_i$, is valid if:

**1)** Each connected subgraph,$\widehat{G}_L^{S_{1_j}}(A,\gamma_i), j = 1,2,\cdots,l_1, \widehat{G}_L^{S_{2_j}}(A,\gamma_i)$, $j = 1,2\cdots,l_2$, generated by the cut-set,$\gamma_i$ includes at least one A-node.

**2)** There exists a collision free path between the cluster of parts, $P|A$, corresponding to $\widehat{G}_L^{S_1}(A,\gamma_i)$ and $A-P|A$, corresponding to $\widehat{G}_L^{S_2}(A,\gamma_i)$ for their mating.

The first condition ensures that $P|A$ and $A-P|A$, can be handled by a tool for assembly whereas the second condition ensures that $S_i^d|A$ can be placed in its mating position. For the test of the second condition, we first test the feasibility of the local mating motion between $P|A$ and $A-P|A$, so that the computational complexity involved in the path verification can be reduced. To be more specific on this point, let us define the following:

**Definition:** *Local Mating Motion, LM*

The predicate $LM(P|A, A-P|A)$ is true if all the liaisons between $P|A$ and $A-P|A$ has at least one common axis of separation.

**Definition:** *Path Existence, PE*

The predicate $PE(P|A, A-P|A)$ is true if there exists a path along which the cluster of parts $P|A$ can be brought to its mating position without colliding with the rest of the assembly.
Since the premise that $LM(P|A, A-P|A)$ is not true is a sufficient condition that $PE(P|A, A-P|A)$ is not true, the test of $LM(P|A, A-P|A)$ preceding the test of $PE(P|A, A-P|A)$ can provide a considerable reduction in the number of the costly $PE(P|A, A-P|A)$ test.

**Algorithm:** Identification of Direct Subassemblies

**Input:** $\widehat{G}_L(A)$, A-Node Set $=$ {all the A-nodes of $\widehat{G}_L(A)$}, Tested Cut-set List={$\emptyset$}, Valid Cut-set List={$\emptyset$}.

**Output:** A list of direct subassemblies specified in Valid Cut-set List.

**Method:**

**Step 1.** If A-node Set is empty, stop.

**Step 2.** Select an A-node from A-node Set, and remove it from A-node Set.

**Step 3.** If there exists a cut-set $\gamma_i, \gamma_i \notin$ Tested Cut-set List, such that $\widehat{G}_L^{S_1}(A, \gamma_i)$ includes the A-node selected in Step 2 and $\widehat{G}_L^{S_2}(A, \gamma_i)$ includes at least one A-node of $\widehat{G}_L(A)$, then continue. Otherwise, go to Step 1.

**Step 4.** Put $\gamma_i$ into Tested Cut-set List.

**Step 5.** Test $\gamma_i$ for $LM$. If it is false, go to Step 3.

**Step 6.** Test $\gamma_i$ for $PE$. If it is true, put $\gamma_i$ in Valid Cut-set List. Otherwise, go to Step 3.

The above algorithm results in the list of all valid direct subassemblies of $A$ contained in Valid Cut-set List. Table 14.1 illustrates the directed subassemblies generated for the ROS optical assembly.

Table 14.1: Valid Cut-Sets or Direct Subassemblies generated from $\widehat{G}_L(A)$ for the ROS Optical Assembly

| Cut-set | $P\|A$ | $A-P\|A$ | Result |
|---|---|---|---|
| $\{e_1\}$ | MG | S1-S2-IOU | Valid |
| | S1-S2-IOU | MG | Valid |
| $\{e_2, e_3\}$ | MG-S1 | S2-IOU | Valid |
| | S2-IOU | MG-S1 | Valid |
| $\{e_3, e_4\}$ | S2 | MG-S1-IOU | Valid |
| | MG-S1-IOU | S2 | Valid |
| $\{e_1, e_2, e_3\}$ | S1 | MG,S2-IOU | Failed in the LM test |
| $\{e_1, e_3, e_4\}$ | S1-IOU | MG,S2 | Failed in the LM test |
| $\{e_2, e_4\}$ | MG-S1-S2 | IOU | Failed in the A-node test |
| $\{e_1, e_2, e_4\}$ | S1-S2 | MG,IOU | Failed in the A-node test |

## 14.5 Subassembly Evaluation Criteria

The direct subassemblies identified from $\widehat{G}_L(A)$ are subject to further evaluation for the selection of a few best direct subassemblies. This serves to maintain a manageable number of assembly sequences, out of a potentially explosive number of possible assembly sequences in assembly planning.

The evaluation of a subassembly is based on the following criteria: 1) Stability, 2) Directionality, 3) Manipulability, 4) Process Planning(Processing Cost), and 5) Parallelism. Stability, directionality, and manipulability provide an indirect measure of the assembly cost involved in local assembly operations that can be directly used for the analysis of Design for Assembly (DFA). Processing planning and parallelism are concerned about the optimality associated with the order of special processes and the adaptability to flexible assembly environments. Note that the optimal selection of a direct assembly at each stage of backward assembly planning based on local criteria may not yield a globally optimal plan. As will be shown later, this problem is handled by the $AO^*$ algorithm with its cost and heuristic functions defined in terms of the above criteria.

### 14.5.1 Stability

To analyze the stability of a subassembly, let us first define the following:

**Definition:** A *floating cluster* of parts of $S_i^d|A$

A cluster of parts of $S_i^d|A$, $\tilde{P}_k|(S_i^d|A)$, is said to be floating if it is connected to the rest of $S_i^d|A$ only by floating liaisons. $\tilde{P}_k|(S_i^d|A)$ corresponds to a subgraph of $G_L(S_i^d|A)$ that can be separated from $G_L(S_i^d|A)$ by a cut-set consisting only of floating liaisons, called a floating cut-set.

**Definition:** A *disconnected cluster* of parts of $S_i^d|A$

A cluster of parts of $S_i^d|A$, $\dot{P}_k|(S_i^d|A)$, is said to be disconnected if it has no liaison connected to the rest of $S_i^d|A$. $\dot{P}_k|(S_i^d|A)$ corresponds to a separate subgraph of $G_L(S_i^d|A)$.

The stability of a subassembly, $S_i^d|A$, can be defined based on a set of floating clusters of parts, $\tilde{P}_k|(S_i^d|A)$, and a set of disconnected clusters of parts, $\dot{P}_k|(S_i^d|A)$, included in $S_i^d|A$.

A floating cut-set, $\tilde{r}_k$, of $G_L(S_i^d|A)$ decomposes $S_i^d|A$ into $\tilde{P}_k|(S_i^d|A)$ and $S_i^d|A-\tilde{P}_k|(S_i^d|A)$. The local freedom of motion of $\tilde{r}_k$, $LFM(\tilde{r}_k)$, can be defined as the local freedom of motion of $\tilde{P}_k|(S_i^d|A)$ against $S_i^d|A - \tilde{P}_k|(S_i^d|A)$:

$$\begin{aligned} LFM(\tilde{r}_k) &\triangleq LFM(\tilde{P}_k|(S_i^d|A), S_i^d|A - \tilde{P}_k|(S_i^d|A)) \\ &= \bigcap LFM(l_j; \tilde{P}_k|(S_i^d|A), S_i^d|A - \tilde{P}_k|(S_i^d|A)), \quad \text{for all } l_j, l_j \in \tilde{r}_k. \end{aligned}$$

Note that $\tilde{P}_k|(S_i^d|A)$ is chosen under the constraint that $S_i^d|A - \tilde{P}_k|(S_i^d|A)$ includes one of the A-nodes of the assembly $A$ that will be used for holding $S_i^d|A$ during the mating of $S_i^d|A$ with $A - S_i^d|A$.

**Definition:** Internal Freedom of Motion of $S_i^d|A$, $IFM(S_i^d|A)$

The internal freedom of motion of $S_i^d|A$, $IFM(S_i^d|A)$, is defined as a collection of assembly directions to which $S_i^d|A$ can be broken apart. $IFM(S_i^d|A)$ can be calculated by the following rules:

1) $\nexists\ \dot{P}_k|(S_i^d|A)$ and $\tilde{P}_k|(S_i^d|A) \Longrightarrow IFM(S_i^d|A) = \emptyset$;
2) $\nexists \dot{P}_k|(S_i^d|A)$ but $\exists \tilde{P}_k|(S_i^d|A) \Longrightarrow IFM(S_i^d|A) = \cup\, LFM(\tilde{r}_k),\ \forall \tilde{r}_k$;
3) $\exists \dot{P}_k|(S_i^d|A) \Longrightarrow IFM(S_i^d|A) = \{\pm x,\ \pm y,\ \pm z,\ \pm \psi,\ \pm \theta,\ \pm \phi\}$.

As an example, let us consider a simple 2-D assembly shown in Figure 14.8. Since $l_1$, $l_2$ and $l_3$ are floating liaisons, we have that

$$\tilde{r}_1 = \{l_1, l_2\}, \qquad \tilde{r}_2 = \{l_1, l_3\}, \qquad \text{and} \qquad \tilde{r}_3 = \{l_2, l_3\}.$$

Assuming that $S_i^d|A$ is oriented with reference to the assembly pose of $A$ as shown in Figure 14.8, and that $P_1$, an $A$-node of $A$, is selected for grasping of $S_i^d|A$, we have

$$\begin{aligned} LFM(\tilde{r}_1) &= LFM(P_2 \cup P_3, P_1) = \{+x, +z\}, \\ LFM(\tilde{r}_2) &= LFM(P_2, P_1 \cup P_3) = \{+z\}, \\ LFM(\tilde{r}_3) &= LFM(P_3, P_1 \cup P_2) = \{+x, +z\}. \end{aligned}$$

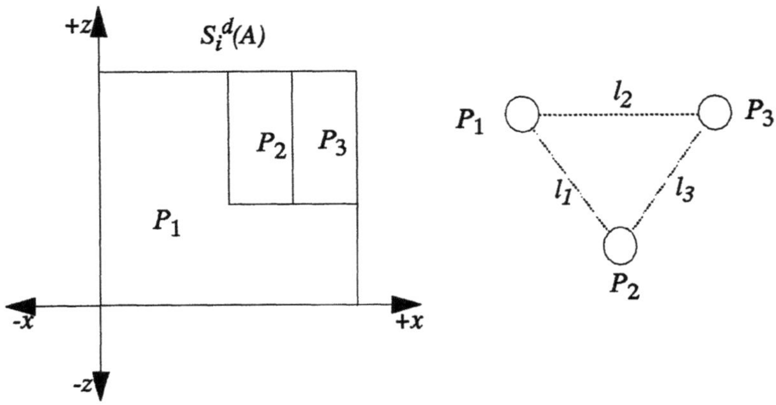


Figure 14.8: An Example to show the Calculation of Internal Freedom of Motion of $S_i^d|A, IFM(S_i^d|A)$

Therefore,

$$IFM(S_i^d|A) = \{+x, +z\}.$$

Based on the definition of $IFM(S_i^d|A)$, we can establish the stability conditi for $S_i^d|A$, as follows:

1) $S_i^d|A$ is said to be self-stable or stable without the assistance of holdi devices, if $IFM(S_i^d|A)$ is null (i.e. $S_i^d|A$ contains neither $\dot{P}_k|(S_i^d|A)$ n $\tilde{P}_k|(S_i^d|A)$ of non-null $LFM$); or $IFM(S_i^d|A)$ has at most a single trar lational freedom of motion, possibly with a rotational freedom of moti about the axis of translation (i.e. $S_i^d|A$ contains a single peg-and-hc type of $\tilde{P}_k|(S_i^d|A)$, e.g., $IFM(S_i^d|A) = \{+z, \pm\phi\}$).

2) $S_i^d|A$ is said to be stable with the assistance of holding devices, if each $\dot{P}_k|(S_i^d|A)$ or $\tilde{P}_k|(S_i^d|A)$ with more than a single translational freedom of motion contains an $A$-node of the assembly $A$. This implies that the mating operation of $S_i^d|A$ can be stabilized and completed with the assistance of external devices holding $\dot{P}_k|(S_i^d|A)$ and $\tilde{P}_k|(S_i^d|A)$ of more than a peg-and-hole type of motion freedom.

3) Otherwise, $S_i^d|A$ is said to be unstable.

A stable $S_i^d|A$, whether it requires a holding device or not, has one or more stable assembly poses, where an assembly pose is represented by assembly coordinate aligned with the direction of part stacking against gravity. For instance, $S_i^d|A$ with null $IFM^h$ (denoting $IFM$ after the incorporation of necessary holding devices) can have assembly pose of $\pm x$, $\pm y$ and $\pm z$. $S_i^d|A$ with $IFM^h$ of $\{+x, \pm\psi\}$ can have an assembly pose of $+x$, requiring a reorientation of $S_i^d|A$ to align $+x$ with the stacking direction (against gravity).

Let us now consider the stability associated with an assembly operation:

**Definition:** *Stable Assembly Operation*

> The assembly operation between $S_i^d|A$ and $A - S_i^d|A$ is said to be stable, having a stable assembly direction, if $S_i^d|A$ and $A - S_i^d|A$ have at least one common stable assembly pose.

The evaluation of $S_i^d|A$ in terms of stability is based on the stability of $S_i^d|A$ and $A - S_i^d|A$ and the stability of the assembly operation between $S_i^d|A$ and $A - S_i^d|A$, as follows:

1) If either $S_i^d|A$ is unstable or $A - S_i^d|A$ is unstable, $S_i^d|A$ can not be selected for a direct subassembly of $A$.

2) When $S_i^d|A$ and $A - S_i^d|A$ have no common stable assembly pose, $S_i^d|A$ can not be selected for a direct subassembly of $A$.

3) Otherwise we evaluate the assembly cost incurred by the need to stabilize $S_i^d|A$ and $A - S_i^d|A$ as well as the assembly operation between $S_i^d|A$ and $A - S_i^d|A$.

The assembly cost is directly related to the number of holding devices required for stabilizing $S_i^d|A$ and $A - S_i^d|A$, and the necessity of reorienting $S_i^d|A$ and $A - S_i^d|A$ for a stable assembly operation. The latter will be analyzed in more detail in the next section in relation to the directionality in assembly and the determination of best assembly poses. Table 14.2 summarizes how to evaluate the relative assembly cost of $S_i^d|A$ due to stability.

## 14.5.2 Directionality and Assembly Pose

The directionality in assembly is another important factor affecting assembly cost. Locally, a stacking operation is considered more cost-effective than a non-stacking operation. Globally, a single direction of assembly is preferred to multiple directions of assembly. Therefore, the evaluation of directionality in assembly should be based on both the local assembly direction between $S_i^d|A$ and $A - S_i^d|A$, and the uniformity of assembly directions embedded in $S_i^d|A$ and $A - S_i^d|A$.

It should be noted that whether or not the local assembly direction between $S_i^d|A$ and $A - S_i^d|A$ can be a stacking direction depends on the choice of the mating pose (as one of the stable assembly poses common to $S_i^d|A$ and $A - S_i^d|A$). However, the selection of a mating pose between $S_i^d|A$ and $A - S_i^d|A$ based

Table 14.2: The Relative Assembly Cost due to Stability associated with $S_i^d|A$ and $A - S_i^d|A$.

| Evaluation | Relative Weight |
|---|---|
| 1. Unstable $S_i^d|A$ or $A - S_i^d|A$ | $\infty$ (Design fault) |
| 2. No common stable assembly pose between $S_i^d|A$ and $A - S_i^d|A$ | $\infty$ (Design fault) |
| 3. The number of holding devices required for stabilizing $\dot{P}_k|(S_i^d|A)$ and $\dot{P}_k|(A-S_i^d|A)$, as represented by the number of $\dot{P}_k|(S_i^d|A)$ and $\dot{P}_k|(A - S_i^d|A)$ | 15 / device |
| 4. The number of holding devices to stabilize $\tilde{P}_k|(S_i^d|A)$ and $\tilde{P}_k|(A - S_i^d|A)$ with more than a single translational freedom of motion. This number of holding devices can be computed by counting the A-nodes of the assembly A, which are included in those $\tilde{P}_k|(S_i^d|A)$ and $\tilde{P}_k|(A - S_i^d|A)$ that require stabilization | 15 / device |
| 5. The reorientation of $S_i^d|A$ and $A - S_i^d|A$ | 10 / reorientation |

Note: the above relative assembly costs are made compatible with the relative weights used in [6].

solely on implementing a stacking operation may incur the need to reorient the assembly of $S_i^d|A$ and $A - S_i^d|A$, so that the assembly of $S_i^d|A$ and $A - S_i^d|A$ can be brought into the, previously selected, best assembly pose for $A$. Furthermore, the assembly pose of $S_i^d|A$ or $A - S_i^d|A$ should be chosen from the set of stable assembly poses of $S_i^d|A$ or $A - S_i^d|A$, which may differ from the selected mating pose between $S_i^d|A$ and $A - S_i^d|A$. This also incurs the need to reorient $S_i^d|A$ or $A - S_i^d|A$, so that $S_i^d|A$ or $A - S_i^d|A$ is brought into its pose. This implies that the determination of an assembly pose and an assembly direction should consider the trade-off between maximizing the directionality in assembly and minimizing the reorientation of assembly pose. In principle, it is desirable to avoid a costly reorientation, unless the reorientation is required to allow many local stacking operations in the subsequent backward assembly planning, thus justifing the cost of reorientation.

Now, let us first introduce the following notational conventions to be used in the algorithm for selecting the best assembly poses for $S_i^d|A$ and $A - S_i^d|A$, and for evaluating the relative assembly cost of $S_i^d|A$ and $A - S_i^d|A$ in terms of the directionality in assembly:

1) $t_1, t_2 \triangleq$ an assembly pose of $S_i^d|A$ and $A - S_i^d|A$ represented with reference to the previously determined assembly pose of $A$, $t^*$.

2) $\{t_1^s\}, \{t_2^s\} \triangleq$ a set of stable assembly poses for $S_i^d|A$ and $A - S_i^d|A$.

3) $\{t_{12}^s\} \triangleq$ a set of stable assembly poses common to $S_i^d|A$ and $A - S_i^d|A$, i.e., $\{t_{12}^s\} = \{t_1^s\} \cap \{t_2^s\}$.

Then, we can associate each pair $(t_1^s, t_2^s)$, where $t_1^s \in \{t_1^s\}$ and $t_2^s \in \{t_2^s\}$, with the relative assembly cost, $L$, involved in a local mating operation.

The relative assembly cost involved in a local mating operation, $L$, can be determined based on the need of reorientations and the directionality of mating operations (whether it is a stacking operation or a non-stacking operation), as well as the difficulty of handling the related subassemblies (which is analyzed in detail in the next section in terms of manipulability).

The reorientation of the assembly poses of $S_i^d|A$ and $A - S_i^d|A$ during assembly becomes necessary due to:

1) The need to transform $t_1^s$ and/or $t_2^s$ into a mating pose, $t_{12}^s, t_{12}^s \in \{t_{12}^s\}$, in the case where $t_1^s \neq t_{12}^s$ or $t_2^s \neq t_{12}^s$.

2) The need to transform the selected mating pose, $t_{12}^s$, into the assembly pose of $A$, $t^*$, in the case where $t_{12}^s \neq t^*$.

Table 14.3 shows the reorientations required for the mating between $S_i^d|A$ and $(A - S_i^d|A)$, under various conditions on $t_1^s$ and $t_2^s$. The directionality of the

Table 14.3: The Reorientations required for the Local Mating Operation between $S_i^d|A$ and $A - S_i^d|A$

| Conditions | | | The Required Reorientations |
|---|---|---|---|
| $t^* \notin \{t_{12}^s\}$ | $t_1^s = t_2^s$ | $t_1^s \in \{t_{12}^s\}$ | $1(t_{12}^s \rightarrow t^*)$ |
| | | $t_1^s \notin \{t_{12}^s\}$ | $3(t_1^s \rightarrow t_{12}^s, t_2^s \rightarrow t_{12}^s, t_{12}^s \rightarrow t^*)$ |
| | $t_1^s \neq t_2^s$ | $t_1^s \in \{t_{12}^s\}$ or $t_2^s \in \{t_{12}^s\}$ | $2(t_1^s \rightarrow t_{12}^s$ or $t_2^s \rightarrow t_{12}^s, t_{12}^s \rightarrow t^*)$ |
| | | $t_1^s \notin \{t_{12}^s\}$ and $t_2^s \notin \{t_{12}^s\}$ | $3(t_1^s \rightarrow t_{12}^s, t_2^s \rightarrow t_{12}^s, t_{12}^s \rightarrow t^*)$ |
| $t^* \in \{t_{12}^s\}$ | $t_1^s = t_2^s$ | $t_1^s = t^*$ | 0 |
| | | $t_1^s \neq t^*, t_1^s \in \{t_{12}^s\}$ | $1(t_{12}^s \rightarrow t^*)$ |
| | | $t_1^s \neq t^*, t_1^s \notin \{t_{12}^s\}$ | $2(t_1^s \rightarrow t^*, t_2^s \rightarrow t^*)$ |
| | $t_1^s \neq t_2^s$ | $t_1^s = t^*$ or $t_2^s = t^*$ | $1(t_2^s \rightarrow t^*$ or $t_1^s \rightarrow t^*)$ |
| | | $t_1^s \neq t^*$ and $t_2^s \neq t^*$ | $2(t_1^s \rightarrow t^*, t_2^s \rightarrow t^*)$ |

mating operation can be tested by transforming the mating directions (between $S_i^d|A$ and $A - S_i^d|A$) in terms of $t^*$ into the mating directions in terms of $t_{12}^s$, where the mating directions in terms of $t^*$ are identified during the verification of the path existence ($PE$).

The relative assembly cost, $L$, involved in a local mating operation can now be calculated for individual $(t_1^s, t_2^s)$, by

$$L = [\gamma_1 REO(t_1^s) + \gamma_2 REO(t_2^s) + \gamma_3 REO(t_{12}^s)]\alpha_0 + \gamma_4 \beta_0,$$

where $REO(t_1^s), REO(t_2^s)$, and $REO(t_{12}^s)$ are binary functions of either 1 (when the reorientation of the corresponding assembly pose is required) or 0; $\alpha_0$ and $\beta_0$, represent respectively the normal relative assembly cost due to a reorientation ($\alpha_0 = 10$) and a mating motion ($\beta_0 \triangleq \beta_{01} = 1$ for a stacking operation and $\beta_0 \triangleq \beta_{02} = 5$ for a nonstacking operation); $\gamma_1, \gamma_2, \gamma_3$ and $\gamma_4$ represent the effect of part manipulability on the relative assembly cost (refer to the manipulability section for more detail).

The best assembly poses of $S_i^d|A, t^*(S_i^d|A)$, and $A - S_i^d|A, t^*(A - S_i^d|A)$ can then determined based on achieving the minimum relative assembly cost, $L$, due to the local mating operation.

Let us consider the relative assembly cost, $R(S_i^d|A)$, involved in the assembly of $S_i^d|A$ as the global estimation of the relative assembly cost associated with $S_i^d|A$. Since the exact evaluation of $R(S_i^d|A)$ can only be obtained after a complete assembly plan is formulated, we indirectly estimate $R(S_i^d|A)$ based on the following two major factors contributing to $R(S_i^d|A)$:

**1)** The estimated relative assembly cost, $R_o(S_i^d|A)$, due to the number of re-orientations involved in the assembly of $S_i^d|A$.

**2)** The estimated relative assembly cost, $R_s(S_i^d|A)$, due to the number of stacking and non-stacking operations involved in the assembly of $S_i^d|A$.

To obtain $R_o(S_i^d|A)$ and $R_s(S_i^d|A)$, let us define the following:

**Definition:** *Directionality of $S_i^d|A$, $D_\xi(S_i^d|A)$*

$S_i^d|A$ is said to have $m$ degrees of directionality in $\xi$ ($\xi = x, y, \text{or } z$), denoted as $D_\xi(S_i^d|A) = m$, in the case where the number of $+\xi$ or $-\xi$ included in the list of $\{LFM(l_i), \forall l_i, l_i \in G_L(S_i^d|A)\}$ is $m$.

Note that in defining the directionality of $S_i^d|A$, $x$, $y$, and $z$ are referenced in terms of the assembly pose of $A$, and $LFM(l_i)$, $l_i \sim (P_1, P_2)$, can be computed either by $LFM(l_i; P_1, P_2)$ or $LFM(l_i; P_2, P_1)$, since $LFM(l_i; P_1, P_2) = -LFM(l_i; P_2, P_1)$.

**Definition:** *Directional Uniformity of $S_i^d|A$, $U_\xi(S_i^d|A)$*

$S_i^d|A$ is said to have the directional uniformity of $r$ in $\xi$, denoted as $U_\xi(S_i^d|A) = r$, in the case where $D_\xi(S_i^d|A)/\text{Card}\{l_i, l_i \in G_L(S_i^d|A)\} = r$.

$S_i^d|A$ has the maximum directional uniformity in $\xi$, if $S_i^d|A$ has the maximum directionality in $\xi$.

**Definition:** *Directionality of a Base Node, $n_B$,* of $S_i^d|A$, $D_{n_B}(S_i^d|A)$

The directionality of a base node, $n_B$, is defined by the independent directions involoved in $\{LFM(l_i; P_i, Base), \forall l_i$: $l_i$ is associated with $n_B$ representing $Base\}$, where an $n_B$ is a node of $G_L(S_i^d|A)$ which has the degree far greater than the average degree of a node of $G_L(S_i^d|A)$, i.e., the degree of $n_B \geq k \cdot$ Average Degree of $G_L(S_i^d|A)$ with $k$ cosntant, $k >> 1$.

We can select $U_\xi(S_i^d|A)$ from $\{U_\xi(S_i^d|A),\ \xi = x, y, \text{ or } z\}$ in a decreasing order until the accumulation of the selected $U_\xi(S_i^d|A)$'s becomes greater than or equal to unity. Let us define $\{\xi^*(S_i^d|A)\}$ as a set containing $\xi$'s which are associated with the selected $U_\xi(S_i^d|A)$'s.

Then, $R_o(S_i^d|A)$ can be estimated by the following equation:

$$R_o(S_i^d|A) \quad = \quad \alpha_o[(\sum_{\xi=x,y,z} a_\xi \cdot b_\xi - 1) + c_\xi],$$

where

$$a_\xi = \begin{cases} 2, & \text{if } \pm\xi \in D_{n_B}(S_i^d|A),\ n_B \in \{n_B\}: \\ & \qquad \text{A set of base nodes in } G_L(S_i^d|A) \\ 1, & \text{otherwise} \end{cases}$$
$$b_\xi = \begin{cases} 1, & \text{if } \xi \in \{\xi^*(S_i^d|A)\} \\ 0, & \text{otherwise} \end{cases}$$
$$c_\xi = \begin{cases} 1, & \text{if the selected best pose of } S_i^d|A \notin \{\xi^*(S_i^d|A)\} \\ 0, & \text{otherwise.} \end{cases}$$

$R_s(S_i^d|A)$ can be estimated based on the following equation:

$$R_s(S_i^d|A) \quad = \quad N \cdot [\beta_{01} U_{\xi_{max}}(S_i^d|A) + \beta_{02}(1 - U_{\xi_{max}}(S_i^d|A))],$$

where $N$ represents the number of parts included in $S_i^d|A$, and

$$U_{\xi_{max}} \triangleq \max_\xi\{U_\xi(S_i^d|A), \xi = \text{x, y and z}\}.$$

Now, the relative assembly cost, $R_s(S_i^d|A)$, representing a global estimation of the relative assembly cost associated with $S_i^d|A$ becomes

$$R(S_i^d|A) \quad = \quad R_o(S_i^d|A) + R_s(S_i^d|A).$$

Finally, the global estimation of the relative assembly cost, $R$, due to directionality, can simply be obtained by $R = R(S_i^d|A) + R(A - S_i^d|A)$.

### 14.5.3 Manipulability

A subassembly subject to either a reorientation and/or a translation for mating should be easily manipulable by tools or hands. The term *manipulability of* $S_i^d|A$ is used to quantify the efficiency in orienting $S_i^d|A$ and in handling of $S_i^d|A$. The manipulability of $S_i^d|A$ is closely linked to the size, shape and weight of $S_i^d|A$. More specifically[6], the orientation efficiency can be measured based on the symmetry and marked polarity in the geometry and weight of $S_i^d|A$, whereas the handling efficiency can be measured based on the regularity in the size, weight and shape of $S_i^d|A$, and the flexibility and fragility of $S_i^d|A$, which determine the need for special tooling, as shown in Table 14.4.

The manipulabilities of $S_i^d|A$ and $A - S_i^d|A$ affect the relative assembly cost of the local mating operation between $S_i^d|A$ and $A - S_i^d|A$, since they directly influence the relative assembly cost for the required reorientations as well as the mating motion.

To take this into consideration, in the previous section, the relative assembly cost for a reorientation, $\alpha$, as well as the relative cost for a mating motion,

Table 14.4: The Criteria for Measuring Manipulability of a Part or a Subassembly[6]

| Orientation Efficiency | Relative Assembly Cost |
|---|---|
| Part tangles, nests or shingles | 5 |
| Asymmetric part without marked polarities of weight or geometry | 5 |
| Asymmetric part with marked polarities of weight or geometry | 3 |
| Symmetric part | 1 |
| Part delivered to the assembly station with a known orientation | 1 |

| Handling Efficiency | Relative Assembly Cost |
|---|---|
| Large off center weight potentially causing loss of orientation | 5 |
| Very large parts | 5 |
| Very small parts | 5 |
| Fragile | 3 |
| Flexible | 3 |
| Irregular shaped part requiring special tooling | 3 |
| Easily handled part with standard tooling (tooling can handle more than 1 part) | 1 |

$\beta$ were determined by multiplying the manipulability coefficient, $\gamma$, to their nominal values, $\alpha_0$ and $\beta_0$.

The manipulability coefficient, $\gamma$, can be determined for $s_i^d|A$ as follows:

$$\gamma(S_i^d|A) = \frac{\sum \text{the scores of orientation and of handling efficiency for } S_i^d|A}{\sum \text{the nominal scores of orientation and of handling efficiency}}$$

### 14.5.4 Process Planning

The special process forest, $G_s(A)$, represents the precedence relationship among special processes associated with $A$, where a child process should precede its parent process. The association of $G_s(A)$ with $A$ implies that the processes included in $G_s(A)$ should be accomplished during the assembly of $A$. The backward assembly planning of $A$, which decomposes $A$ into $S_i^d|A$ and $A-S_i^d|A$, also requires the decomposition of the processes included in $G_s(A)$.

1) The processes that should be accomplished with $A$ prior to the decomposition of $A$ into $S_i^d|A$ and $A - S_i^d|A$.

2) The processes that should be left for $S_i^d|A$.

3) The processes that should be left for $A - S_i^d|A$.

Note that the processes to be accomplished with $A$ should be selected in the top-down order starting from the root nodes of $G_s(A)$.

In general, the decomposition of special processes associated with $A$ into the above 3 categories is not unique. For instance, let us assume that $A$ is associated with $G_s(A)$ consisting of 3 trees, $\{T_1,T_2,T_3\}$, as shown in Figure 14.9 (a). By selecting $\{P_A\}$, $\{P_A\} = \{P_0^1, P_{11}^1, P_0^2\}$, as a set of special processes to be accomplished with $A$, the remaining forest representing the special processes that should be accomplished with $S_i^d|A$ and $A - S_i^d|A$ consists of 5 trees, $\{T_1'$, $T_2'$, $T_3'$, $T_4'$, $T_5'\}$, as shown in Figure 14.9 (b).

Now, the selection of $S_i^d|A$ dictates a particular decomposition of $\{T_1'$, $T_2'$, $T_3'$, $T_4'$, $T_5'\}$ into two disjoint sets of trees to be associated with $S_i^d|A$ and $A-S_i^d|A$. For instance, $S_i^d|A \sim \{T_1',T_2'\}$ and $(A - S_i^d|A) \sim \{T_3',T_4',T_5'\}$, as shown in Figure 14.9 (b).

It should be noted that the selection of the special processes, $\{P_A\}$, from $G_s(A)$ for $A$, and the decomposition of $G_s(A) - \{P_A\}$ into the two disjoint sets, $G_s(S_i^d|A)$ and $G_s(A - S_i^d|A)$, to be associated with $S_i^d|A$ and $A - S_i^d|A$ impose additional precedence constraints among special processes. That is, a different selection of special processes, $\{P_A\}$, for $A$, and/or a different selection of $S_i^d|A$ may result in a different partial order among special processes due to the different $G_s(S_i^d|A)$ and $G_s(A - S_i^d|A)$, resulting in a difference in

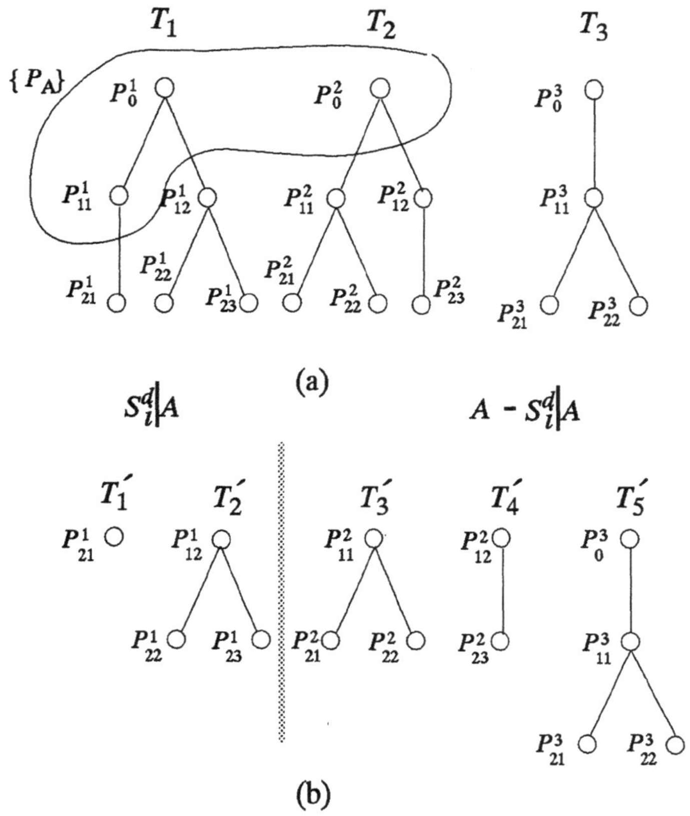


Figure 14.9: (a) The Special Process Forest, $G_s(A)$, associated with an Assembly $A$; (b) The Special Process Forests that should be assembled with $S_i^d|A$ and with $A - S_i^d|A$, in the case where a set of Special Processes, $\{P_A\}$, $\{P_A\}$ $= \{P_0^1, P_{11}^1, P_0^2\}$, are selected for $A$.

processing cost. This prompts the need to incorporate process planning, which concerns the minimization of the cost involved in special processes, into the selection of $S_i^d|A$. Planning of special processes requires the estimation of the cost in accomplishing a special process at a different stage of assembly, as well as the estimation of the overall cost of the special processes associated with $S_i^d|A$ and the overall cost of the special processes associated with $A - S_i^d|A$. The details of process planning including the estimation of processing costs are highly domain dependent and will not be elaborated here.

### 14.5.5 Parallelism

The total parallelism in an assembly sequence differs depending on the selected subassemblies. Parallelism can shorten the assembly time, although it is not necessarily linear due to the increased material transfer time. Parallelism may require additional resources such as workstations, fixtures, manpower, and material transfer facilities for the implementation. Therefore, an assembly plan with the maximum parallelism may not be always desirable. However, parallelism can be a useful feature for flexible assembly systems.

Parallelism can be measured approximately by the depth of an assembly partial order graph. However, the exact depth can be obtained only by generating the whole assembly order. Therefore, we consider the estimation of a particular decomposition ($S_i^d|A$ and $A - S_i^d|A$) on the parallelism of assembly, based on the number of parts in $S_i^d|A$ and $A - S_i^d|A$, as follows:

$$w_p(S_i^d|A) = |N_1 - N_2|/(N_1 + N_2),$$

where $w_p(S_i^d|A)$ represents the effect of selecting $S_i^d|A$ on the parallelism in assembly, and $N_1$ and $N_2$ represent the number of parts in $S_i^d|A$ and $A - S_i^d|A$, respectively.

## 14.6 Selection of Best Subassemblies based on AO* Algorithm

As indicated in the beginning of this section, the selection of $S_i^d|A$ based solely on the relative assembly costs involved in the local mating operation between $S_i^d|A$ and $A - S_i^d|A$ may not produce a globally optimal assembly plan. Therefore, we adopt the AO* algorithm with a properly defined evaluation function to search for a globally optimal or suboptimal plan.

The search space to which the AO* algorithm is applied can be represented by an AND/OR tree. The decomposition of an assembly $A$ in backward assembly planning implies the expansion of an AND node (representing an assembly $A$) into its OR children representing the alternative decompositions of

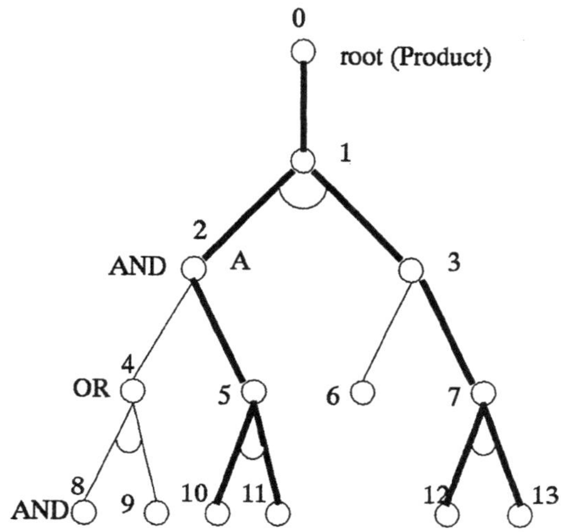


Figure 14.10: The AND/OR Tree representing the Search Space for AO* Algorithm, where $2 \sim A$ implies that $4 \sim \{S_1^d|A, A - S_1^d|A\}, 5 \sim \{S_2^d|A, A - S_2^d|A\}$, $8 \sim S_1^d|A, 9 \sim (A - S_1^d|A), 10 \sim S_2^d|A$, and $11 \sim (A - S_2^d|A)$.

$A$, $\{(S_i^d|A, A - S_i^d|A), i = 1, \cdots, l\}$, and its AND grandchildren $\{S_i^d|A$ and $A - S_i^d|A$, for $i = 1, \cdots, l\}$ attached to indivisual OR children, as shown in Figure 14.10. The AO* algorithm searches for an optimal solution tree by expanding those AND nodes of the current potential solution tree that are open to expansion, and by evaluating the next alternatives based on an evaluation function.

A potential solution tree is an AND tree[2] having the minimum value for the evaluation function at the current stage of search, whereas a solution tree is an AND tree with leaves consisting of only single parts.

To formulate the evaluation function, $e_f$, for the AO* algorithm, let us introduce the following definitions:

**Definition:** The *Local Cost, $c_l(n_i^0)$, associated with an OR node, $n_i^0$*

$c_l(n_i^0), n_i^0 \sim (S_i^d|A, A - S_i^d|A)$, represents the relative assembly cost incurred by the local mating operation between $S_i^d|A$ and $A - S_i^d|A$. $c_l(n_i^0)$ can be computed by the weighted sum of the following three components:

**1)** The relative assembly cost due to the stabilization of $S_i^d|A$ and $A - S_i^d|A$ by using holding devices and/or reorientations, as described in

[2]An AND tree is an AND/OR tree every AND node of which has no more than one OR child.

Table 14.3.

2) The relative assembly cost due to the reorientations and translations required for mating between $S_i^d|A$ and $A - S_i^d|A$, as described in Table 14.4 and Table 14.5. Note that this cost is linked to the directionalities and best assembly poses for $S_i^d|A$ and $A - S_i^d|A$, as well as the manipulabilities of $S_i^d|A$ and $A - S_i^d|A$.

3) The relative cost of the special processes assigned to $A$, the parent node of $n_i^0$.

**Definition:** The *Accumulated Cost, $c_a(T_i^a)$, associated with an AND tree, $T_i^a$*

$c_a(T_i^a)$ represents the weighted sum of the following two components:

1) The sum of $c_l(n_i^0)$ for all $n_i^0$, $n_i^0 \in T_i^a$.

2) The depth of $T_i^a$ defined by the maximum depth of $n_i^0$ for all $n_i^0$, $n_i^0 \in T_i^a$, where the depth of an OR node is measured in terms of the depth among OR nodes without considering AND nodes.

**Definition:** The *Local Heuristic Estimate, $h_e(n_i^0)$, associated with an OR node, $n_i^0$*

$h_e(n_i^0)$, $n_i^0 \sim (S_i^d|A, A - S_i^d|A)$, represents an estimate of the optimal relative assembly cost to assemble $S_i^d|A$, and can be computed by the weighted sum of the following components:

1) The relative assembly cost, $R$, associated with the directional uniformity of $S_i^d|A$ and $A - S_i^d|A$, as defined in the previous section.

2) The relative assembly cost, $S$, associated with the internal stability of $S_i^d|A$ and $A - S_i^d|A$ :

$$S = \delta[\chi(S_i^d|A) + \chi(A - S_i^d|A)]$$

where $\chi(S_i^d|A)$ and $\chi(A - S_i^d|A)$ represent the internal stability of $S_i^d|A$ and $A - S_i^d|A$, respectively, and are defined by

$$\chi(S_i^d|A) = \frac{\text{the number of floating liaisons in } G_L(S_i^d|A)}{\text{the average degree of a node in } G_L(S_i^d|A)}$$

$$\chi(A - S_i^d|A) = \frac{\text{the number of floating liaisons in } G_L(A - S_i^d|A)}{\text{the average degree of a node in } G_L(A - S_i^d|A)}$$

and $\delta$ represents the relative assembly cost due to a holding device.

3) The effect of parallelism, $w_p(S_i^d|A)$, as defined in the previous section.

**Definition:** The *Accumulated Heuristic Estimate, $h_a(T_i^a)$, associated with an AND tree, $T_i^a$*

$h_a(T_i^a)$ represents the sum of $h_e(n_i^0)$ for all $n_i^0$, $n_i^0 \in T_i^a$.

Table 14.5: A DFA Analysis Table for an OR Node, $n_i^0$, $n_i^0 \sim \{S_i^d \mid A, A - S_i^d \mid A\}$

| DFA Analysis Category | | DFA Criteria Details | |
|---|---|---|---|
| Stability | Total relative cost due to the need to stabilize $S_i^d \mid A$ and/or $A - S_i^d \mid A$ | The number of holding devices required for the stabilization of $S_i^d \mid A$ and/or $A - S_i^d \mid A$ | The relative cost due to the required holding devices |
| | | The number of reorientations required for the stabilization of $S_i^d \mid A$ and/or $A - S_i^d \mid A$ | The relative cost due to the required reorientaion |
| Manipulability and Directionality | Total relative cost involved in mating between $S_i^d \mid A$ and $A - S_i^d \mid A$ due to manipulability and directionality | The manipulability factors (Refer to Table 14.4 for more details) | $\gamma(S_i^d \mid A)$ and $\gamma(A - S_i^d \mid A)$ |
| | | The best assembly poses for $S_i^d \mid A$ and/or $A - S_i^d \mid A$ | |
| | | The number of reorientations required for mating between $S_i^d \mid A$ and/or $A - S_i^d \mid A$ | The ralative cost due to the required reorientations |
| | | The translatinal motion during mating between $S_i^d \mid A$ and/or $A - S_i^d \mid A$ | The ralative cost due to the required translatinal motion for mating |
| Process | Total relative cost for the special processes assigned to $A$ | The list of special processes assigned to $A$ | The relative cost of individual special processes |
| | Total cost at $n_i^0$ | | |

Then, the evaluation function, $e_f(T_i^a)$, associated with an AND tree $T_i^a$ simply becomes

$$e_f(T_i^a) = c_a(T_i^a) + \eta h_a(T_i^a),$$

where $\eta$ adjusts the contribution of $h_a(T_i^a)$ to $e_f(T_i^a)$ in relation with $c_a(T_i^a)$.

## 14.7 Assembly Planning with DFA Analysis

As shown in the previous section, the evaluation of the local cost, $c_l(n_i^0)$, at an OR node, $n_i^0$, is based on the detailed analysis of $c_l(n_i^0)$ in terms of the stability, the directionality, the assembly pose and the manipulability associated with the assembly of the children of $n_i^0$, as well as the cost of special processes assigned to the parent of $n_i^0$.

The result of this analysis at each OR node of the search tree can directly be used for the identification of the assemblability of a product and for the evaluation of DFA criteria, which can be fed back to the designer for proper design evaluation and modification. The assembly planner developed here has both the capability of selecting an optimal assembly partial order as well as the capability of conducting DFA analysis, serving as a powerful tool for automating the DFA evaluation and modification cycle in concurrent engineering.

DFA analysis performed during the process of computing the local cost, $c_l(n_i^0)$, associated with an OR node, $n_i^0$, $n_i^0 \sim \{S_i^d|A, A - S_i^d|A\}$, is summurized into the DFA analysis table for $n_i^0$, as illustrated in Table 14.5.

Now, the analysis of DFA for a given product can be accomplished based on the DFA tables associated with all of the OR nodes of the solution tree.

**Example:** The ROS Optical Assembly

The AO* algorithm with the cost and heuristic functions defined in the previous section is applied to the ROS optical assembly for finding an optimal solution tree and performing DFA analysis. Figure 14.11 illustrates first several nodes of the AND/OR search tree formed by the AO* algorithm, where DFA analysis tables are attached to individual OR nodes.

At Node 1, the system identifies two alternative direct subassemlies, MG and S2, of the product through the generation of the abstract liason graph (refer to Figure 14.7) and the identification of the valid cut-sets based on the generated abstract liason graph. These two alternatives are represented as the OR nodes 2 and 3 in Figure 14.11, while the two OR nodes are expanded to their AND children,(4,5) and (6,7), respectively.

The system then calculates the evaluation function at Node 2, $e_f$(Node 2), and the evaluation function at Node 3, $e_f$(Node 3), based on the local costs,

$c_l$(Node 2) and $c_l$(Node 3), and the local heuristic estimates, $h_e$(Node 2) and $h_e$(Node 3), as follows:

At Node 2, the system identifies that

1) MG and S1+S2+IOU are self-stable.
2) The best assembly pose of S1+S2+IOU is determined to be the same as the assembly pose of the product in order to avoid a costly reorientation. The assembly pose of the product is initially given in such a way that MG is located on top of the product.
3) MG can be stacked onto S1+S2+IOU.
4) MG is symmetric and easy to handle.

This implies that the decomposition represented by Node 2 requires one stacking operation ($\beta_0 = 1$) without the need of holding devices and reorientations, and that the manipulability coefficient of MG is low(0.4), incurring low assembly cost. As a result, we have that $c_l(\text{Node 2}) = 1 \times 0.4 = 0.4$. $c_a$(Node 2) can then be obtained directly from $c_l$(Node 2) by adding the depth(1) of Node 2: $c_a(\text{Node 2}) = 1.4$. For the calculation of the local heuristic estimate at Node 2, the system identifies the following:

1) The estimates of the relative assembly cost due to the directional uniformity, $R$, of MG and the internal stability, $S$, of MG are zero, since MG is a single part.
2) S1+S2+IOU consists of 15 parts with the maximum uniform directionality of 1 in $z$. However, it has a base node which has directionality of $+z$ and $-z$. Therefore, $R_0(\text{S1+S2+IOU}) = 10 \cdot 1 = 10$, and $R_s(\text{S1+S2+IOU}) = 1 \cdot 15 = 15$, given the selected best assembly pose of S1+S2+IOU. As a result, the estimate of the relative assembly cost due to the directional uniformity, $R$, of S1+S2+IOU is 25.
3) S1+S2+IOU has 4 floating liaisons and has the average degree of node of 33/14. Therefore, the estimate of the relative assembly cost due to the internal stability, $S$, of S1+S2+IOU is 25.5 ($S = 15 \times 4/(33/14) = 25.5$), where 15 is used for the relative assembly cost for a holding device.
4) The effect of Node 2 on assembly parallelism,$w_p$, can be estimated as 8.67 ($w_p = 10 \times 13/15 = 8.67$ with 10 assigned as a weight).

As a result, we have that $h_e(\text{Node 2}) = 59.17$. Furthermore, $h_a(\text{Node 2}) = h_e(\text{Node 2})$, since no other OR node exists at the current potential solution

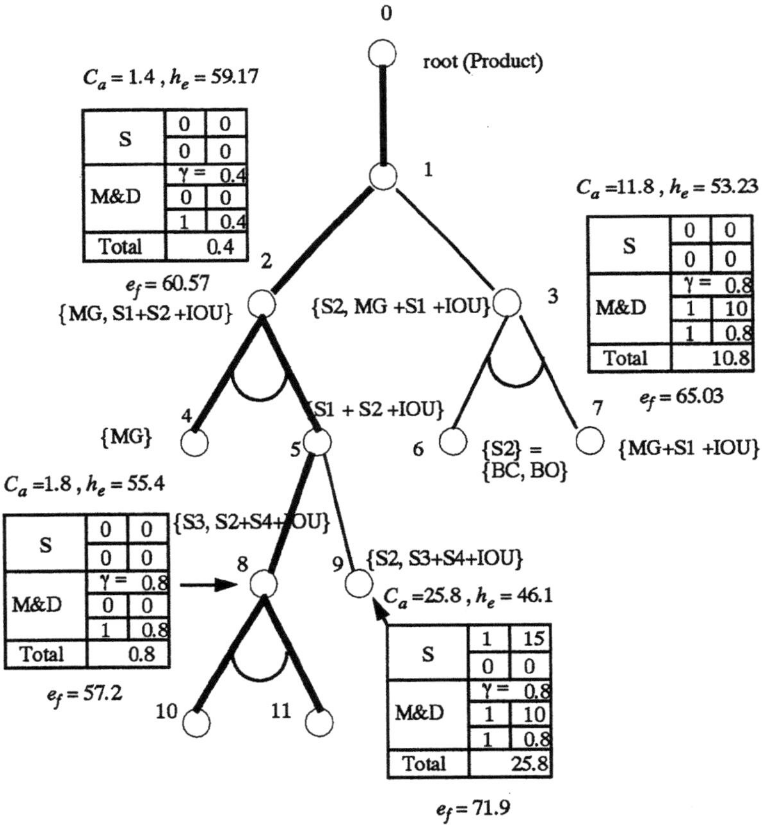


Figure 14.11: The AO* Search of an Optimal Plan for the ROS Optical Assembly, where the generated DFA analysis tables are attached to individual OR nodes. S, M, and D inside the tables represent respectively stability, manipulability and directionality.

tree candidate.

Therefore, we have that

$$
\begin{aligned}
e_f(\text{Node } 2) &= c_a(\text{Node } 2) + \eta h_a(\text{Node } 2) \\
&= 60.57 \text{ (with } \eta = 1.0).
\end{aligned}
$$

At Node 3, the system follows the same steps that are used for calculating $e_f(\text{Node } 2)$, and results in the following:

$$
\begin{aligned}
c_l(\text{Node } 3) = c_a(\text{Node } 3) &= 11.8, \\
h_e(\text{Node } 3) = h_a(\text{Node } 3) &= 53.23, \\
\text{with} \quad R = 25, S = 20.9, &\text{ and } w_p = 7.33.
\end{aligned}
$$

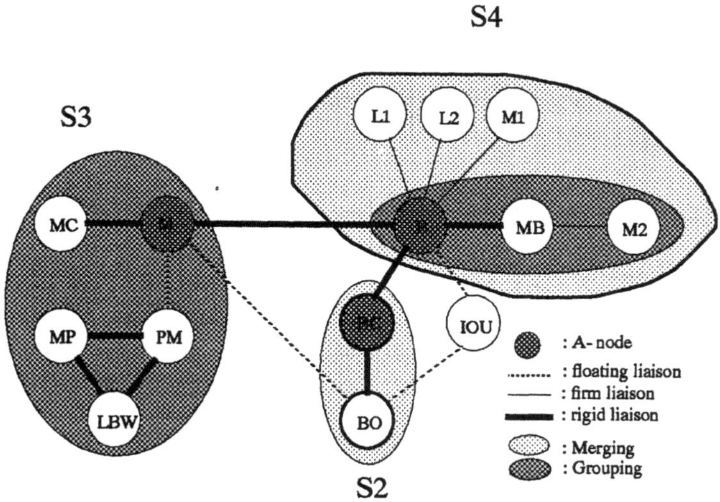


Figure 14.12: The merging and grouping operations applied to the liaison graph generated at Node 5 for the construction of the abstract liason graph.

Therefore, we have that

$$\begin{aligned} e_f(\text{Node 3}) &= c_a(\text{Node 3}) + \eta h_a(\text{Node 3}) \\ &= 65.03 \text{ (with } \eta = 1.0). \end{aligned}$$

Finally, comparing $e_f$(Node 2) and $e_f$(Node 3), the system selects Node 2 for further expansion. The result of such an expansion is shown in Figure 14.11. It is noted that, at Node 5, $G_L$(S1+S2+IOU) should be generated first, so that the process of constructing $\hat{G_L}$(S1+S2+IOU) and identifying the valid cut-sets from $\hat{G_L}$(S1+S2+IOU) can start. Figure 14.12 illustarates the merging and grouping operations applied to the $G_L$(S1+S2+IOU) to construct $\hat{G_L}$(S1+S2+IOU).

## 14.8 Conclusion

This chapter contributes to bringing automatic assembly planning closer to reality by

**1)** Developing an efficient backward assembly planner which handles the case where an assembly sequence is not same as the reverse of a disassembly sequence.

**2)** Achieving the efficiency in planning with the reduction of search space not only by merging parts based on interconnection feasibility constraints but also by grouping parts based on the special process precedence constraints.

3) Extending assembly planning into assembly process planning by incorporating special assembly processes such as testing, cleaning, etc, in planning.

4) Establishing and evaluating the subassembly selection criteria with a direct connection to assembly cost.

5) Developing the AO* algorithm for the search of a globally optimal assembly plan.

6) Developing an automatic DFA analysis tool for concurrent engineering by combining automatic assembly planning with DFA analysis.

However, there still remain many problems to overcome in turning automatic assembly planning into practice. Further research on the more powerful geometric and physical reasoners for assembly planning should follow, in order to have a direct connection to CAD database, to handle more complicated products, to achieve greater efficiency in assembly planning, and to solidify methods that evaluate an assembly plan in terms of assembly cost and DFA.

# Acknowledgements

This research was supported in part by the National Science Foundation under grants CDR-87-17322, and in part by the industrial members of the Institute for Manufacturing and Automation Research (IMAR).

# References

[1] G. Boothroyd and P. Dewhurt, *Design for Assembly*, Pendon/IPC, Inc., 1984.

[2] T. L. De Fazio and D. E. Whitney, "Simplified Generation of All Mechanical Assembly Sequences," *IEEE J. Robotics Automata*, RA-3(6):640-658, December 1987. Corrections *ibid* RA-4(6):705-708, December 1988.

[3] Sukhan Lee and Yeong G. Shin, "Automatic Construction of Assembly Partial-Order Graph," *Proceedings of the 1988 International Conference on Computer Integrated Manufacturing*, RPI, Troy, New York May 1988, pp. 383-392.

[4] L. S. Homem de Mello and A. C. Sanderson, " Automatic Generation of Mechanical Assembly Sequences," *Carnegie-Mellon Univ.*, CMU-RI-TR-88-19 1988.

[5] Sukhan Lee, "Disassembly Planning by Subassembly Extraction," *The Proceedings of the Third ORSA/TIMS Conference on Flexible Manufacturing Systems,* Elsevier Science, MIT, MA. Aug., 1989. pp. 383-388.

[6] Robert L. Hoekstra, "Design for Automated Assembly: An Axiomatic and Analytical Method," *SME Technical Paper,* Detroit, Michigan, May 1989.

[7] J. D. Wolter, "On the Automatic Generation of Assembly Plans", *Proceedings of IEEE Conference on Robotics and Automation*, pp. 62-68, 1989.

[8] L. S. Homem de Mello and A. C. Sanderson, " Evaluation and Selection of Assembly Plans," *Proceedings of the 1990 IEEE Conference on Robotics and Automation*, Cincinnati, OH. May 1990.

[9] Sukhan Lee and Yeong G. Shin, "Assembly Planning Based on Subassembly Extraction",*Proceeding of the 1990 IEEE Conference on Robotics and Automation*, Cincinnati, OH. May 1990. pp. 1606-1611.

[10] Sukhan Lee and Yeong G. Shin, "A Cooperative Planning System for Flexible Assembly," *Proceedings of the 2nd International Conference on Computer Integrated Manufacturing*, Troy, NY., May 1990.

[11] Sukhan Lee and Yeong G. Shin, "Assembly Planning Based on Geometric Reasoning," *Computers & Graphics, an International Journal of Applications in Computer Graphics*, vol. 14, No. 2, 1990. pp. 237-250.

[12] Richard L. Hoffman, "Automated Assembly in a CSG Domain," *IEEE International Conference on Robotics and Automation*, pp. 210-215, May 1989.

[13] L. De Floriani and G. Nagy, "A graph-based model for face-to-face assembly," *Proceedings IEEE International Conference on Robotics and Automation*, Scottdale, pp 75-78, 1989.

[14] A. A. G. Requicha and H. B. Voelcker, "Boolean Operation in Solid Modeling: Boundary Evaluation and Merging Algorithms," *Proceedings IEEE*, vol. 73, No. 1, Jan., 1985.

[15] Nils J. Nilsson, *Problem-Solving Methods in Artificial Intelligence*, McGraw-Hill, Inc., 1971.

## Chapter 15

# Computer aids for finding, representing, choosing amongst, and evaluating the assembly sequences of mechanical products

Thomas E. Abell, Guillaume P. Amblard,
Daniel F. Baldwin, Thomas L. De Fazio,
Man-Cheung Max Lui, Daniel E. Whitney

Sequence of assembly of a set of parts plays a key role in determining important characteristics of the tasks of assembly and of the finished assembly. Matters such as difficulty of assembly steps, needs for fixturing, potential for damage during assembly, ability to do in-process testing, occurrence of need for rework, and cost of assembly, are all affected by assembly sequence choice. The rational exploration and choice of assembly sequence is then an important task for a production engineer.

Exploring assembly sequence choice is difficult for two reasons: the number of assembly sequences can be large even at a small parts count, and can rise rapidly with increasing parts-count; and seemingly minor design changes can drastically modify the available choices of assembly sequences. Production engineers seldom consider all assembly sequences before choosing a sequence to be used. That this is so is due only in part to the potentially large number of sequences involved in most assemblies. Means for generating the complete set of assembly sequences have been few, not well-known, and not always convenient. Past techniques for exploring the choices of assembly sequence have been informal or incomplete.

Our interest in choosing good assembly sequences dates from a 1977 demonstration of robot assembly of automobile alternators[23]. Our past work used parts-trees and connection diagrams to represent assemblies[11,27].

Our current algorithms are rooted in the work of Bourjault and his colleagues[6, 7,8,18] and Homem de Mello and Sanderson[19,21,20,26]. Bourjault used a parts-connection diagram or liaison diagram to generate yes-no questions to be addressed by the designer. "Yes" or "no" represents the ability or inability to assemble a part to a subassembly, which depends on whether a clear approach path without geometric interference exists for that part and subassembly. Answers to the questions are processed to generate a list of the possible sequences. Henrioud and Bourjault[18] use the same information but process the connection diagram differently, posing far fewer questions for the same result. De Fazio and Whitney[12] altered the form of the Bourjault[6] yes-no questions, asking the user fewer questions; theirs are not yes-no questions and require geometric reasoning and anticipation by the user. They showed how to represent assembly sequences as paths through a network of assembly states (nodes) and assembly moves (arcs) such as that shown in Figure 15.4. This compact representation is called the assembly sequence diagram and forms a basis for the following evaluation and editing methods. An implementation similar to Henrioud and Bourjault's is described in De Fazio et al.[13]. Homem de Mello[19] uses cut-sets of the connection diagram of the assembly and subassemblies as bases for disassembly questions. An approach combining aspects of the techniques of Homem de Mello and Sanderson, and of Henrioud and Bourjault is used here to generate precedence relations and is referred to as the "cut-set method." In this method, the ability or inability to disassemble each cut-set of the assembly and each subassembly is inferred where possible, or answered by the user where not. Frommherz and Hornberger[14] describe a similar approach.

To make benefits of assembly sequence consideration widely available, we have developed a set of computer aids to generate possible assembly sequences and provide an environment that allows the designer to select a good sequence. Figure 15.1 illustrates the process schematically. It shows generation of possible assembly sequences followed by user sequence choice according to such criteria as: ease or reliability of assembly, fixturing, or gripping[2,4]; least assembly unit-cost or least fixed, variable, or total assembly system cost[9,15,17]; best product-testing strategy for on-line testing[25]; assembly-line layout, or other production-related criteria.

## 15.1 Background

Earlier work used concepts from Bourjault[6] to also generate a complete set of assembly sequences, though certain simplifications allowed practical application of the technique to assemblies with higher parts counts. Bourjault begins by using information contained in a parts list and an assembly drawing to characterize an assembly by a network where nodes represent parts and lines between nodes represent user-defined relations between parts called "liaisons." A definition of "liaison" follows the principal literal definition[1], "a close bond or connection," and generally includes physical contact between parts. A liaison exists between two parts if the two parts may be assembled together alone.

Once the assembly is characterized by a network of nodes (parts) and lines (liaisons), names are associated with these two sets of elements: parts names with the nodes; and liaison numbers with the lines. Subsequently any assembly step is characterized by the establishment of one or more of the liaisons of the assembly. Completion of assembly from start can then be characterized by a punctuated string of numbers representing, in some sequence, all of the liaisons of the assembly. Bourjault's and our simplified technique correspond up to this point.

Bourjault derives rules that permit algorithmic generation of (only) valid number strings representing assembly sequences from the answers to a series of questions about individual mates. Each question is answered with "yes" or "no." For assemblies consisting of rigid parts alone, the questions are of the single form ($L_i$ is read "the liaison numbered $i$"):

Is it true that $L_i$ cannot be done after $L_j$ and $L_k$ have been done?

The method of Bourjault involves asking and answering a large number of questions arranged in sets, each of which involves imagining some set of liaisons first completed, then not yet completed.

### 15.1.1 Simple technique for generating assembly sequences

Points differentiating our simple technique from that of Bourjault are: a smaller set of questions regarding conditions of liaison establishment, a set whose size increases in proportion to liaison count rather than faster than the square of liaison count; and algorithmic generation of liaison sequences from a more compact and evocative form of the liaison-sequence rules. Both techniques

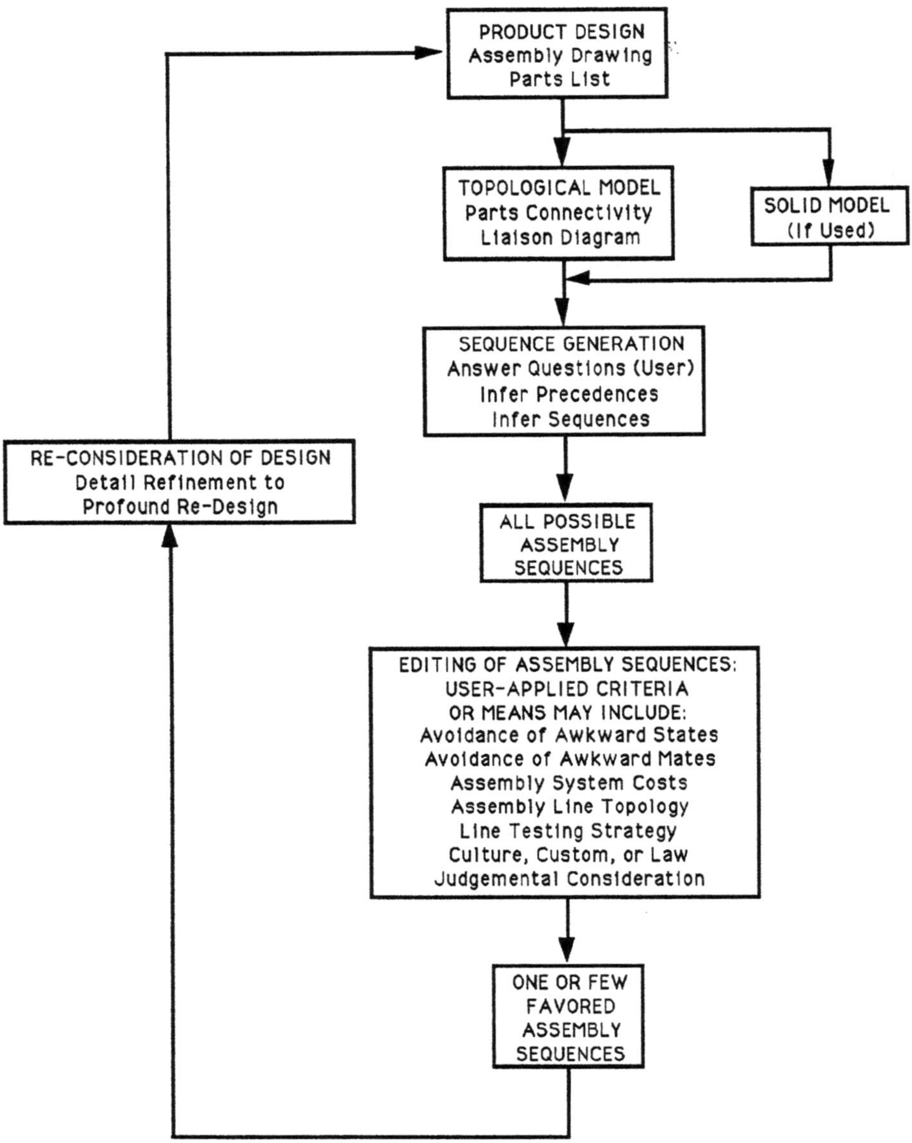


Figure 15.1: The roles of assembly sequence design and evaluation in early product design

share opening moves, based on information in an assembly drawing and parts list or in a prototype or sample assembly. Figure 15.2 is an idealized assembly drawing with parts list representing an assembly from industry (AFI), the final assembly of an automotive automatic transmission.

One begins by representing the assembly as a network of nodes and lines. Each part is represented by a node bearing the name of the part. Liaisons, as relationships between parts, are represented as numbered lines connecting related parts. Figure 15.3 is an example network representation of assembly.

Once the assembly is characterized as a network of parts and liaisons, the user must answer a question for each liaison. If each component part of the assembly is rigid, then the questions to be answered, for each liaison $i$, $i = 1$ to $l$, are:

Q1: What liaisons must be done prior to doing liaison i?

Answers are to be expressed as precedence relationships between liaisons or logical combinations of liaisons. Example answers may be of the form:

$$\begin{aligned}
(L_j \text{ or } (L_k \text{ and } L_m)) &\rightarrow L_i \\
L_i &\rightarrow (L_s \text{ or } (L_t \text{ and } L_u)) \\
(L_j \text{ or } (L_k \text{ and } Lm)) &\rightarrow (L_s \text{ or } (L_t \text{ and} Lu))
\end{aligned}$$

The symbol "→" is read "must precede."

The user must seek and anticipate all the alternatives which permit each liaison to be done. Doing so results in a close knowledge of the design details of the assembly. Overlooking alternatives falsely constrains assembly sequence count. Overlooking precedence rules yields spurious assembly moves. A false move is exposed by trying to practice it, if not earlier by consideration of the assembly drawing.

One may be concerned that question count reduction is accompanied by a staggering increase in the difficulty of answering each question. This is a subjective matter, but empirically an increase in difficulty does not seem to occur in many cases, and where it does occur, it is a reasonable increase in difficulty. A difference is that a question of the simpler technique must evoke an answer that contains the same logical relations implied a (large) set of yes/no answers to questions of the technique of Bourjault. Perhaps surprisingly, most of the questions of the simpler technique have answers that are at once simple and accessible, and easily expressed. A complicated design can result in an answer that is complicated or difficult to express. It is often useful and possible to answer difficult questions in prose. The prose may be translated into symbolic logical form, and logical technique may be used to reduce answers to simpler form. Examples of prose response to assembly questions appear in subsection 15.1.3.

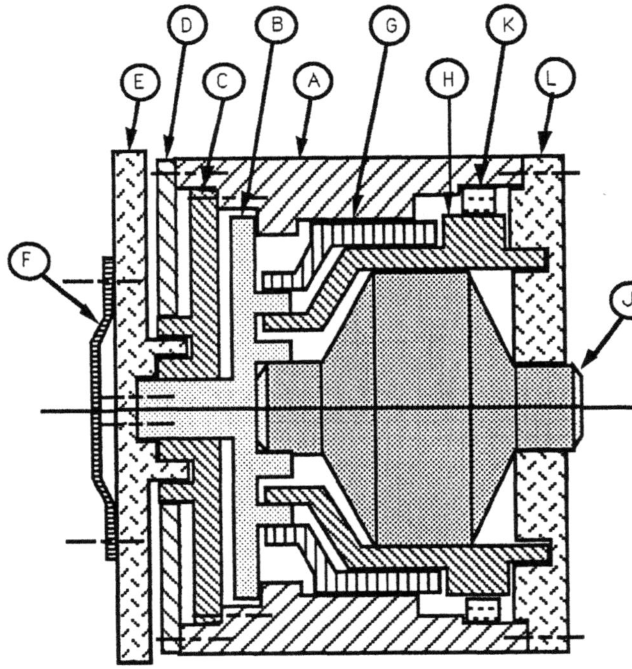


Figure 15.2: Assembled parts of example assembly from industry (AFI)

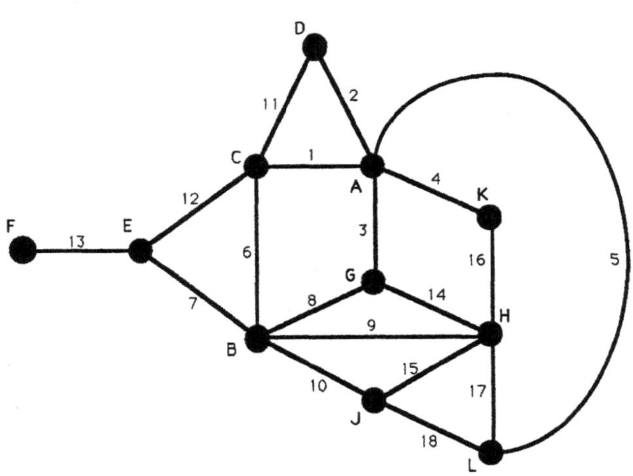


Figure 15.3: Liaison diagram of example assembly from industry (AFI)

### 15.1.2 Assembly state and assembly move representation of assembly

The symbolically stated answers to the l questions are in form of precedence relations between liaisons or logical combinations of liaisons. Liaison sequences may be generated directly from the answers. The initial state is disassembly, no liaison is established. "State" refers to the state of establishment of liaisons. An explicit list of which liaisons are and which are not established represents the state of assembly. Assembly proceeds from state to state by adding a part or a subassembly to another part or subassembly until all liaisons are established. The imaginary path associated with the attachment of a part or subassembly is called an assembly state transition, a state transition, or an assembly move. Each state may be represented by a box with a list of numbers representing established liaisons, and assembly moves may be represented as lines connecting states. The starting state's list has no entries.

To generate liaison sequences, begin by scanning the liaison list and the answers for liaisons which are not precedented. Any of these may serve as the first liaison to be established. Line up representations of each first possible state across a rank and connect each with the starting state by a line. For each possible first liaison, explore for all possible subsequent states by again scanning the liaison list, the precedence relations (answers) and any other constraints imposed on the assembly, thereby generating another rank. It is convenient to show no state more than once, so if it occurs that there are two or three ways of getting to a state in the second rank, its representation will have two or three assembly moves (lines) entering it. In this way one proceeds algorithmically to the end state where all liaisons are established. State and assembly-move diagrams are seen as Figures 15.4 and 15.5.

Name the ranks ordinally, zeroth for the unassembled state, first for the prospective first liaisons, and so forth. Note that there are as many ranks as parts. Since $l \geq (n-1)$, a single liaison per assembly move is the rule only for assemblies where $l = (n-1)$. For assemblies where $l > (n-1)$, some assembly moves involve establishing two or more liaisons. One may consider that an assembly move involves placing a part or a subassembly, but the bookkeeping is not by part name but by liaison number. But it is already known that parts count and liaison count can differ by more than one. Another manifestation of the same matter is noted on the liaison diagram where closed figures (triangles, quadrilaterals, pentagons, etc.) may occur with parts at the vertices. If a last part is placed in a set that makes a closed figure, two liaisons (lines) are established. If a part placement closes two figures, three liaisons are established in the assembly move, and so on. The AFI example, Figure 15.2, has a liaison diagram, Figure 15.3, with multiple closed figures.

Even though liaison-sequence generation is algorithmic and can be arranged to be done by a computer, it is useful that a graphical form of a state and assembly-move diagram be available. This diagram may be arranged in (inverted) tree form such that a state is shown as many times as there are paths to reach that state and so that the final assembled state is shown as many times as there are liaison sequences to completion[6]. Alternately, the diagram may be arranged so that no state is shown more than once and some states show a plurality of assembly moves entering or leaving[12]. The latter representation is chosen for convenience. The following example involves 818 liaison-sequences. Their representation, Figure 15.4, would be huge if it were in tree-form rather than the more compact "diamond" form.

### 15.1.3 Development of an example

An example is developed representing an assembly from industry (AFI), final assembly of an automotive automatic transmission. Its geometry is represented by circular symmetry about the axial centerline. Excepting the axial centerline, the centerline segments in Figure 15.2 represent bolts on bolt circles. Since these fasteners are transparent to the liaison sequence process when not represented by nodes, one is obliged to respond to questions Q1 in a way that considers the placement and securing of fasteners as part of the liaison of the parts being joined.

The liaison diagram of AFI, Figure 15.3, follows almost directly from the assembly drawing, Figure 15.2. One feature calls for comment. Part K has axial freedom and may come into contact with part L. No liaison is shown between parts K and L since no subassembly of parts K and L alone is anticipated. The assembly is highly integrated; there is a large number of liaisons for the number of parts. This is manifest on the liaison diagram as a relatively large number of closed figures or network loops. Parts count and liaison count are: $n = 11$, $l = 18$, $l > (n - 1)$.

Each assembly sequence will have some assembly moves with multiple liaison establishment. The liaison-sequence diagram has 11 ranks including fully-disassembled and fully-assembled states.

The liaison diagram established, the next step is to ask and answer a group of questions. Answers to Q1 for some liaisons are shown below:

$i = 1$: Once L1 is done, part B cannot be installed. Part B must be installed prior to L1, either into part C (L6) or into part A. Were part B to be placed in part A there must be something to receive it, a jig (not characterized in the example) or part G (L8). But if part G is in place in part A, then L3 is done. So: (6 or (3 and 8)) $\rightarrow$ 1.

$i = 3$: Nothing need be done prior to doing L3.

$i = 4$: Nothing need be done prior to doing L4 .

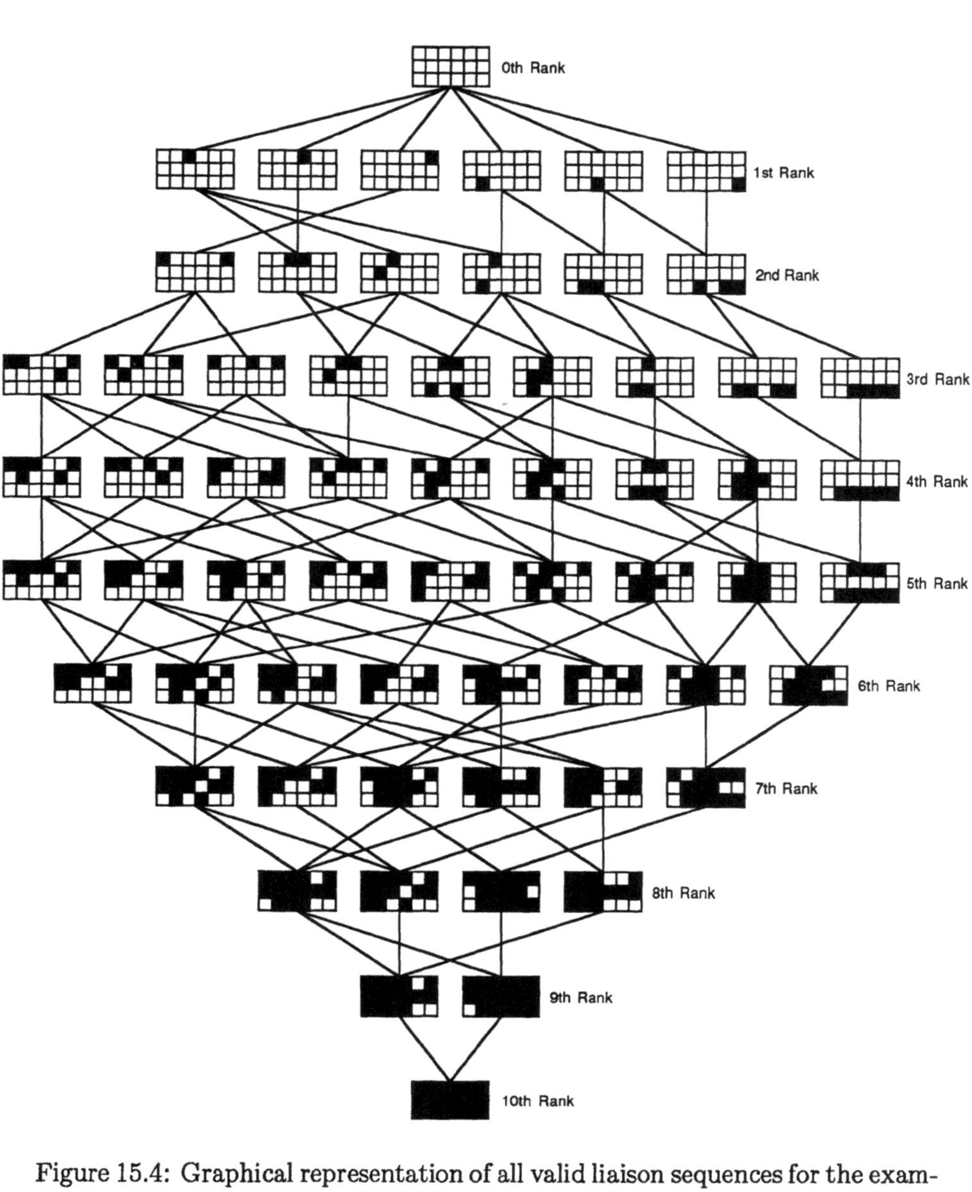


Figure 15.4: Graphical representation of all valid liaison sequences for the example assembly from industry (A.F.I.) under the additional constraint precluding a plurality of unconnected subassemblies in an assembly state

$i = 5$: Once parts A and L are mated (L5), access for the internal parts G, H, J, and K is denied. These internal parts must be placed before the L5 mate. The needs and alternatives are implied by the following: Part G must be in part A or on part H (3 or 14); Part H must be in part G or on part L (14 or 17); Part J must be in part H or in part L (15 or 18); Part K must be in part A or on part H with part H on part L (4 or (16 and 17)) Also (so that parts G and H cannot be assembled together (L14) but externally to parts A and L) Part G must be in part A or part H must be on part L (3 or 17). So: ((3 or 14) and (14 or 17) and (3 or 17) and (15 or 18) and (4 or (16 and 17))) → 5.

$i = 8$: Parts B and G may only be usefully mated while they are inside part A. To insure this, either part, B or G, must initially be placed inside part A. For part B to be secure, part C must be fastened to part A (L1) . For part G to be placed, L3 is completed. So: (1 or 3) → 8

$i = 16$: There is no physical constraint requiring a precedent for liaison 16, but part K alone on part H (L16) is unstable and difficult to support. One may choose to constrain the assembly so that part K is stably supported. Part K is axially free on part H (L16) and needs a backstop before the liaison is stable. Part K can be supported on part L after part H mates to part L (L17) or on part A. For parts K and H to be mated (L16) with part K in part A (L4), part G must be in part A or mated with part H. So: (17 or (3 and 14) or (4 and (3 or 14))) → 16

$i = 18$: Nothing need be done prior to doing L18.

The following is a summary of the precedence constraints that follow from the design, geometry, and dimensions of AFI when all questions are answered:

| | | |
|---|---|---|
| (6 or (3 and 8)) | → | 1 |
| 1 | → | 2 |
| ((3 or 14) and (14 or 17) and (3 or 17) and (15 or 18) and (4 or (16 and 17))) | → | 5 |
| 2 | → | 7 |
| (1 or 3) | → | 8 |
| (8 or ((1 or 3) and 14)) | → | 9 |
| (9 or (8 and 15)) or ((1 or 3) and 14 and 15)) | → | 10 |
| 1 | → | 11 |
| 2 | → | 12 |
| 7 | → | 13 |
| (17 or (3 and 14) or (4 and (3 or 14))) | → | 16 |
| (15 or 18) | → | 17 |

It remains to generate the valid liaison sequences. First an additional constraint is imposed to reduce the count of liaison sequences. The additional constraint is a preclusion of a plurality of unconnected subassemblies. This is done for convenience here, though it can be done by design. The constraint is equivalent to imposing a sequential assembly line, without branches. All liaison sequences for AFI, subject to the precedence constraints based on design, dimensions, and geometry, and to the avoidance of any plurality of subassemblies, are represented in Figure 15.4. There are 818 sequences. Assembly states are represented by boxes. In this case each box has 18 cells in a three row by six column array corresponding to 18 liaisons: liaisons one through six, left to right across the top row; seven through 12 left to right across the middle row; and 13 through 18, left-to-right across the bottom row. A blank cell denotes a liaison not established and a marked cell denotes an established liaison.

For the technique to be industrially useful, an engineer must be able to encompass output as that represented in Figure 15.4 and reduce it to a relatively small number of choices by deleting the awkward and retaining the favorable. Such sweeping reductions are possible in this case.

Consider this characterization of assembly of AFI: Part A is "filled" with parts from two ends and the "fill" at each end is independently secured. Parts B,C,D,E, and F fill the front, and parts G,H,J,K, and L fill the back. Consider that during assembly the axis will be vertically oriented using gravity to keep parts in place until they can be secured. If orientation is front-up, parts B,C,D,E, and F can be placed but G,H,J, and K would fall out before L is secured; a similar but opposite situation obtains for rear-up orientation. This suggests constraining assembly further, so that if a front fill is begun, nothing is put in the back until the front fill is finished, and so that if a rear-fill is begun, nothing is put in the front until the rear fill is finished. This additional constraint is easily expressed by writing another menu classifying liaisons as front-fill liaisons, rear-fill liaisons, and front-to-rear association liaisons and; once any single liaison is completed from either the front-fill or the rear-fill liaison category, that no liaison be completed from the opposite category until all liaisons from the first-used category are completed. The menu is presented in Table 15.1.

Table 15.1: Menu for an Additional Constraint Precluding Simultaneous or Mixed (from front and rear) Filling of Part A for the Assembly from Industry

| | |
|---|---|
| Front-fill Liaisons | 1, 2, 6, 7, 11, 12, 13 |
| Front-to-Rear Association Liaisons | 8, 9, 10 |
| Rear-fill Liaisons | 3, 4, 5, 14, 15, 16, 17, 18 |

All liaison sequences for AFI, subject to precedence constraints based on design, dimensions, and geometry, and the additional constraints, are displayed in Figure 15.5. There are now but 18 assembly path choices, down from 818.

The new constraint eliminates states that include incomplete sets of liaisons from both front-fill and rear-fill categories. In Figure 15.5, note a channel, uncrossed by any state transition, internal to the state and state-transition diagram, that persists from the 1st rank through the 9th rank inclusively. To the left of the channel lie all the liaison sequences involving first the front-fill liaisons complete, followed then and only then by the rear-fill liaisons. To the right of the channel lie all the liaison sequences involving first the complete set of rear-fill liaisons, followed then and only then by the front-fill liaisons. That this channel may exist, uncrossed by state transitions, is a predictable manifestation of the last constraint.

Further reductions in liaison-sequence path count remain. They result from recognition and removal of particularly awkward assembly moves, which occur in some but not all liaison sequences. Two such assembly moves have been recognized and are marked on Figure 15.5. They are, first, the installation of part K between parts A and H simultaneously, characterized as liaisons 4 and 16 occurring simultaneously, marked in four places by circles; and, second, mating part B simultaneously with parts G, H, and J, characterized as the simultaneous establishment of liaisons 8, 9, and 10, marked by a triangle. The former involves three simultaneous splined-shaft to splined-bore or toothed-bore mates and some journal to bore mates; the latter represents feeding a loose stack of clutch discs between the splined female cage and the splined male part.

If the decision is now made to avoid all these (five) awkward assembly moves, there remain but two liaison-sequence paths. In particular, avoiding simultaneous establishment of liaisons 8, 9, and 10 disqualified all of Figure 15.5 which lay to the right of the previously identified channel. Following the injunctive decision and the earlier constraints implies that assembly of AFI properly begins by filling the front and securing the last front element, followed by reorienting the assembly and filling parts into the rear.

The preceding material explains basic technique of simple liaison sequence analysis by description and example. The topology, geometry, and dimensions of an assembly determine the necessary and inviolable constraints on assembly sequence which are expressed as precedence relations. Subsequently the user considers other constraints, not dictated by geometry and dimensions but rather optionally imposed, to simplify the field of choice of assembly sequences. The question of optional constraints is a delicate one. On one hand, each optional constraint represents a potentially significant reduction in the complexity of the liaison-sequence diagram; on the other hand, carelessness in applying constraints may preclude important assembly-sequence options. The editing effort is explained in detail in the last section of this chapter.

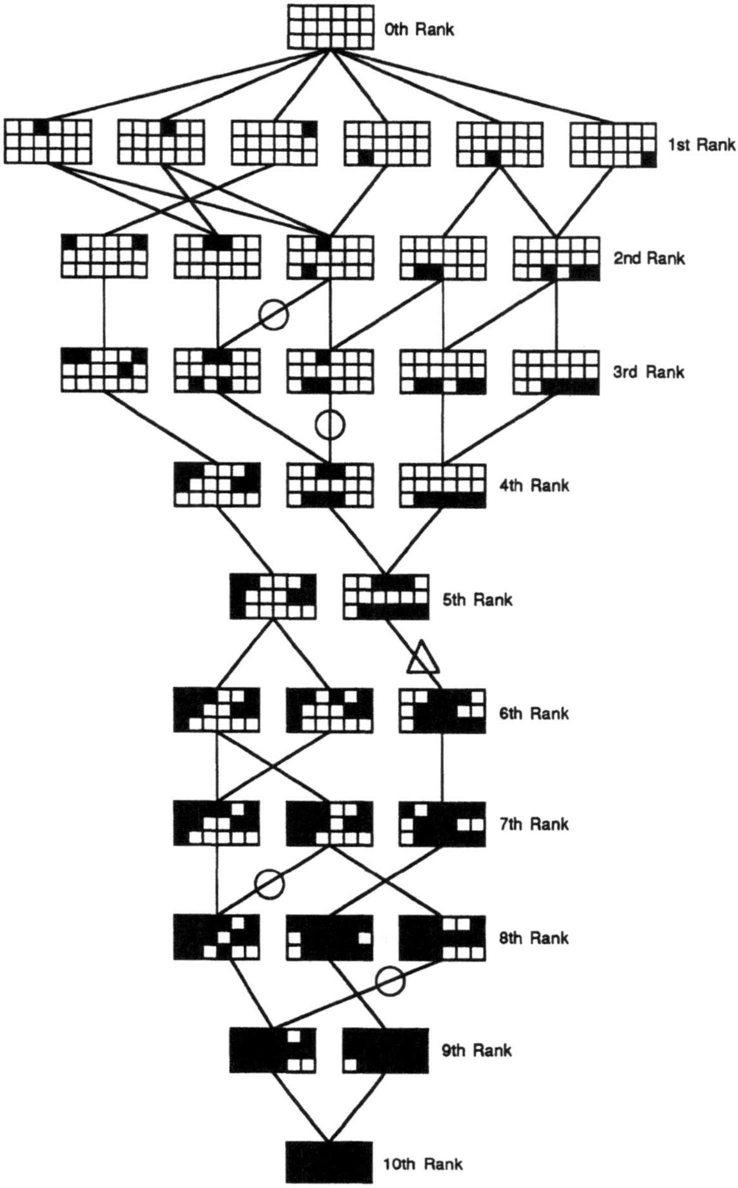


Figure 15.5: Graphical representation of all valid liaison sequences for the example assembly from industry (A.F.I.) under two additional constraints precluding a plurality of unconnected subassemblies and precluding the mixing of front-fill and rear-fill liaisons. Moves marked with a circle involve the difficult simultaneous establishment of liaisons 4 and 16; that marked with a triangle involves the difficult simultaneous establishment of liaisons 8, 9, and 10. Two liaison sequence paths remain if both of these assembly paradigms are avoided.

## 15.2 Interactive program for finding and editing assembly sequences

There are disadvantages associated with the simplified assembly sequence analysis: the technique requires mental analysis of assembly alternatives; in complicated assemblies these alternatives may be occasionally many and difficult to consider; and it is not algorithmic. Our user-interactive assembly-sequence analysis and editing programs avoid such difficulties by combining the disassembly analysis based on assembly cut-sets pioneered by Homem de Mello and Sanderson[21], with simplifications based on liaison diagram loop closure recognized by De Fazio and Whitney[12], and those based on assembly of subassembly subsets and supersets recognized by Bourjault et al.[8]. Using disassembly analysis precludes trial generation of any "dead-end" assembly states, but requires answering a disassembly question for each assembly cut-set. (A "dead-end" assembly state is one from which full assembly cannot be reached). Using loop-closure, subassembly subset, and subassembly superset considerations often permits many answers to cut-set disassembly questions to be inferred from the answer to one question.

The method assumes that the parts to be assembled and their mates are rigid. Many commercially significant products fit these assumptions. Each liaison must be accomplished once and once only. The assembly sequence generating algorithm includes the following "rules:"

**Loop-Closure Rule:** A cycle in a liaison diagram implies a need to simultaneously complete two of the liaisons in the cycle. (This rule follows from assuming rigid parts and liaisons.)

**Superset Rule:** If two parts or subassemblies cannot mate due to interference in the approach path, then adding a part to either set will not change this situation.

**Subset Rule:** If two parts or subassemblies can mate, then removing any part, itself not associated with the mating liaison(s), from either subassembly, will not change this situation.

These rules let the algorithm make heavy use of previously-answered questions to deduce answers to others. In typical "hard" problems we have addressed, the algorithm uses the rules to infer answers to over 95%[4] of the questions. The interactive aids described below make answering the rest so easy that we have postponed attempts to use solid modeling to automatically answer approach-path interference questions. Nonetheless we have created a fast, easily-used tool that addresses usefully complicated assemblies.

Answering the questions yields a set of precedence relations connecting liaisons with ordering operators. The Bourjault and cut-set methods both create precedence relations of the forms (three examples):

$$1 \;>=\; 3$$
$$1 \;>=\; (3\&4)$$
$$(1\&2) \;>=\; 3.$$

The operator ">=" means "must precede or concur with." The simplified[12] method generates statements containing "&," "or," or "not," and the operator "$\rightarrow$," meaning "must precede." The numbers represent liaisons. For example, the statement:

$$(1\&2) \;>=\; (3\&4)$$

reads that both liaisons 1 and 2 must be completed, before or concurrently with, completion of (both) liaisons 3 and 4, but not necessarily before or concurrently with either liaison 3 or 4.

Precedence relations of any of the above forms enter a program written by Lui[22] to generate an assembly sequence diagram[12] representing valid assembly sequences. The assembly sequence diagram is generated by forward-chaining through the relations, starting with liaisons having no precedents. Lui's program lets the user edit resulting sequences by removing individual states and moves from a displayed representation of sequences, using a graphic and mouse-menu interface.

Henrioud and Bourjault, Homem de Mello and Sanderson, and Bourjault have shown that all possible assembly sequences can be generated without explicit precedence relations. However, we have found it useful to be able to add other constraints by expressing them as logical relations. Examples include: addressing cases that violate rigidity assumptions, as with liquid or compliant parts in an assembly; and enforcing early assembly of a particular subassembly to support an in-process test. For this reason we have chosen to generate explicit precedence relations.

Lui's program has been extended to the noted user capabilities: to choose desirable assembly states, moves, or partial sequences; avoid awkward assembly states or moves; represent part or subassembly fixturing and orientation opportunities; consider fixture-change and re-orientation counts; or choose good sequences based on desirable fixturing or orientation sequences. Editing capabilities include: edit all redundant full or partial assembly sequences associated with separate subassemblies made at parallel sites[3]; impose constraints such as: "avoid simultaneous completion of a set of designated liaisons;" or "liaison $k$ must immediately follow liaison $j$;" or eliminate of states that involve a plurality of subassemblies, or that involve a branched assembly line or parallel assembly operations; or eliminate, or incorporate, a particular state or move.

The implementation on SUN 3/60 workstations integrates the processes of answering questions and editing and evaluating sequences into one seamless activity in which the user is aided by graphic representations of parts, subassemblies, and assembly states, moves, and sequences.

### 15.2.1 Assembly sequence generation

A cut-set method[4] partly based on the work of Homem de Mello and Sanderson[20] is used to find and represent all geometric or mechanical assembly constraints as precedence relations. The method is similar in its use of graph cut-sets to analyze the liaison diagram. Our method also uses the superset and subset rules developed by Bourjault. The Lui[22] program uses the precedence relations and the liaison diagram to generate the assembly sequence diagram.

The cut-set method includes three elements to increase efficiency or utility not fully utilized by either Bourjault and Henrioud or Homem de Mello and Sanderson. They are incorporation of the subset and superset rules; implementation of an efficient precedence relation search algorithm; and generation of explicit precedence relations.

#### Overview of the cut-set method

A flow chart for the cut-set method implementation is shown as Figure 15.6. First, part assembly topology is entered as a liaison diagram. Next, all subassemblies are found by generating all possible part combinations, and testing connectivity of the subgraph formed by the combination of parts or nodes in the liaison diagram. Connected subgraphs are subassemblies. Assembly cut-sets, generated next, are defined by two part sets (i.e. subassemblies) $N_i$ and $N_j$ and the connecting liaison set S. The assembly cut-sets are used to generate all the questions needed to determine the precedence relations for an assembly. The questions, represented by "$R(N_i; N_j)$?," take the form: "Can the subassembly of parts of $N_i$ be disassembled from the subassembly of parts $N_j$?"

Next, the questions are answered by logical inference if possible or by the user if not. A check for invalid assembly states, the superset rule, the subset rule, a precedence relation search algorithm, and user input are invoked to do so. Starting from the largest assembly cut-sets formed by the full liaison diagram, the question $R(N_i; N_j)$? is checked against all previously-obtained precedence relations to insure that the assembly state $N_i \cup N_j$ is not invalid. If it is not invalid, then $R(N_i; N_j)$? is checked against all previous question answers using the subset rule. If $R(N_i; N_j)$? passes the subset rule in that $N_i$ and $N_j$ are subsets of $N_i''$ and $N_j''$ for which $R(N_i''; N_j'')$? has already been answered YES, then $R(N_i; N_j)$? is YES. If $R(N_i; N_j)$? fails the subset rule, it is checked against all previous question answers using the superset rule. If $R(N_i; N_j)$? passes the superset rule in that $N_i$ and $N_j$ are supersets of $N_i''$ and $N_j''$ for which $R(N_i''; N_j'')$? has already been answered NO, then $R(N_i; N_j)$? is NO. If $R(N_i; N_j)$? fails both the subset and superset rule checks, the user is asked the question, and the answer is stored.

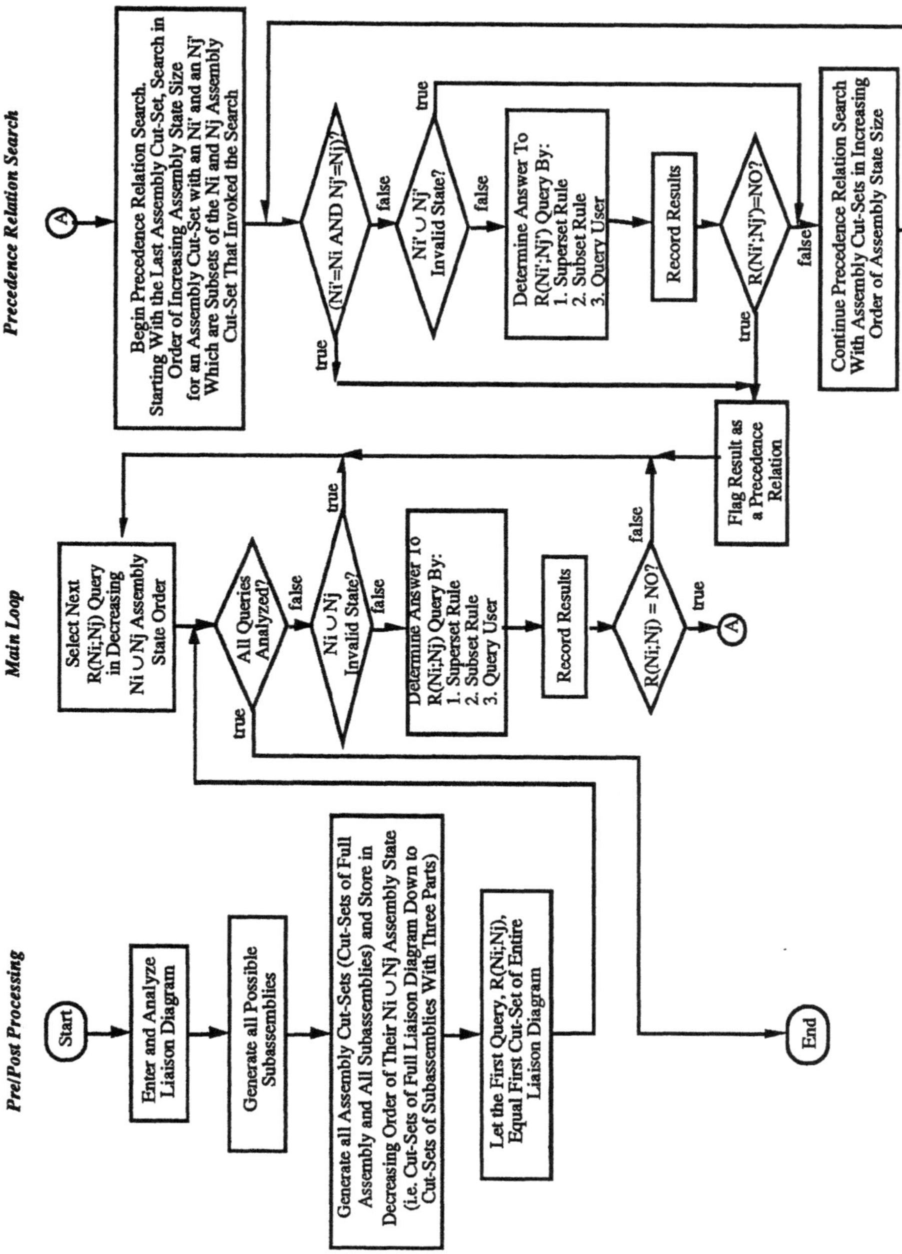


Figure 15.6: Flow chart of the cut-set method software implementation

If $R(N_i; N_j)$? is NO, then the program seeks to write a precedence relation. The search algorithm looks for the smallest $N_i'$, $N_j'$ assembly cut-set (where $N_i'$ and $N_j'$ are subsets of $N_i$ and $N_j$ respectively) for which $R(N_i'; N_j')$ is NO. Answers for the $R(N_i'; N_j')$? questions are determined in the same manner as above. The resulting precedence relation takes the form:

Liaisons in $S$ >= (Liaisons in $N_i'$ and Liaisons in $N_j'$).

A more detailed description of the cut-set method and an example follow.

### Assembly cut-sets

A cut-set of a connected graph is typically defined as a minimum set of edges in the graph which leave two disjoint connected subgraphs if removed[4]. That set of edges is called the edge-set S, and the two connected subgraphs are called node-sets $N_i$ and $N_j$. If any edge in S remains, the graph remains connected. An alternate definition of a cut-set is used here: cut-set is defined by the node-sets $N_i$ and $N_j$ such that each node-set forms a connected subgraph, $N_i \cup N_j$ is the node-set of the full graph, and $N_i \cap N_j$ is the empty set $\emptyset$. Here a cut-set is referenced by any two node-sets (subassemblies) that meet the above definition; and the corresponding cut liaisons joining $N_i$ to $N_j$ is referenced by the edge set (or liaisons) $S$. A single node is a connected subgraph.

Assembly cut-sets are cut-sets of the full assembly and all subassemblies. Equivalently they are the cut-sets of the liaison diagram and of all connected subgraphs of the liaison diagram. Assembly cut-sets omit cut-sets formed by single part pairs since any pair of rigid mating parts can be assembled or disassembled. For example, the assembly cut-sets of the liaison diagram of Figure 15.7 are given in Table 15.2.

### Cut-sets as questions

Assembly cut-sets are posed as questions to determine the precedence relations. The question can be presented as either an assembly or a disassembly operation since these operations are assumed equivalent and reversible. We chose to use the disassembly form as follows:

$R(N_i; N_j)$?: If assembled together, can subassembly $N_i$ be separated from subassembly $N_j$?

$N_i$ and $N_j$ are the subassemblies or subgraph node-sets created by breaking the liaisons in the cut-set $S$. The above question asks whether the cut liaisons $S$ can be established if the liaisons of the connected subgraphs $N_i$ and $N_j$ have been previously established. Answers are based solely on the assembly cut-set information given, and not on whether a final completely assembled or disassembled state can be reached. Equivalently, determination of a cut-set's feasibility is based solely on geometric and mechanical aspects of the subassemblies given in the question.

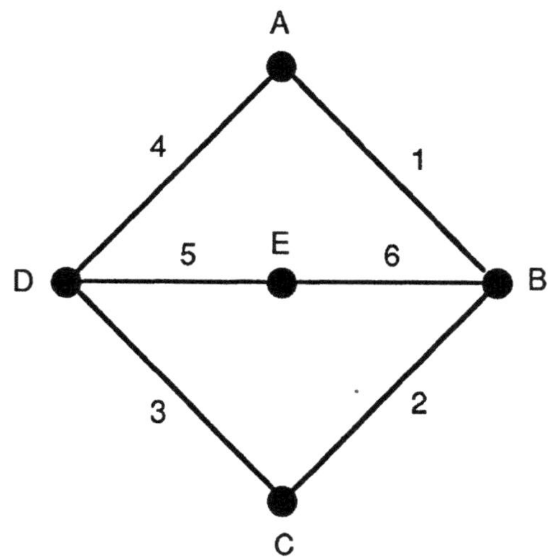


Figure 15.7: Example liaison diagram

### Assembly cut-set generation

The algorithm to generate assembly cut-sets takes as input the liaison diagram, and all possible subassembly subgraphs arranged in order of decreasing parts-count in assembly state $N_i \cup N_j$, with the smallest subgraphs having three nodes. Assembly cut-sets are generated by looping through each subgraph starting with the full liaison diagram and ending with the last subgraph with three nodes. The cut-sets of each input graph are generated by determining all possible subassemblies (node sets of connected subgraphs) $N_i$ and $N_j$ such that $N_i \cup N_j$ is the node-set of the input graph and $N_i \cap N_j$ is the empty set.

### Simplification rules

The assembly cut-sets represent a sufficient question set to determine precedence relations for an assembly. If knowledge from earlier answers is not exploited, the interference question count needed to determine the precedence relations equals the assembly cut-set count, a quite large number for even moderately-sized assemblies. As questions are formed, several techniques are used to answer most of them by inference from previous answers. First, invalid $N_i \cup N_j$ assembly states are not considered. Suppose we are faced with $R(N_i'; N_j')$? and have already determined the precedence relation "Liaisons in $S >=$ (Liaisons in $N_i$ and Liaisons in $N_j$)." If the liaisons in $N_i' \cup N_j'$ comprise a superset of the liaisons in $N_i$ and $N_j$ and do not comprise a superset of $S$, then $N_i' \cup N_j'$ is an invalid state. Cut-sets of invalid assembly states need not be posed as questions.

Other uses of prior information are simple extensions of the superset and subset

Table 15.2: Assembly cut-sets for the example in figure 15.7

| Graph or Subgraph | Cut-Sets $N_i$ | Cut-Sets $N_j$ |
|---|---|---|
| A,B,C,D,E | A | B,C,D,E |
| | B | A,C,D,E |
| | C | A,B,D,E |
| | D | A,B,C,E |
| | E | A,B,C,D |
| | A,B | C,D,E |
| | A,D | B,C,E |
| | B,C | A,D,E |
| | B,E | A,C,D |
| | C,D | A,B,E |
| | D,E | A,B,C |
| B,C,D,E | B | C,D,E |
| | C | B,D,E |
| | D | B,C,E |
| | E | B,C,D |
| | B,C | D,E |
| | B,E | C,D |
| A,C,D,E | A | C,D,E |
| | C | A,D,E |
| | E | A,C,D |
| A,B,D,E | A | B,D,E |
| | B | A,D,E |
| | D | A,B,E |
| | E | A,B,D |
| | A,D | B,E |
| | A,B | D,E |

| Graph or Subgraph | Cut-Sets $N_i$ | Cut-Sets $N_j$ |
|---|---|---|
| A,B,C,E | A | B,C,E |
| | C | A,B,E |
| | E | A,B,C |
| A,B,C,D | A | B,C,D |
| | B | A,C,D |
| | C | A,B,D |
| | D | A,B,C |
| | A,B | C,D |
| | A,D | B,C |
| C,D,E | C | D,E |
| | E | C,D |
| B,C,E | C | B,E |
| | E | C,B |
| A,D,E | A | D,E |
| | E | A,D |
| A,C,D | A | C,D |
| | C | A,D |
| A,B,E | A | B,E |
| | E | A,B |
| A,B,C | A | B,C |
| | C | A,B |
| A,B,D | B | A,D |
| | D | A,B |
| B,D,E | B | D,E |
| | D | B,E |
| B,C,D | B | C,D |
| | D | B,C |

rules of Bourjault et al.[8] and are stated in terms of parts, but they also have a dual form for liaisons:

If $R(N_i; N_j)? = \text{YES}$, then $R(N_i'; N_j')?$ is also YES, where $N_i'$ and $N_j'$ are subsets of the parts in $N_i$ and $N_j$ respectively (Subset Rule).

If $R(N_i; N_j)? = \text{NO}$, then $R(N_i'; N_j')?$ is also NO, where $N_i'$ and $N_j'$ are supersets of the parts in $N_i$ and $N_j$ respectively (Superset Rule).

The last two rules indicate that the most information can be gained from YES answers to $R(N_i; N_j)?$ with large $N_i$ and $N_j$ (subassemblies with many parts) and from NO answers to $R(N_i; N_j)?$ with small $N_i$ and $N_j$ (subassemblies with few parts). This implies advantages for a logical ordering of cut-set questions. Using the heuristic that, for most industrial assemblies, precedence relations spring from a small subset of parts which cause geometric or mechanical interferences, the initial questions are posed using cut-sets in decreasing order of the number of parts in assembly state $N_i \cup N_j$ because the probability of quickly obtaining YES answers is high. As soon as a NO is obtained, the precedence relation search algorithm is entered and is pursued via questions that use cut-sets in increasing order of the number of parts in assembly state $N_i \cup N_j$ because the probability of quickly obtaining NO answers is high.

### Precedence relation search algorithm

The precedence relation search algorithm, shown on the right side of Figure 15.6, is invoked when $R(N_i; N_j)? = \text{NO}$. Search starts with assembly cut-sets of subassemblies with three parts and continues in order of increasing parts count in the assembly state $N_i \cup N_j$. Nested within the precedence relation search is a series of question searches. A question search also starts with the assembly cut-sets of subassemblies with three parts and continues in order of increasing number of parts in the assembly state $N_i \cup N_j$. The question search stops once an assembly cut-set $N_i'$, $N_j'$ is found where $N_i'$ and $N_j'$ are subsets of, or ultimately equal, the $N_i$, $N_j$ assembly cut-set that invoked the precedence relation search. If $N_i'$ equals $N_i$ and $N_j'$ equals $N_j$, then the precedence relation search ends, and the original $R(N_i; N_j) = \text{NO}$ forms a precedence relation. An answer for the new question $R(N_i'; N_j')?$ is found by using the subset rule, the superset rule, or asking the user. If $R(N_i'; N_j')?$ is YES, the precedence relation search continues with another question search. If $R(N_i'; N_j')?$ is NO, the search ends and $R(N_i'; N_j')$=NO forms a precedence relation.

### Formation of explicit precedence relations

The precedence relations used to generate assembly sequences follow from $R(N_i; N_j) = \text{NO}$ results found during a precedence relation search. The general form is:

(Set of Cut Liaisons $S$) >= (Liaisons forming Subassemblies $N_i$ and $N_j$).

$R(N_i; N_j)?$ asks whether all liaisons of cut-set $S$ can be established once the liaisons forming $N_i$ and $N_j$ are established. The precedence relation is the logical conclusion from $R(N_i; N_j)? = \text{NO}$. For example, suppose the assembly cut-set $N_i = \text{A,B}$; $N_j = \text{C,D}$ of the liaison diagram of Figure 15.7 forms a precedence relation. The precedence relation's explicit form is:

(2&4) >= (1&3)

or that liaisons 2 and 4 must be established prior to or concurrently with the liaisons 1 and 3.

### An example illustrating the use of the program

The algorithms, programmed in "C," run interactively on a SUN 3/60 workstation. The software is shown in use on an example product, shown with its liaison diagram in figures 15.2 and 15.3, AFI, an automatic transmission[12]. AFI has 8221 assembly cut-sets, found in about 40 seconds on a SUN 3/60 work station.

The user can display the parts in a window. Part drawings are made using SUN's drawing software, SUNDRAW, and are stored and displayed as raster images. The subassemblies $N_i$ and $N_j$ are shaded white and grey respectively in a parts display window. The user answers questions posed based on geometric or mechanical constraints. Figure 15.8 shows the first question. Geometric constraints preclude removal of part A from the entire assembly. The NO answer starts a precedence relation search ending with the question "R(C,D; A)?", Figure 15.9. Removal of part A from parts C and D is impossible as part D blocks access to bolts connecting parts C and A. The user continues until all necessary questions are answered, responding to most questions quickly and easily simply by looking at the displayed subassemblies. The precedence relations resulting are printed to the screen, Figure 15.10, and to a file for use by Lui's LSG program. Of the potential 8221 questions in this case, 111 are asked of the user. An engineer familiar with the design answered them in about 14 minutes.

The algorithm has been extended to request local-mating-condition information about liaisons in the question-answering process. The user describes the local separation direction of each liaison, represented as a Cartesian coordinate vector. A computer routine inspects separation directions of all liaisons connecting subassembly $N_i$ to subassembly $N_j$ to determine if a common direction exists. If not, the answer to $R(N_i; N_j)$ is NO. If so, the user must be questioned to determine if any global interferences exist. Applying this feature to the AFI transmission example reduces the number of interference questions a user must answer to 55. Adding local separation information has greatest potential for reducing user effort for more complex assemblies, since separation information enters in proportion to liaison-count, not cut-set count.

In comparison, with neither local direction information nor a check for validity of states, the transmission example asks the user 142 questions; the Bourjault

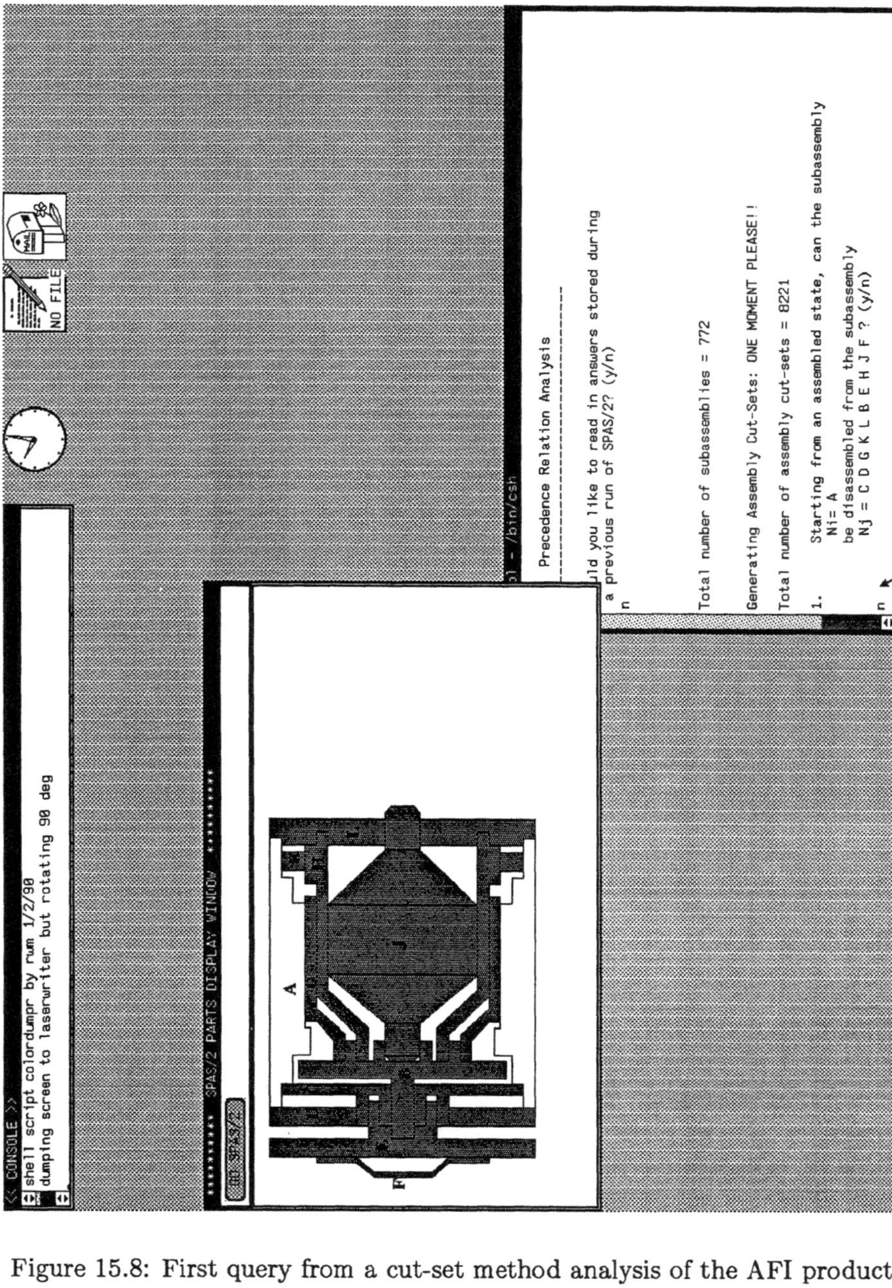


Figure 15.8: First query from a cut-set method analysis of the AFI product

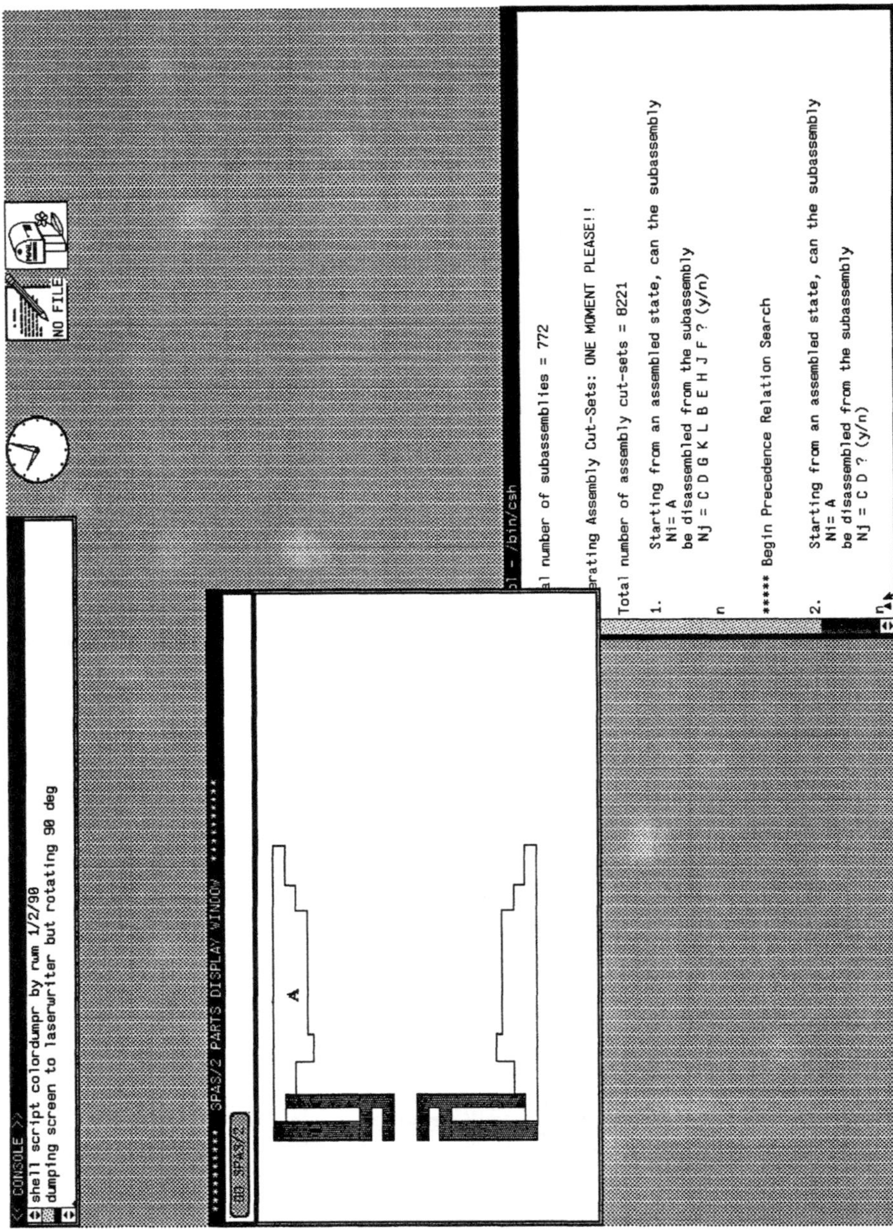


Figure 15.9: Second query from a cut-set method analysis of the AFI product

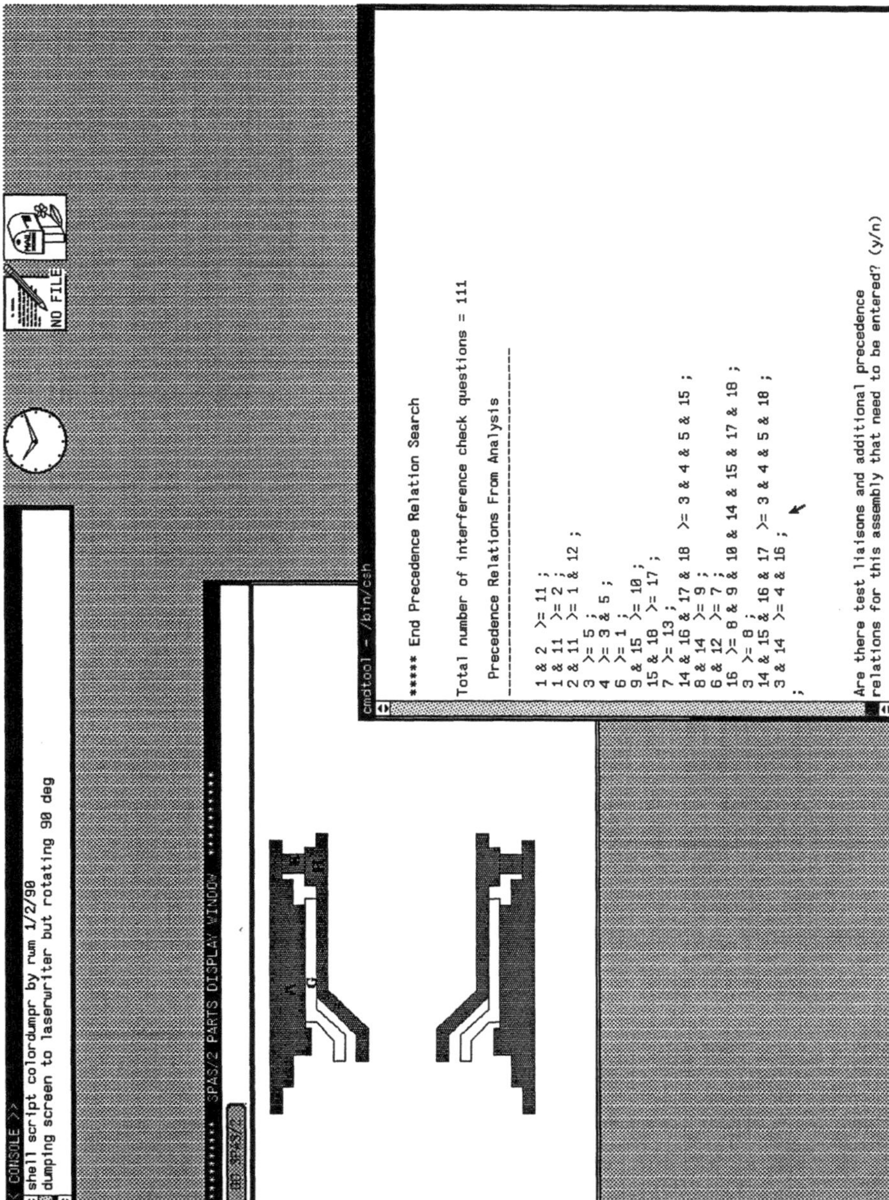


Figure 15.10: Precedence relations from a cut-set method analysis of the AFI product

method applied to the same example, as implemented in De Fazio et al.[13], asks the user 482 questions. Table 15.3 presents statistics for four assemblies.

Table 15.3: Question-Count statistics for the cut-set method for four example assemblies. Total question count equals the assembly cut-set count as earlier defined. Geometric interference count is the number of questions requiring a geometric interference check more extensive than a check of local separation constraints. The major factors in reducing the user-referred question count are the superset rule, the subset rule, and the invalid state check.

| ASSEMBLY | Part Count | Liaison Count | User-Referred Question Count | Total Question Count (Cut-Set Count) |
|---|---|---|---|---|
| Gimbal | 11 | 18 | 32 | 6,693 |
| AFI | 11 | 18 | 55 | 8,221 |
| Viewfinder | 16 | 21 | 403 | 313,530 |
| Seeker Head | 17 | 26 | 58 | 203,754 |

#### Generating the assembly sequences

Precedence relations are now passed to Lui's program "LSG" [22] which converts them into an assembly sequence diagram, as shown in Figure 15.4. Each box in the diagram is an assembly state; each line is an assembly move. Each path from top to bottom is an assembly sequence that builds the entire product. Editing and evaluation can now begin by calling the editing program.

## 15.3 Editing Means to Select Favorable Sequences

Once a product's possible assembly sequences are found, the user must narrow the choice to a few "good" sequences. These few sequences can then be evaluated more rigorously to find any best sequence. Typical products have thousands of possible assembly sequences, and many factors influence the quality of assembly sequences; so the designer needs a tool that represents assembly sequences and allows a variety of editing options.

Physical or judgmental characterizations of the states (boxes) and assembly moves (lines) of the liaison-sequence diagram provide rational bases for the choice of the "best" assembly sequence. Characterizations may be either qualitative or quantitative, and will generally suggest to the user the appropriate evaluation and choice technique.

An important qualitative state characterization is stability of the assembly state. Briefly, stable subassemblies are welcome states and unstable assembly states are best avoided if possible since they call for the complication and expense of stabilizing jigging. Conditionally stable assembly states are welcome only to the extent that the assembly sequence does not call for any move that violates the condition of stability. For example, if a state is stable front-up and unstable rear-up, an assembly move calling for a front-up to rear-up flip, prior to realization of an unconditionally stable state, should be avoided if possible.

Assembly moves can be qualitatively characterized in terms of the ease or difficulty of the parts mate implied, or in terms of the skill-level required for completing the assembly move or mate, or in terms of whether the assembly move threatens part damage.

Qualitative characterizations of states and assembly moves were quite sufficient to reduce 818 liaison-sequence choices to two in the example, noting that an applied optional constraint was invoked to preclude unstable states and undue flips for access. The nature of a network such as the liaison-sequence diagram is that a few forbidden states and moves may force the abandonment of other preceding or subsequent states and assembly moves, very effectively pruning the diagram. Reduction of this sort can be done very quickly on a graphical representation of the liaison-sequence diagram.

Quantitative characterizations may include times for various technologies to accomplish assembly moves, costs of the hardware associated with the technologies appropriate to assembly moves, probabilities of failure for particular technologies addressing particular assembly moves, costs of fixturing or tooling needed to secure unstable states, and so forth. The user can apply a simple criterion to choice of assembly sequence, such as a shortest time path through the liaison-sequence diagram. Alternately, subsequent to extensive quantitative economic and durational characterization of the states and assembly moves of the liaison-sequence diagram, the user can extend a synthesis routine[16] to simultaneously suggest both assembly sequence and assembly technology.

Editing assembly sequences to find a preferred sequence is done on four bases: editing representationally redundant sequences; pruning away any difficult, awkward, or unwanted elements while maintaining any preferred elements in the network of assembly possibilities; minimizing non-productive assembly tasks such as reorientation of a subassembly; and choosing among candidate assembly sequences on a calculated economic basis. The middle bases are the design or assembly engineer's provinces, thus seem the richest and most interesting. Editing bases, moves, and paradigms include the following.

### Purge of representationally-redundant partially-parallel sequences

Editing redundant sequences applies only to sequences with parallel assembly operations, branched work flow, and one or more states with two or more subassemblies. The redundancy is implicit in the representation of assembly. Only one assembly operation may be represented at a time on an assembly sequence diagram. If two operations, A and B, are to be done in parallel and nominally simultaneously, they are represented twice and in both possible sequences: as (A, then B); and as (B, then A). The assembly operations A and B may be done in parallel and the order of completion is unimportant (in lieu of any applicable constraints) though each order is representated in the analysis. Three operations done in parallel are represented six times, N are represented N! times. All but one of parallel-operation sequences is eliminated by this purge, described in detail in Amblard[3] and in Abell[2]. Elimination of redundant representations is done as a default option in our sequence editing software. This purge may be undone or redone at will with single keystrokes. Assembly graphs that admit to many parallel operations are massively pruned by this purge, as the simple combinatoric examples suggest. An AND/OR graph representation lacks the exhaustive list of parallel-operation assembly sequences explicit in a liaison sequence diagram before removal by a purge of redundant sequences. After purge of a liaison sequence diagram, each parallel-operation line layout is represented by an arbitrary single sequence, rather than the implicit choice of the AND/OR graph.

### Choice of Assembly-Line Topology

In assembly line design, an engineer often has reason or need to impose an assembly line topology. Final assembly lines are often sequential, withou branches. Branched assembly lines may be difficult to implement or to suppl with parts, and worker access may be awkward. Supply lines may have t cross branches of a branched assembly line, while a sequential line may be fe from one side and manned on the other. Branched assembly lines are suited t some products and sequences: where several stable subassemblies are created or where sequential line-balance is difficult, and where line-supply and acces are easily accomodated, for example.

A branched assembly line with one or more parallel operations implies an as sembly state with two or more unconnected subassemblies. Imposing a sequen tial assembly line is done by purging all assembly states containing a pluralit of unconnected subassemblies, and is a two-keystroke operation if using our editing software.

An engineer wanting a branched assembly line must identify the state or states that represent the subassemblies that are to be processed on parallel lines; then impose that each assembly sequence pass through one of the identified states.

### Avoid an assembly state

Avoiding or excising a particular assembly state is one of a few primitive editing actions. Reasons to invoke include to avoid subassembly instability or fixturing difficulty. State avoidance typically is a least powerful editing move (erases fewest sequences) at mid-assembly; and most powerful (erases most sequences) when assembly is just beginning or just ending. Avoiding a state is done by using a mouse to denote, and highlight, the state on the assembly sequence diagram, and calling "delete" from a command menu.

### Avoiding an assembly paradigm

Every connection between parts must be made at some time between start and finish of assembly, none is avoided. If connection count exceeds parts-count less one, then one or more assembly steps include making a multiple connection. A "difficult assembly move" that may be avoided excludes single parts-pair connections since each is unavoidable; and includes only multiple simultaneous connections among parts, some combinations of which may be avoided.

In any complicated assembly a particular difficult assembly move consisting of the simultaneous establishment of two or more particular connections may occur in several locations on an assembly sequence diagram, either alone or as subset of a more complicated set of connections. Such a move is called an assembly paradigm. Engineers responsible for product design and assembly can identify difficult assembly paradigms and can easily characterize them in terms of their simultaneous connections. With the editing software, any such paradigm is then excluded with a few keystrokes. This is typically a moderately strong editing action.

### A particular state is to occur

It is occasionally important that one or more particular assembly states occur during assembly; reasons may include a need to do a production test, take a measurement, or to take advantage of particular stable subassembly for a needed refixturing or reorientation move. Assuring the occurrence of a state is a primitive move and the complement of avoiding a state. It is a powerful editing action, especially so midway in the assembly sequence. That this is so is attested to by the fact that making subassembly decisions, deciding that certain subassemblies will be built as part of the assembly sequence, is generally the first basis of any unaided choice of assembly sequence. Passing through a state is assured without additional constraint by deleting all other states in the same rank of the assembly sequence diagram.

### A particular subassembly is to occur

Any particular subassembly will occur only once in an assembly sequence diagram that is limited to sequential assembly lines, but may occur in two or more states of an assembly sequence diagram that includes branched assembly lines as well. The reasons for invocation match those of assuring occurrence of a particular state. The complementary editing action, that a particular unconnected subassembly not occur, is used to avoid subassemblies that are hard to fixture except as part of a greater subassembly.

### A particular partial assembly sequence is to occur

Assuring occurrence of a particular partial assembly sequence is a useful and powerful editing action. This move is logically represented by one or a chain of statements of the form "task B is to immediately follow task A" and is invoked with a few keystrokes. Reasons include to immediately secure an unstable subassembly following a latest assembly move.

### Conditional precedences amongst states

Conditional assembly, deferring completion of a particular subassembly until a result, possibly measured, of another, possibly subset, subassembly is known, is occasionally needed, and is not evoked by the means of finding possible sequences. Such conditional subassembly pairs must be recognized during editing sequences. Such constraints are usually easily stated logically and can be quite powerful editing actions. In branched assembly line cases, if a conditional precedence constraint is needed, it must be applied prior to purge of representationally redundant sequences.

### Path editing based on assembly state fixturing and orientation hypotheses

Refixturing and reorienting subassemblies are subassembly operations that involve no part mating and that add no value to the product. Minimizing the count of such moves across an assembly sequence is a goal. Fixturing and orientation options are state-related data that can be, but are not yet represented on an assembly sequence diagram. The data must be synthesised and provided, say by the user. Many earlier described editing actions consider individual assembly states or assembly moves, or partial assembly sequences. Minimizing subassembly operations that add no product value requires a view of entire assembly sequences.

The editing program leads a user state-by-state through supplying the needed information. The user is asked to hypothesize fixturing opportunities for subassemblies of each assembly state, and characterize each opportunity in terms of parts and surfaces of parts supported by fixture, and corresponding stable orientations. When the characterizations are complete, assembly sequences are searched and grouped according to refixturing and reorientation counts. This stage of editing is represented in Figure 15.11. The engineer may choose any one of the groups and step through its assembly sequences, reviewing assembly moves and states, fixturing needs, and subassembly orientations. This provides a basis for further screening and sequence choice.

### Economic editing basis

A goal in assembly system design is to produce an assembly system that is economical at around the anticipated production rate. Assembly cost can be dominated by scrapping costs and direct and rework labor costs associated with an ill-chosen assembly sequence. Such sequences can usually be found and avoided by applying the editing techniques. Where avoidance is impossible, product redesign may be appropriate. Similarly, costs of unneeded operations

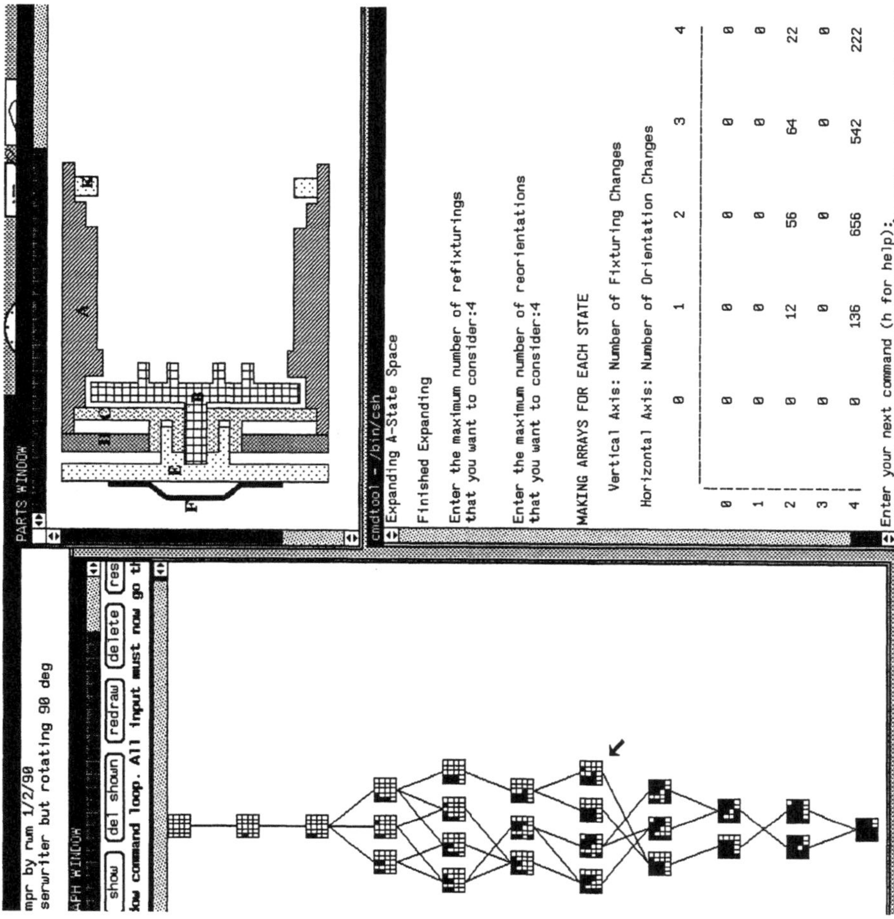


Figure 15.11: The assembly sequence editing environment. The left panel represents a partially-edited assembly sequence diagram. The "show(state)" menu command is active and the assembly state denoted by the cursor is represented in the upper right. Several editing actions have already been taken: Purge of representationally redundant sequences; assurance of a sequential assembly line; excision of two difficult part mating paradigms; excision of several assembly states due to fixturing difficulty. Data about fixturing opportunities and corresponding stable subassembly orientations has been entered and the remaining sequences have been examined for refixturing and reorientation counts. There are 12 combinations of sequence, fixturing, and state orientation that yield an assembly with 2 fixturing changes and 1 subassembly reorientation.

that add no value can dominate any savings from careful choice of assembly line technology, and such operations also can be anticipated and controlled by applying editing techniques.

Once some physically acceptable assembly sequences are known, an engineer may do comparative economic studies of the candidate sequences (Gustavson[17], Cooprider[9], Graves and Holmes-Redfield[15]). Assembly System Design Program (ASDP) is an example of a computer based assembly system design optimization routine that explores possibilities of assembly resources addressing multiple tasks (Gustavson[17]). Economic screening of assembly system designs often requires supplying information characterizing the costs and times of assembly steps. This effort is appropriate for comparisons amongst a few candidate sequences. Under some circumstances, for example in cases of products for which there are no problem assembly moves or difficult fits, data such as operation times, costs, and candidate technologies may be provided directly from a data base and without engineering consideration.

### 15.3.1 Assembly Sequence Editing

A program called EDIT has been written that allows display and editing of assembly sequences in this manner. EDIT is written in the programming language "C" using SUN's graphics package, SUNVIEW. EDIT reads assembly sequences from input files written by LSG[22] and creates a data structure of the sequence graph[2]. Most of the information in this data structure is stored under the states of the assembly sequence diagram. The data structure for assembly states is shown in Figure 15.12. As shown, the state structure holds information about moves that lead to and from the state, the status of the state, and the state's subassemblies.

The delete_status element in the data structure notes whether a state is on, highlighted, or removed from the display. The software is flexible about invoking and undoing editing moves. When editing moves are made, the program changes the status of the state instead of removing the state from the structure.

EDIT reads as input the assembly sequences generated by LSG and accepts raster images of SUNDRAW part drawings. The assembly sequence diagram, part drawings, and information about states of assembly are combined into a screen display of possible assembly sequences. Figure 15.13 is an example. Two part sets shown on the right represent two subassemblies coexisting in a designated state, marked with the cursor in Figure 15.13. At bottom left is an editing choice menu, representing the set of user commands. Above is part of the AFI assembly sequence diagram.

Two classes of editing have been implemented and are described below: editing states and moves; and editing based on fixturing and refixturing and reorientation counts.

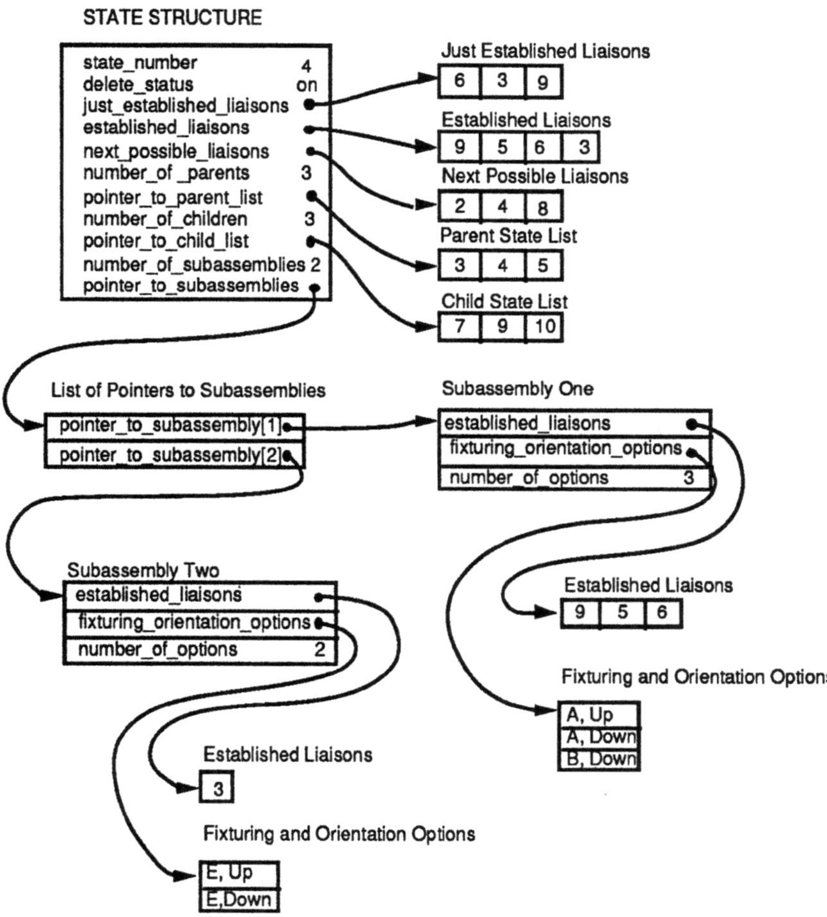


Figure 15.12: Assembly state data structure

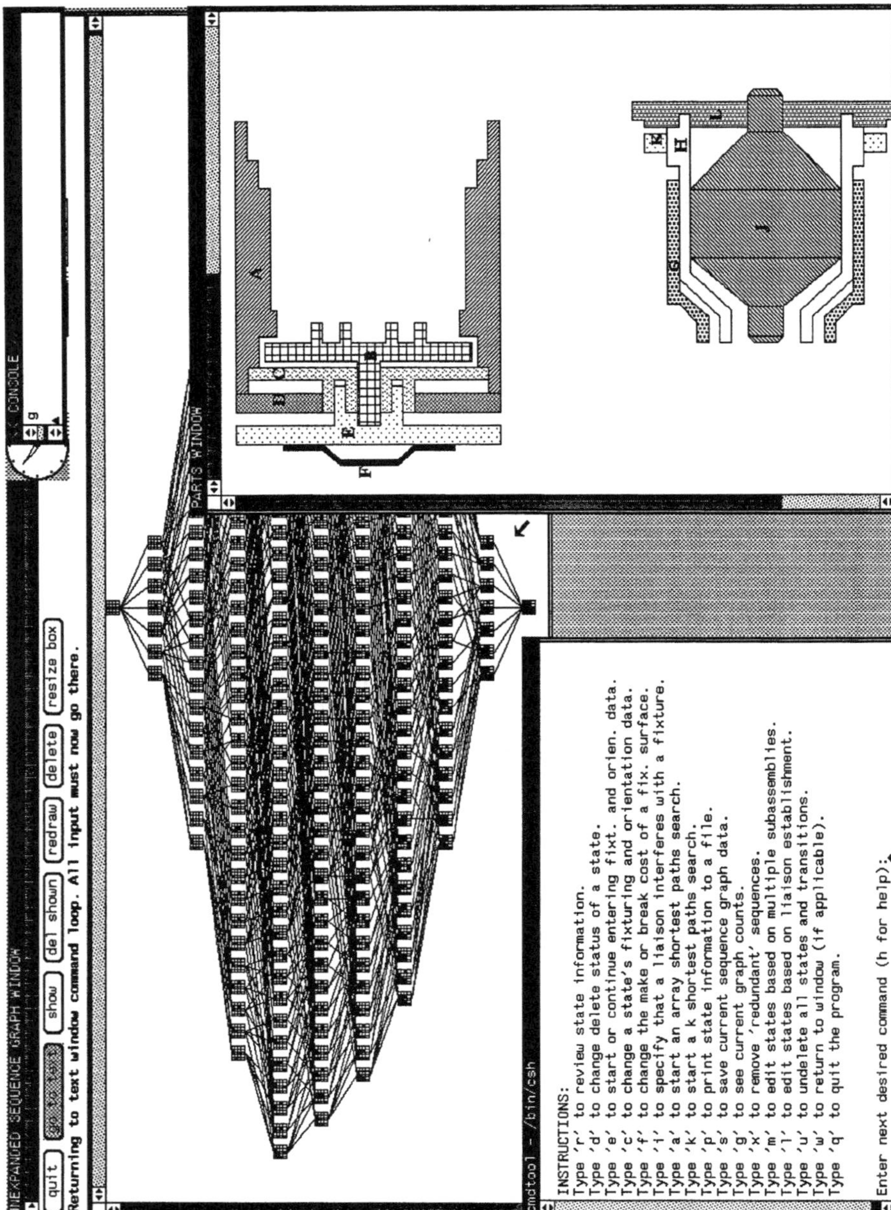


Figure 15.13: Screen display of the assembly sequence editing program EDIT

**Editing States and Moves**

A variety of editing choices is based on assembly states and moves. Eliminating individual states or moves is the most basic. An edit menu allows us to invoke several editing features. The "SHOW" option gives a view of the particular state or move designated by the mouse. Choosing "SHOW STATE" and designating a state in the assembly sequence diagram shows the following state data: (1) its part drawings concatenated as one or more subassemblies, (2) a text description in the text window, and (3) the state and all its parent and child states highlighted in color on the diagram. Choosing "SHOW TRANSITION" and designating the two states associated with a move highlights the move in color and brings up part drawings of the states before and after the move. After highlighting a state or move, we can delete it by selecting "DEL SHOWN."

The "DELETE" option deletes a particular state or move without first showing it. The option has four choices: (1) "DELETE STATE," (2) "DELETE TRANSITION," (3) "UNDELETE STATE," and (4) "UNDELETE TRANSITION." Selecting an option activates the mouse to perform the indicated function on selected elements. Deleting an item causes it to be highlighted in red. Invoking "REDRAW" removes all deleted items from the display.

More complex editing paradigms are done by entering text commands in the text window. Editing features allow deletion of assembly states with multiple subassemblies; deletion of moves where a denoted set of simultaneous mates is made; or specification that a particular assembly move must immediately precede another. These are powerful editing tools; invoking one can significantly reduce the original sequence count. Their use is based on knowledge of particular desirable or undesirable states, moves, or partial sequences.

Use of these editing facilities is illustrated on the AFI example of Figure 15.2 as follows:

State and move editing usually begins by removing redundant assembly sequences from among the sequences having branched work flow and one or more states with two or more subassemblies. Redundancy-purge editing is a default option that may be overridden; it uses a single key-stroke ("x" in Figure 15.13) and reduces the assembly sequence count in this example from 50,748 to 3319.

Editing continues by removing two awkward assembly paradigms, simultaneous establishment of liaisons 8, 9, and 10; and of 4 and 16. 1607 sequences remain after invoking this editing option, shown in Figure 15.14. Note that none of the five states in the next-to-last rank lack liaisons 8, 9, and 10 or 4 and 16. Note also invocation of the "SHOW (State)" option. In Figure 15.15, "SHOW (State Transition)" is invoked. The assembly move between assembly states in the 3rd-to-last and next-to-last ranks, both on the left edge, is shown.

Finally, editing command "m" is used to eliminate all states containing more than one subassembly, eliminating branched assembly lines.

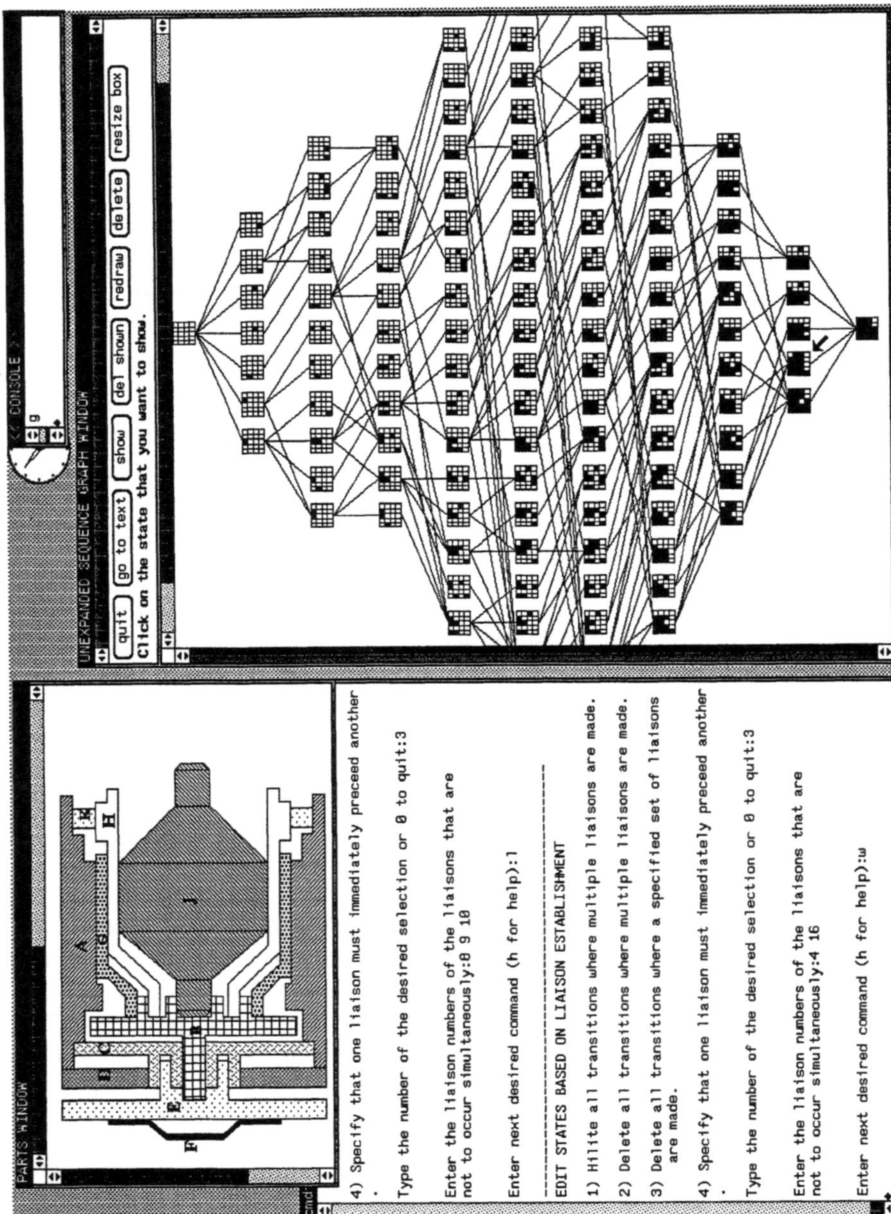


Figure 15.14: Editing awkward assembly paradigms

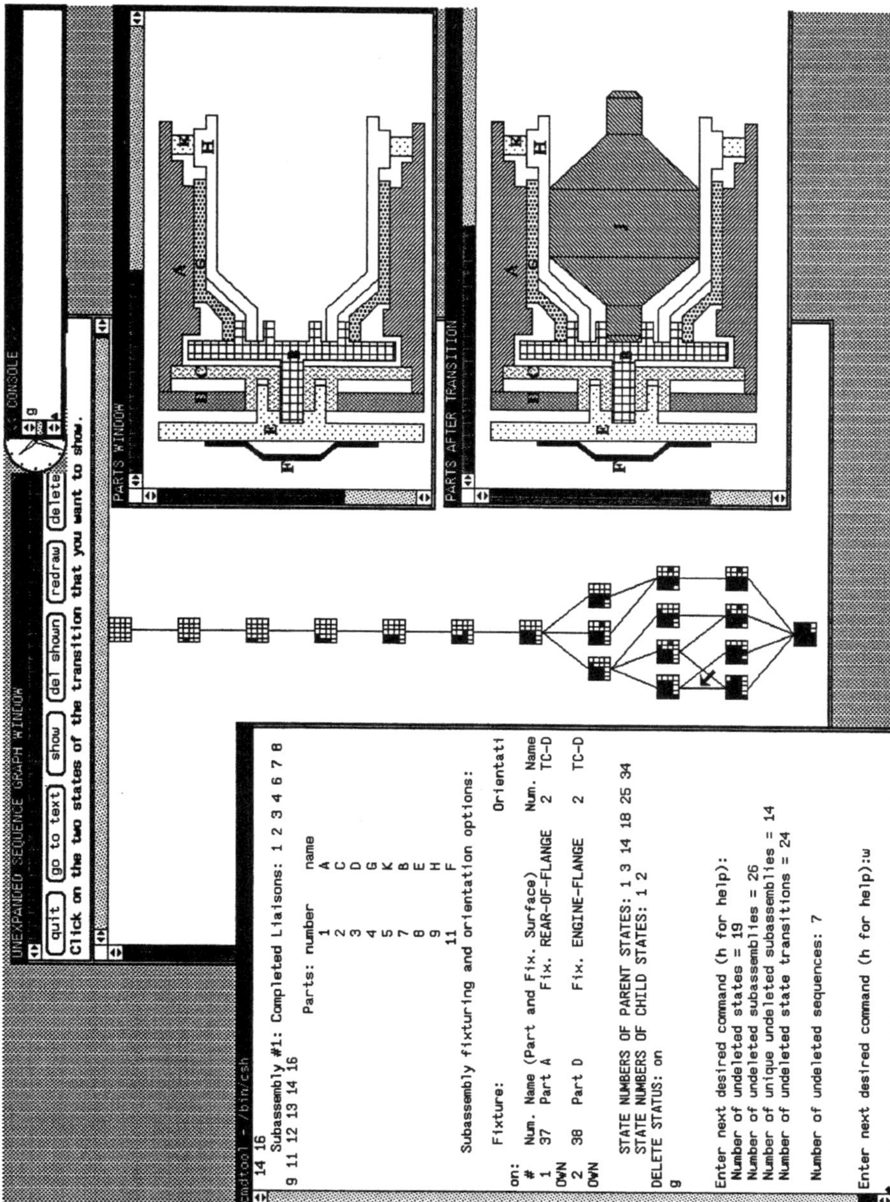


Figure 15.15: Editing option to show an assembly move or state transition

This sort of editing, manually-applied progressive invocation of a sequence of logically-described editing paradigms, is very rapid and effective for a fair range of products. For the AFI example, one progression is shown in Table 15.4.

Table 15.4: An editing progression

| EDITING PARADIGM (CUMULATIVE) | NUMBER OF SEQUENCES REMAINING |
|---|---|
| Unedited Sequence Count | 50,748 |
| Deletion of Redundant Sequences[3] | 3,319 |
| Liaisons 8, 9 and 10, and liaisons 4 and 16 not Done Concurrently[12] | 1,607 |
| Constrain Against Branched Assembly Lines | 312 |
| Load Case (Part A) from Either End before Starting Other End[12] | 2 |

Screens representing the last two of these cumulative editing paradigms are shown as Figures 15.16 and 15.17. The text in Figure 15.16 shows the keystroke entries to invoke the middle three editing paradigms. The display of Figure 15.17, invoking the "SHOW (State)" option, shows the right branch of the two remaining assembly sequences. The last editing paradigm was done manually by repeated use of the "SHOW (State)" option and appropriate use of the "Delete (State)" option; it takes a few minutes.

### Editing Based on Refixturing and Reorientation

Another editing means allows evaluating and editing all individual assembly sequences based on fixturing, refixturing, orientation, and re-orientation issues. Refixturing and re-orientation are production moves that cost but add no value; usually it pays to avoid them during assembly. The user must supply substantial fixturing and orientation information associated with each assembly state to use this evaluation option. Because of the state-associated need for substantial additional information, editing based on fixturing and orientation issues is generally left until initial editing based on assembly states, moves, and line topology has been completed and state- and move-counts have been reduced.

To evaluate sequences this way, information about fixturing and orientation possibilities for each remaining state is entered. Each original state and assembly move is replaced by a set of states and moves, allowing each possible fixture and orientation state combination and refixturing and reorientation move to be represented. The result is an expanded assembly sequence diagram. Each path through this network represents a sequence of liaisons, fixturing choices, and

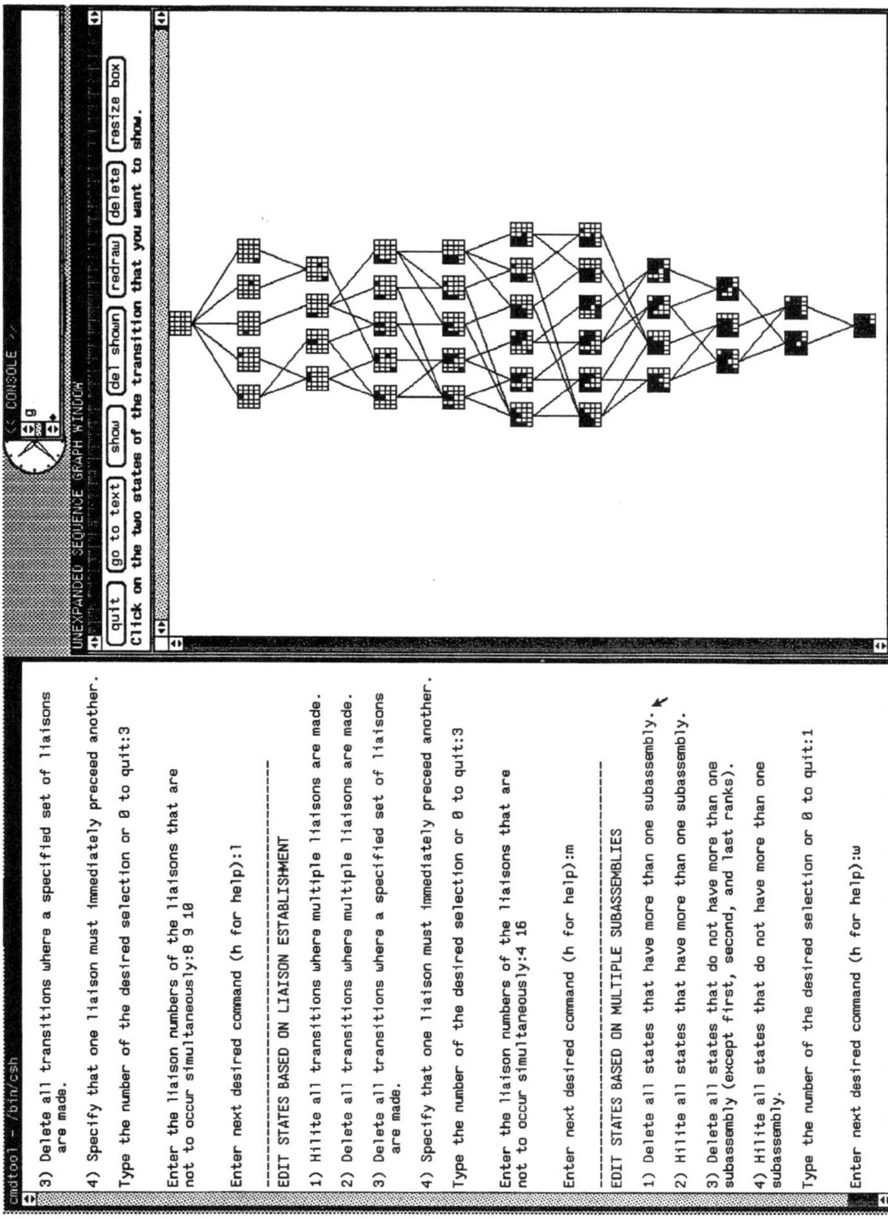


Figure 15.16: Surviving sequences after editing to eliminate branched assembly lines.

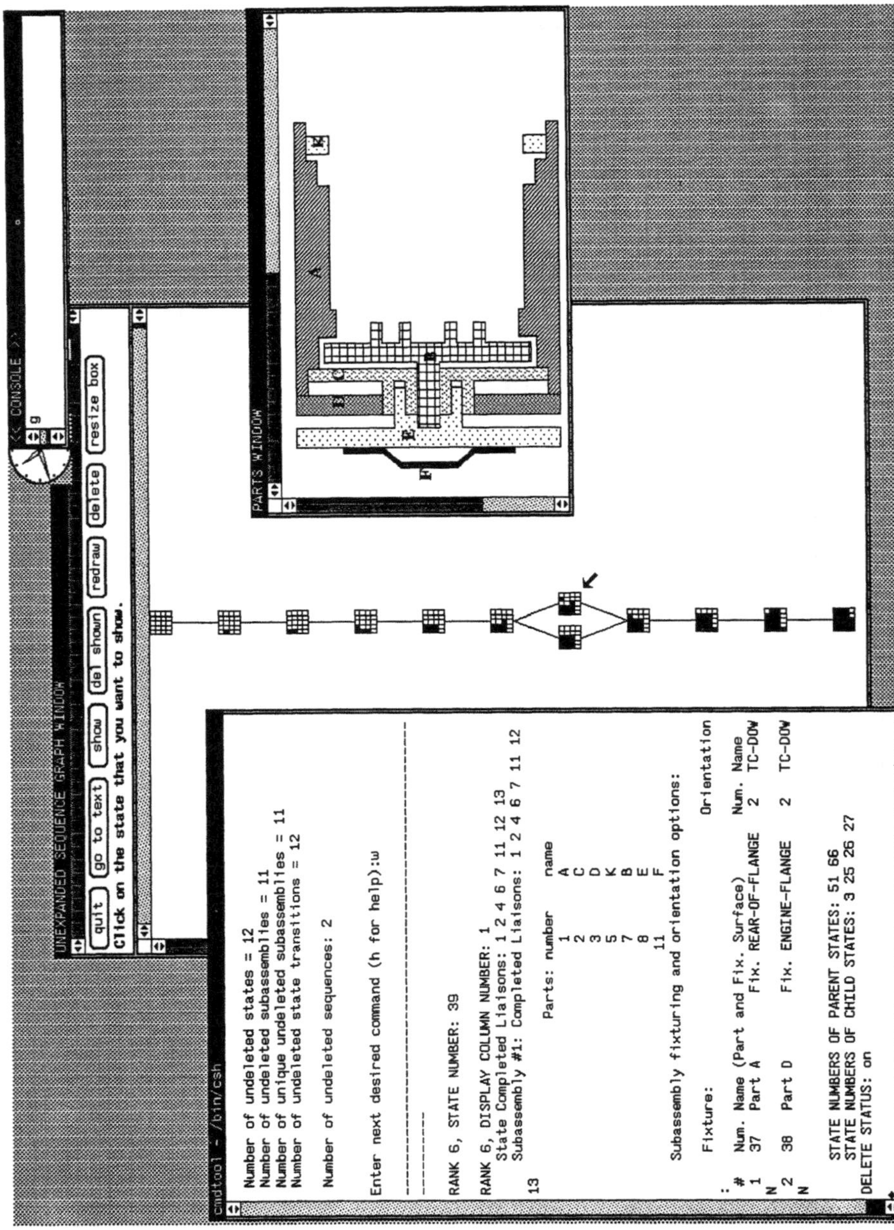


Figure 15.17: Two surviving sequences remain after invoking constraint: load case (A) from either end completely before starting other end.

orientations. The software then uses an N-th shortest-path algorithm to rank assembly sequences according to how many fixturings and orientations they require, allowing the user to consider those sequences that require relatively few fixturing and orientation changes.

A shortest path approach is used so as to find paths that minimize fixturing and orientation change counts, and since the sequence graph is a well-defined directed network from disassembly to the fully assembled state. Fixturing and orientation changes occur during assembly moves and costs are associated with arcs. Associating fixturing and orientation options with assembly states (nodes) determines the number of changes needed in each assembly move (arc). These changes provide the costs for the shortest path algorithm. Operating the shortest path algorithm associates paths with a matrix of re-orientation count and refixturing count.

Fixturing is represented by a user-named user-defined part-surface or subassembly-surface to which a fixture is attached; and orientation as any of several user-specified orientations for which fixturing to the designated surface is stable. Fixturing and orientation are represented as pairs since they are interdependent. Each subassembly may have several of these pairs, each representing a surface and possible orientations to be associated with each fixture.

Evaluating assembly sequences on the bases of reorientation and refixturing costs starts by entering names of mating surfaces and possible fixturing surfaces for all the parts in the assembly. One specifies all orientations to be considered for assembly. Figure 15.18 shows the surface and orientation information for the AFI transmission example.

Possible fixturing and orientation pairs for each unique subassembly in the sequence graph are found next, and involves two steps. First, the user specifies whether any of the subassembly's surfaces are blocked. An inaccessible surface will not become accessible again as more parts are added, so it is eliminated from further consideration. The program then presents a list of fixturing-surface and orientation pairs that consist of all available surfaces paired with all orientations. The user eliminates all surface-orientation pairs considered infeasible, leaving acceptable fixturing-surface and orientation options for further consideration. The AFI assembly has 64 unique subassemblies, and it takes about 2 hours to enter fixturing and orientation data. Entering these data constitutes preliminary design consideration of possible fixturing for each subassembly of each state. The user-engineer is asked "Can, and how can each subassembly of each state be fixtured? (answer for each considered orientation)."

As there is usually a plurality of fixture and orientation options for each state, entering these data represents significant expansion of the possibly pre-edited assembly-sequence diagram. For example, consider two states in adjacent ranks on the assembly sequence diagram; the first with two fixture-orientation options, the second with three. To encompass the new data, the first "state"

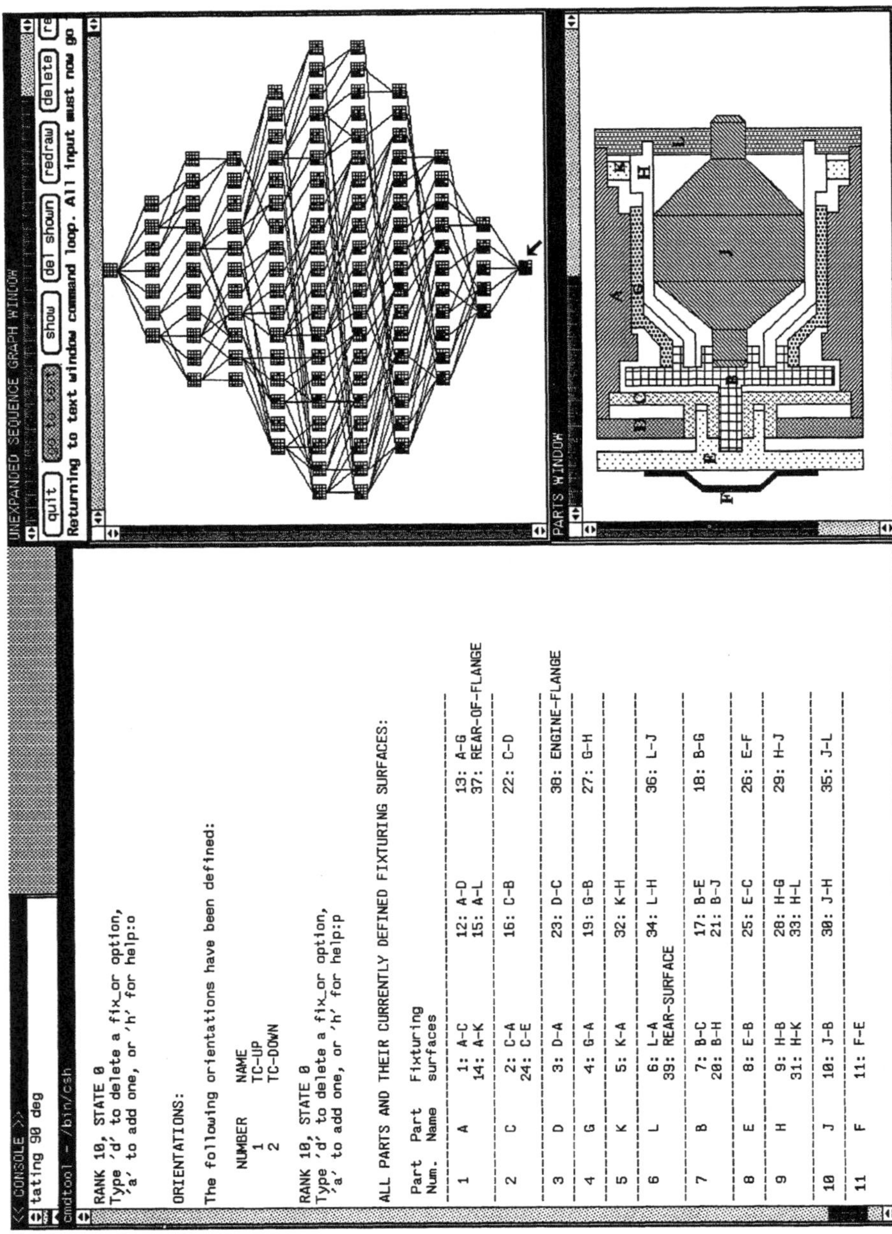


Figure 15.18: Listing of fixturing surface and orientation information for the AFI product.

now must be represented as two states, the second as three. Additionally, since each (of two) state/option combination in the earlier rank has access to each (of three) state/option combination in the later rank, the single state transition or assembly move becomes six moves. Only two of these moves can represent an assembly move without orientation or fixturing change.

Once the data have been entered, one may invoke the shortest path algorithm, showing the sequences requiring the fewest fixturing and orientation changes. One enters the command "a" and specifies the maximum number of fixturing changes and orientation changes to be considered. See Figure 15.11. The shortest path algorithm yields a matrix which indicates how many sequences are associated with each fixturing and orientation change pair. In the example there are 12 sequences that require two fixturing and one orientation changes. Ask to see all the paths associated with a particular change count, and the program will list the paths and plot them on the screen. It can list and plot any of these sequences individually, and step through partial assembly drawings of the states in this sequence. Figure 15.19 shows this for the AFI, where the fifth state in the sequence is shown. Having considered the paths with fewest fixturing and orientation changes, one may choose an assembly sequence from them.

This edit is very valuable as fixtures can be very costly and reorientation takes assembly-line time, length, and resources. Substantial time is needed to enter fixturing and orientation data, but the time is well-spent in that the program structure prompts the user-engineer to do preliminary fixturing design, records his thoughts, and then allows concurrent consideration of assembly-sequence choice and fixture and orientation option choice. For an engineer familiar with a design and details of part-geometry, editing based on states and moves can be completed in a few to many minutes; while considering, generating, and entering fixturing and orientation data can take a few hours. We are currently examining ways to automate some of this process.

### 15.3.2 Editing Strategies - Order of Application of Editing Means

Preceding material presents editing moves and actions available to reduce a large selection of assembly sequences to a workable few for assembly system design. Editing is hierarchical, so that the sequence of applying editing criteria and actions, called an "editing strategy," is an issue. Part of a rationale for an editing strategy is implicit in the preceding material. A rationale for an editing strategy is made more explicit now.

Several criteria are considered to evolve an editing strategy as follow:

1. *Interference between editing bases.* If application of editing basis "A" adversely affects the possibility of applying editing basis "B" but not vice versa,

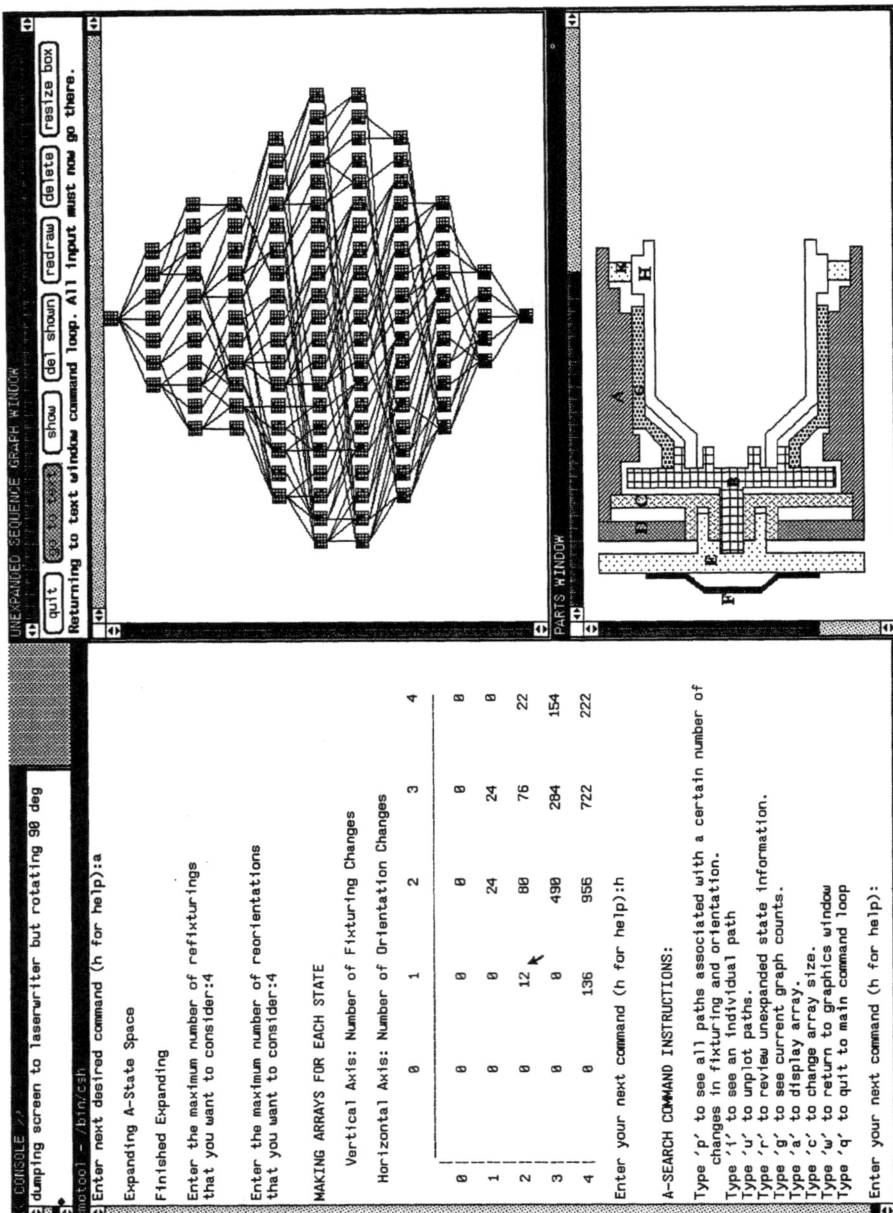


Figure 15.19: Illustration of the editing program stepping through a particular assembly sequence for the AFI product, showing fixtured surfaces, subassembly orientations, and assembly state.

then application of basis B should precede application of basis A.

2. *Expected economic impact of an editing basis.* An editing basis with a high expected economic impact should precede one with a lower expected economic impact. The economic impact of a poor assembly sequence choice, one that leaves a product unneccesarily difficult to assemble, is high, as it extends beyond assembly labor cost and tooling and fixturing costs, to rework, repair, scrap, and warranty costs.

3. *Ease of application of an editing basis.* The easier-applied of two editing bases should be applied earlier. One measure of application ease of an editing basis is the amount of data, if any, that must be user-supplied to allow editing. Data, if needed, are often needed on a state-by-state or move-by-move basis, so that state and move counts are crude multipliers of data need. This suggests a strategy that leaves editing actions requiring added data to follow actions needing no added data.

There is a single issue of potential conflict between a pair of editing bases. Applying the purge of redundant branched-assembly-line sequences constrains the possibility of applying conditional precedences among assembly states. See Table 15.5. Any needed conditional precedences among assembly states must be assured prior to exercising the purge of redundant sequences.

Consider the three remaining broad editing bases for expected economic impact. They are: excising difficult or awkward elements while maintaining any needed elements in the network of assembly possibilities; minimizing non-productive assembly tasks such as reorientation of a subassembly; and choosing among candidate assembly sequences on a calculated economic basis.

Matters of assembly ease and of non-productive task count often have large economic effects on both assembly and rework costs, even of the order of 50% to 100%, and 10% to 20% of assembly cost, respectively. Choice of assembly sequence candidates on a calculated economic basis is a sensitive editing means, but extensive data must be supplied. Difficult assembly states or moves often require extensive and expensive assembly labor or tooling, or even incur repair, rework, and scrapping costs, wasting the discrimination of an economic analysis. Sequence choice on an economic basis is thus best left for when few acceptable candidates, already screened for assembly ease and convenience, fixturing needs, and subassembly orientation, are available.

Assemby line tasks fall into two categories: non-productive tasks and productive or value-adding tasks. Productive tasks involve mating of parts; are represented on an assembly sequence diagram; and their count is invariant for any assembly sequence of a product and equal to the parts count. Non-productive tasks involve no part mating; are often fixture-related; require additional data to be represented on an assembly sequence diagram; and their count is arbitrary and path-dependent. Example non-productive moves are: moving a subassembly from one conveyance to another; testing or measuring; reorienting

Table 15.5: Effects of editing moves on assembly sequence choice and the assembly sequence diagram

**State-conditional precedence:** Enforces that a specific subassembly exists concurrently with or precedes another specific subassembly. Concurrency implies branched assembly line topology. Precedence is equivalent to a partial assembly sequence enforcement and is applicable to any assembly line topology, branched or sequential. Since the purge of redundant branched sequences chooses an arbitrary sequence from a plurality where sequence is represented but concurrence is the actuality, any state-dependent concurrence must be established before the purge. Any such constraints are at the engineer's bidding.
**Purge of redundant sequences:** The purge of redundant sequences affects only assembly states with a plurality of subassemblies, which are those manifest on parallel branches of a branched assembly line. The connection sequences represent orders to the individual assembly events which occur across the branches of a branched assembly line. Sequence is without meaning for parallel operations which may be concurrent. The purge of redundant sequences arbitrarily chooses one sequence and eliminates all remaining sequences of any set of sequences that share common parallel assembly moves on a branched assembly line topology. The redundant sequences are unneeded and the purge is done routinely as a default option in the editing software.
**Constrain the assembly line topology:** An assembly line with no branches is called sequential and is selected if each assembly state with two or more subassemblies is eliminated. A sequential line is chosen by the engineer with a single software command. Branched assembly "lines" remain as choices if any assembly state or states with a plurality of subassemblies remain on the assembly sequence diagram. The engineer designing a branched assembly line must consider all otherwise not precluded choices individually.
**Enforce a particular state:** If all sequences incorporate a particular state, that state must stand alone in its rank in the assembly sequence diagram; the diagram is waisted to a width of one at that state's rank. The strength of this editing move is evident from its description. The complementary editing move avoids a particular state, removing that state from the assembly sequence diagram. It is generally a weaker move.
**Enforce a particular subassembly:** Incorporating a particular state or a particular subassembly are equivalent for a sequential line. Since a particular subassembly may exist in two or more assembly states that include a plurality of subassemblies, the editing moves are not common if a branched assembly line remains an option. The complementary editing move avoids a particular subassembly by eliminating all states where that subassembly appears unconnected.

Table 15.5 (continued): Effects of editing moves on assembly sequence choice and the assembly sequence diagram

**Enforce a partial assembly sequence:** This editing option is invoked logically by enforcing that one of a pair of assembly moves will immediately follow another. These constraints may be chained. Reduction in assembly sequence options generally increases faster than the chain-length or the count of such constraints. Editing software accomplishes each of these constraints or each link in a chain with a few keyboard strokes.

**Excise Awkward assembly paradigms:** Every connection must be made sooner or later to accomplish full assembly. Thus it is useless to seek to avoid a connection. Assembly move count is the parts count. If the connection count equals or exceeds the parts count, as it does for most designs, then two or more connections must be made simultaneously. Choice of assembly sequence often includes choice of which connections are to be combined. Of these combinations often some choices are clearly more easily done, and some are done with more or even great difficulty. Engineers responsible for product design or assembly can generally call out the difficult connection combinations. Assembly sequence editing software cooperates by accepting commands of a few keystrokes that excise from all choices any noted simultaneous combination of connections.

a subassembly; and refixturing a subassembly.

These simple ideas and an observation about the role of fixturing in assembly suggest that considerations of ease of part mating have a greater expected economic impact than considerations of minimizing non-productive moves, and should be considered earlier. Fixtures are often used when their usage cost is less than that of suffering the part-mating difficulties fixturing avoids. There is no cost associated with using the results of screening assembly sequences to find the physically easiest sequence, and much to be gained. Fixturing does have associated costs, and is often a needed palliative for any remaining difficult assembly moves.

Ease of application of editing basis suggests the same order for the three broad editing bases as does economic impact: screening for assembly ease and convenience first; minimizing non-productive assembly tasks next; and screening on a calculated economic basis, last.

A major determinant of ease of application of editing basis is the quality and quantity of data that must be supplied to accomplish editing on a basis. Data needs for these broad editing bases grow as the bases are traversed in the stated order.

Screening for assembly ease and convenience requires engineering knowledge of the product design and of the mechanics of part mating. The user makes yes/no decisions based on that knowledge but without any need to supply or enter characterizations of that knowledge. Minimizing non-productive assembly tasks across an assembly sequence implies both a knowledge and a counting of those tasks. The common non-productive tasks are subassembly reorientation, refixturing, and conveyance transfer. None of these are represented on the basic assembly sequence diagram. Subassembly fixturing and orientation (as well as assembly tests or measurements) are associated with assembly states; reorientation and refixturing with assembly moves. Assembly system design engineers draw on knowledge of the product design, of mechanics of part mating, and of assembly customs and practices, to supply to a data-base the state-associated information of fixturing and orientation needs and options.

Similarly, information for economic screening of candidate sequences is not represented on a basic assembly sequence diagram; it must be supplied by the user or from a data base. The needed information includes characterizations of appropriate assembly technology, related fixed and variable candidate machine, tooling, and labor costs, and task times; and is associated with assembly moves. Assembly system design engineers must draw on market and vendor cost and performance data, as well as product design and part mating mechanical knowledge, to supply data for economic screening of candidates.

Within the broad editing basis that is pruning the awkward states or difficult moves while retaining the graceful states and easy moves, the following editing sequence is recommended: apply any assembly line topological constraints; assure any needed subassemblies or states; enforce any needed partial assembly sequences; excise any awkward assembly paradigms; and finally excise any individual awkward assembly states.

A model editing strategy based on the combined considerations is presented as Figure 15.20.

## 15.4 Conclusions

Assembly sequence is a major consideration and a component of the issue of finding the most favorable means of assembling a product from parts. Different assembly sequences have different needs for assembly fixturing, for number of orientation changes, for convenience of access, for time of assembly, and for assembly skill level; different sequences have different possibilities and probabilities of part-damage during assembly. The importance of considering these and other consequences of chosen assembly sequence increase with production run size, with product parts-count, and with rising quality standards.

Considerations have so far concerned applying liaison-sequence analysis to established, or stationary, designs. Product design and assembly system requirements are intimately coupled, and product design, materials, manufacturing means, and assembly system design are properly considered concurrently. Product designs typically evolve through many revisions and small changes in prod-

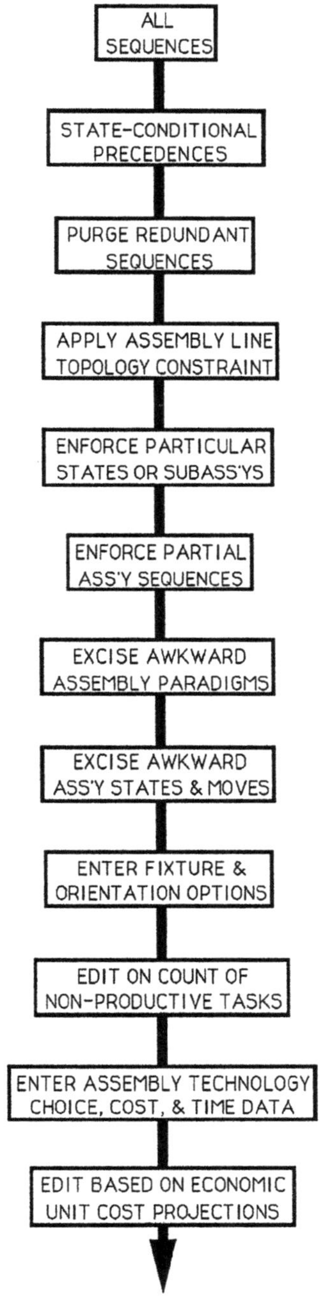


Figure 15.20: An editing strategy

uct design can cause radical changes in assembly sequence. Liaison-sequence analysis as a tool is capable of addressing the issues of interactions between product design and assembly-system design, but of course there is a substantial increase in effort required if many designs are to be analyzed. A liaison-sequence analysis, done before a design is frozen, and showing one or more obviously favorable assembly sequences, can itself be examined to extract the salients allowing the superior assembly sequence choices. The responsible production engineer is then in a position to participate in reviews of the evolving design, now having a knowledge of which design features or changes have little or no effect on ease of assembly, and which features or changes have a major influence on assembly convenience. Liaison-sequence analysis provides a solid and tangible basis for consideration and discussion of production engineering matters during design evolution and at design reviews. See Figure 15.1.

Consideration of assembly sequence issues was informal and heuristic in the past. An assembly engineer devised and compared enough assembly sequences to find one or more acceptable ones, but had no way of knowing whether a better one remained unknown and unconsidered. Formal means for exhaustive consideration of assembly sequence issues exist now. Means involve finding all sequences and reducing that number to the best few candidates by rational editing consideration. Assembly engineers may now consider assembly issues more quickly or more deeply.

The assembly sequence generation and editing software shown does all bookkeeping and algorithmic operations but usurps none of the engineering functions. It is not an expert system which embodies only the parts of the observed behavior of past practitioners, but an engineering aid which responds to all of the skills of the using engineer. All editing decisions of choice are made by cognizant engineers for considered reasons. No editing decision is independently invoked by the computer software. An engineer is involved at all levels of editing, giving complete insight into editing issues. Editing decisions and reasons are easily documented and reconsideration of editing actions is conveniently possible.

Application of editing bases is hierarchical, and the heirarchical sequence may be chosen by the engineer for technical and convenience reasons. There are rational bases for choice of editing sequence. Experience in use indicates that for many carefully-considered designs there are few good sequences and that they are usually revealed quickly by any rational editing sequence.

The newly found speed of technique provided by the editing software allows rapid response to design changes in a concurrent design environment. See Figure 15.1. Past experience includes the example of a product functional designer assessing performance impact of redesign of a complicated assembly in a matter of hours, while assembly system engineers, working without use of the described technique, needed about a week to assess the impact of the same design change on the preliminarily designed assembly system. The observation was made during design, manufacturing, and assembly studies of a conventional automotive automatic transmission. The noted redesign addressed geometry and fastening of the major fixed element within the cases so as to explore casting draft options. There was not a major functional redesign, but there was a major disruption of assembly sequence choices and assembly system consideration and design.

In another instance during a similar study of an automatic automotive transaxle, a redesign addressed a minor functional change affecteing design of the gearbox portion. Taking full advantage of the new assembly sequence determination and editing techniques allowed the product redesign impact assessment on a preliminarily designed assembly system to be completed in about a day. While this period does not quite approach the performance assessment time we have experienced, it is a substantial improvement over past performance and a major enabler of a concurrent engineering environment (Nevins and Whitney[24]).

# References

[1] *Webster's New Collegiate Dictionary.* G. & C. Merriam Co., Springfield MA, 1979.

[2] T. E. Abell. *An Interactive Software Tool for Editing and Evaluating Mechanical Assembly Sequences Based on Fixturing and Orientation Requirements.* S.M. Thesis, M.I.T. Mechanical Engineering Department, Cambridge MA, Aug. 1989.

[3] G. P. Amblard. *Rationale for the Use of Subassemblies in Production Systems: A Comparative Look at Sequential and Arborescent Systems.* S.M. Thesis, M.I.T. Operations Research Center, Cambridge MA, 1989.

[4] D. F. Baldwin. *Algorithmic Methods and Software Tools for the Generation of Mechanical Assembly Sequences* S.M. Thesis, M.I.T. Mechanical Engineering Department, Cambridge MA, Feb. 1990.

[5] D. F. Baldwin, T. E. Abell, M.-C. Lui, T. L. De Fazio and D. E. Whitney. An Integrated Computer Aid for Generating and Evaluating Assembly Sequences for Mechanical Products. *IEEE Trans. on Robotics and Automation* RA-7(1):78–94, Feb. 1991.

[6] A. Bourjault. *Contribution a une Approche Méthodologique de l'Assemblage Automatise: Elaboration Automatique des Séquences Opératoires.* Thesis to obtain Grade de Docteur es Sciences Physiques at L'Universite de Franche-Comte, Nov. 1984.

[7] A. Bourjault. Methodology of Assembly Automation: A New Approach. *Book of Abstracts of Second Conf on Robotics and Factories of the Future.* San Diego CA, Jul. 1987.

[8] A. Bourjault et al. *Outils Methodologiques pour la Definition de Systems d'Assemblage Automatises.* Universite de Franche-Comte, Centre de Recherche Microsystemes et Robotique, Feb. 1987. Also computer implementation of these tools: *SAGA-Systeme d'Elaboration Automatique de Gammes d'Assemblage Version 1.2*, Feb. 1988.

[9] C. B. Cooprider. *Equipment Selection and Assembly System Design under Multiple Cost Constraints.* S.M. Thesis, M.I.T. Sloan School of Management, Cambridge MA, May 1989.

[10] T. L. De Fazio, T. E. Abell, G. P. Amblard, and D. E. Whitney. Computer-Aided Assembly Sequence Editing and Choice: Editing Criteria, Bases, Rules, and Technique. *Proc. 1990 IEEE Int. Conf. on Systems Eng*, pages 416–422. Pittsburgh PA, Aug. 1990.

[11] T. L. De Fazio and D. E. Whitney. *Part and Assembly-Technique Classification.* C. S. Draper Laboratory, Inc., Cambridge MA, Rep. CSDL-R-1643, Apr. 1983.

[12] T. L. De Fazio and D. E. Whitney. Simplified Generation of All Mechanical Assembly Sequences. *IEEE J. Robotics and Automation* RA-3(6):640–658, Dec. 1987.

[13] T. L. De Fazio, D. E. Whitney, M-C Lui, T. E. Abell, and D. F. Baldwin. Aids for the Design or Choice of Assembly Sequences. *Proc IEEE International Conference for Systems, Man, and Cybernetics*, Cambridge MA, Nov. 1989

[14] B. Frommherz and J. Hornberger. Automatic Generation of Precedence Graphs. *Publications 1988 of the University of Karlsruhe Faculty for Informatics, Institute for Real-Time Computer Control and Robotics* (Prof U. Rembold, Prof R. Dillmann).

[15] S. C. Graves and C. Holmes-Redfield. Equipment Selection and Task Assignment for Multiproduct Assembly System Design. *Int. J. Flex. Manuf. Sys.*, 1:31–50, 1988.

[16] R. E. Gustavson. Computer-Aided Synthesis of least-cost Assembly Systems. *Proc. 14th Int'l. Symposium on Industrial Robots*, Gothenburg, Sweden, Oct. 1984.

[17] R. E. Gustavson. *Design of Cost-Effective Assembly Systems.* SME Paper AD88-250, presented at "Successful Planning and Implementation of Flexible Assembly Systems," pages 29–31. Ann Arbor MI, Mar. 1988.

[18] J. M. Henrioud and A. Bourjault. *Determination des Arbres d'Assemblage.* R.A.I.R.O. APII 24:547–564, Nov. 1990.

[19] L. S. Homem de Mello. *Task Sequence Planning for Robotic Assembly.* Ph.D. Thesis, Electrical and Computer Engineering, Carnegie Mellon University, Pittsburgh PA, 1989.

[20] L. S. Homem de Mello and A. C. Sanderson. Representations of Assembly Sequences. *IJCAI-89 Proc. Eleventh Int. Joint Conf. on Artificial Intelligence*, pages 1035–1040, Aug. 1989.

[21] L. S. Homem de Mello and A. C. Sanderson. AND/OR Graph Representation of Assembly Plans. *IEEE Trans. on Robotics and Automation* RA-6(2):188–199, Apr. 1990.

[22] M. M. Lui. *Generation and Evaluation of Mechanical Assembly Sequences Using the Liaison Sequence Method.* M.I.T. Mechanical Engineering Department, S.M. Thesis, Cambridge MA, May 1988.

[23] J. L. Nevins and D. E. Whitney. Computer Controlled Assembly. *Scientific American* 238(2):62–74, Feb. 1978.

[24] J. L. Nevins and D. E. Whitney (editors). *Concurrent Design of Products and Processes - A Strategy for the Next Generation in Manufacturing.* McGraw-Hill Book Co., New York NY, 1989.

[25] S. Pappu. *A Dual Ascent Algorithm for Finding the Optimal Test Strategy for an Assembly Sequence.* S.M. Thesis, M.I.T. Operations Research Center, Cambridge MA, May 1989.

[26] A. C. Sanderson and L. S. Homem de Mello. Task Planning and Control Synthesis for Flexible Assembly Systems. *Machine Intelligence and Knowledge Engineering for Robotic Applications*, Berlin: Springer Verlag, pages 331–353, 1987.

[27] Whitney, D. E. et al. *Design and Control of Adaptable Programmable Assembly Systems.* C. S. Draper Laboratory Report R-1284, Cambridge MA, Aug. 1979.

# Contributors

**Thomas E. Abell**[1]
Ford Motor Company
Dearborn Michigan

**Guillaume P. Amblard**[1]
Sharp Corporation
Nara
Japan

**Daniel F. Balwin**[1]
Massachussetts Institute of Technology
Mechanical Engineering Department
Cambridge Massachusetts

**Alain Bourjault**
Laboratoire d'Automatique de Besançon
URA 822
Institut de Productique - ENSMM
15 Impasse des Saint Martin
25000 - Besançon
France

**Thomas L. De Fazio**
The Charles Stark Draper Laboratory
555 Technology Square
Cambridge Massachusetts 02139

**Leila De Floriani**
Dipartimento di Matematica
Universitá di Genova
Via L. B. Alberti, 4
16132 Genova
Italy

**Jean-Michel Henrioud**
Laboratoire d'Automatique de Besançon
URA 822
Institut de Productique - ENSMM
15 Impasse des Saint Martin
25000 - Besançon
France

**Richard L. Hoffman**
Research and Technology Center
Northrop Corporation
One Research Park
Palos Verdes California 90274

**Luiz S. Homem-de-Mello**
Jet Propulsion Laboratory
California Institute of Technology
Pasadena California 91109-8099

**Sukhan Lee**
Department of Electrical
Engineering – Systems
University of Southern California
Los Angeles California 90089-0782

**Man-Cheung Max Lui**[1]
Oracle Corporation
San Francisco California

**Amitava Maulik**
Electrical, Computer, and Systems
Engineering Department
Rensselaer Polytechnic Institute
Troy New York 12180-3590

[1]Correspondence related to chapter 15 should be sent to Thomas L. De Fazio or Daniel E. Whitney.

**George Nagy**
Electrical, Computer, and Systems
Engineering Department
Rensselaer Polytechnic Institute
Troy New York 12180-3590

**Aristides A. G. Requicha**
Computer Science Department
University of Southern California
Los Angeles California 90089-0782

**Jean-François Rit**
Robotics Laboratory
Computer Science Department
Stanford University
Stanford California 94305

**Arthur C. Sanderson**
Electrical, Computer, and Systems
Engineering Department
Rensselaer Polytechnic Institute
Troy New York 12180-3590

**Yeong Gil Shin**
Computer Science Department
University of Southern California
Los Angeles California 90089-0782

**Federico Thomas**
Instituto de Cibernética
Diagonal, 647, planta 2
08028 Barcelona
Spain

**Joshua U. Turner**
Rensselaer Design Research Center
Rensselaer Polytechnic Institute
Troy New York 12180-3590

**Timothy W. Whalen**
Computer Science Department
University of Southern California
Los Angeles California 90089-0782

**Daniel E. Whitney**
The Charles Stark Draper Laboratory
555 Technology Square
Cambridge Massachusetts 02139

**Randall H. Wilson**
Robotics Laboratory
Computer Science Department
Stanford University
Stanford California 94305

**Jan D. Wolter**
Department of Computer Science
Texas A&M University
College Station Texas 77843-3112

# Index